INFINITY
and the
MIND

INFINITY
and the
MIND

The Science and Philosophy of the Infinite

by

RUDY
RUCKER

BIRKHÄUSER
BOSTON • BASEL • STUTTGART

RUDY RUCKER
Department of Mathematics
Randolph-Macon Woman's College
Lynchburg, Virginia   24503   USA

Library of Congress Cataloging in Publication Data
Rucker, Rudy v. B. (Rudy von Bitter), 1946 —
Infinity and the mind.
Bibliography: p.
Includes index.
1. Logic, Symbolic and mathematical.
2. Set theory.   3. Infinite.   I. Title.
QA9.R79   511.3   81-21756
ISBN 3-7643-3034-1 (Switzerland)   AACR2

CIP — Kurztitelaufnahme der Deutschen Bibliothek

Rucker, Rudy von B.:
Infinity and the mind/Rudy v. B. Rucker. —
Boston; Basel; Stuttgart: Birkhäuser, 1982.
ISBN 3-7643-3034-1
Bitter Rucker, Rudy von

TABLE OF CONTENTS

PREFACE

This book discusses every kind of infinity: potential and actual, mathematical and physical, theological and mundane. Talking about infinity leads to many fascinating paradoxes. By closely examining these paradoxes we learn a great deal about the human mind, its powers, and its limitations.

The study of infinity is much more than a dry, academic game. The intellectual pursuit of the Absolute Infinite is, as Georg Cantor realized, a form of the soul's quest for God. Whether or not the goal is ever reached, an awareness of the process brings enlightenment.

Infinity and the Mind has been written with the average person in mind. Most of the main text should prove digestible, if chewed. By and large, the separate sections are complete in themselves, and the reader should feel free to skip about in the book.

At the end of each chapter there is a section with puzzles and paradoxes; answers are provided. For those who may wish to delve a bit deeper into set theory and logic, I have organized two mathematical excursions which are placed at the end of the book.

Infinity and the Mind was thought out and written over a period of some ten years. I started having ideas for it in that most Sixties of years, 1972. At that time I was writing a doctoral dissertation in set theory for Erik Ellentuck at Rutgers University and attending a logic seminar led by the eminent proof theorist Gaisi Takeuti at the Institute for Advanced Study in Princeton, New Jersey. The first time I met Takeuti I asked him what set theory was really about. "We are trying to get exact description of thoughts of infinite mind," he said. And then he laughed, as if filled with happiness by this impossible task.

The same year I met Kurt Gödel at the Institute for Advanced Study. No one in modern times has thought more logically than Gödel, no one has proved theorems of greater mathematical complexity. Yet the man I met was a joyful, twinkling sage-not some obsessed fossil. What struck me most about Gödel was his intellectual freedom-his ability to move back and forth between frankly mystical insights and utterly precise logical derivations. As I began to study the writings of Georg Cantor, the founder of set theory, I realized that Cantor shared this freedom. Logic and set theory are the tools for an exact metaphysics.

The writing of this book started with a paper I did for a logic colloquium at Oxford University in 1976 and began in earnest with a set of mimeographed lecture notes for an interdepartmental course I taught with my friend William J. Edgar at SUNY Geneseo in 1977. In 1978 I rewrote my notes and reproduced them by photo-offset for an experimental metamathematics course. Those notes make up the present Chapters One and Three and the more technical Excursion I.

I spent the years 1978-1980 at the Mathematics Institute of the University of Heidelberg, a guest of Gert Müller and the Alexander von Humboldt Foundation. While there, I wrote Chapter Four with Excursion II for a course of lectures on the philosophy of mathematics. Chapters Two and Five have been written this winter at Randolph-Macon Woman's College.

Infinity and the Mind is a work of transmission. I dedicate it with love and respect to everyone in the channel.

R.v.B.R.
Lynchburg, Virginia
June 19, 1981

CHAPTER ONE
INFINITY

A SHORT HISTORY OF INFINITY

The symbol for infinity that one sees most often is the lazy eight curve, technically called the lemniscate. This symbol was first used in a seventeenth century treatise on conic sections.[1] It caught on quickly and was soon used to symbolize infinity or eternity in a variety of contexts. For instance, in the 1700s the infinity symbol began appearing on the Tarot card known as the Juggler or the Magus. It is an interesting coincidence that the Qabbalistic symbol associated with this particular Tarot card is the Hebrew letter א, (pronounced alef), for Georg Cantor, the founder of the modern mathematical theory of the infinite, used the symbol $\aleph_0$, (pronounced alef-null), to stand for the first infinite number.

The appropriateness of the symbol ∞ for infinity lies in the fact that one can travel endlessly around such a curve . . . demolition derby style, if you will. Endlessness is, after all, a principal component of one's concept of infinity. Other notions associated with infinity are indefiniteness and inconceivability.

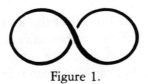

Figure 1.

Figure 2.

Infinity commonly inspires feelings of awe, futility, and fear. Who as a child did not lie in bed filled with a slowly mounting terror while sinking into the idea of a universe that goes on and on, for ever and ever? Blaise Pascal puts this feeling very well: "When I consider the small span of my life absorbed in the eternity of all time, or the small part of space which I can touch or see engulfed by the infinite immensity of spaces that I know not and that know me not, I am frightened and astonished to see myself here instead of there . . . now instead of then."[2]

It is possible to regard the history of the foundations of mathematics as a progressive enlarging of the mathematical universe to include more and more infinities. The Greek word for infinity was *apeiron,* which literally means unbounded, but can also mean infinite, indefinite, or undefined. *Apeiron* was a negative, even pejorative, word. The original chaos out of which the world was formed was *apeiron*. An arbitrary crooked line was *apeiron*. A dirty crumpled handkerchief was *apeiron*. Thus, *apeiron* need not only mean infinitely large, but can also mean totally disordered, infinitely complex, subject to no finite determination. In Aristotle's words, ". . . being infinite is a privation, not a perfection but the absence of a limit. . . ."[3]

There was no place for the *apeiron* in the universe of Pythagoras and Plato. Pythagoras believed that any given aspect of the world could be represented by a finite arrangement of natural numbers, (where "natural number" means "whole number.") Plato believed that even his ultimate form, the Good, must be finite and definite. This was in contradistinction to almost all later metaphysicians, who assumed that the Absolute is necessarily infinite. In the next chapter I will discuss the way in which Greek mathematics was limited by this refusal to accept the *apeiron*, even in the relatively harmless guise of a real number with an infinite decimal expansion.

Aristotle recognized that there are many aspects of the world that seem to point to the actuality of the *apeiron*. For instance, it seems possible that time will go on forever; and it would seem that space is infinitely divisible, so that any line segment contains an infinity of points. In order to avoid these actual infinites that seemed to threaten the orderliness of his *a priori* finite world, Aristotle invented the notion of the *potentially infinite* as opposed to the *actually infinite*. I will describe this distinction in more detail in the next section, but for now let me characterize it as follows. Aristotle would say that the set of natural numbers is potentially infinite, since there is no largest natural number, but he would deny that the set is actually infinite, since it does not exist as one finished thing. This is a doubtful distinction, and I am inclined to agree with Cantor's opinion that ". . . in truth the potentially infinite has only a borrowed reality, insofar as a potentially infinite concept always points towards a logically prior actually infinite concept whose existence it depends on."[4]

Plotinus was the first thinker after Plato to adopt the belief that at least God, or the One, is infinite, stating of the One that, "Absolutely One, it has never known measure and stands outside of number, and so is under no limit either in regard to anything external or internal; for any such determination would bring something of the dual into it."[5]

St. Augustine, who adapted the Platonic philosophy to the Christian religion, believed not only that God was infinite, but also that God could think infinite thoughts. St. Augustine argued that, "Such as say that things infinite are past God's knowledge may just as well leap head-

long into this pit of impiety, and say that God knows not all numbers.
. . . What madman would say so? . . . What are we mean wretches
that dare presume to limit His knowledge?"[6]

This extremely modern position will be returned to in the last section
of this chapter. Later medieval thinkers did not go as far as Augustine
and, although granting the unlimitedness of God, were unwilling to
grant that any of God's creatures could be infinite. In his *Summa Theologiae* St. Thomas Aquinas gives a sort of Aristotelian proof that "although God's power is unlimited, he still cannot make an absolutely unlimited thing, no more than he can make an unmade thing (for this
involves contradictories being true together)."[7] The arguments are elegant, but suffer from the flaw of being circular: it is proved that the notion of an unlimited thing is contradictory by slipping in the premise
that a "thing" is by its very nature limited.

Thus, with the exception of Augustine and a few others, the medieval
thinkers were not prepared to deal with the infinitude of any entities
other than God, be they physical, psychological, or purely abstract. The
famous puzzle of how many angels can dance on the head of a pin can
be viewed as a question about the relationship between the infinite
Creator and the finite world. The crux of this problem is that, on the
one hand, it would seem that since God is infinitely powerful, he should
be able to bid an infinite number of angels to dance on the head of a pin;
on the other hand, it was believed by the medieval thinkers that no actually infinite collection could ever arise in the created world.

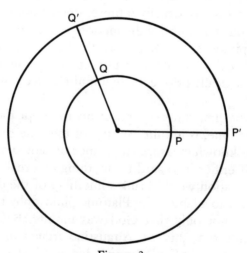

Figure 3.

Their proofs that infinity is somehow a self-contradictory notion were all flawed, but there was at least one interesting paradox involving infinity that the medieval thinkers were aware of. It would seem that any line includes infinitely many points. Since the circumference of a circle with radius two is two times as long as the circumference of a circle with radius one, then the former should include a *larger* infinity of points than the latter. But by drawing radii we can see that each point P on the small circle corresponds to exactly one point P' on the large circle, and each point Q' on the large circle corresponds to exactly one point Q on the small circle. Thus we seem to have two infinities that are simultaneously different and equal.

In the early 1600s Galileo Galilei offered a curious solution to this problem. Galileo proposed that the smaller length could be turned into the longer length by adding an infinite number of infinitely small gaps. He was well aware that such a procedure leads to various difficulties: "These difficulties are real; and they are not the only ones. But let us remember that we are dealing with infinites and indivisibles, both of which transcend our finite understanding, the former on account of their magnitude, the latter because of their smallness. In spite of this, men cannot refrain from discussing them, even though it must be done in a roundabout way."[8]

He resolved some of his difficulties by asserting that problems arise only, "when we attempt, with our finite minds, to discuss the infinite, assigning to it those properties which we give to the finite and limited; but this I think is wrong, for we cannot speak of infinite quantities as being the one greater or less than or equal to another."[9] This last assertion is supported by an example that is sometimes called Galileo's paradox.

1	2	3	4	5	6	7 ...
$\updownarrow$	$\updownarrow$	$\updownarrow$	$\updownarrow$	$\updownarrow$	$\updownarrow$	$\updownarrow$
1	4	9	16	25	36	49 ...

The paradoxical situation arises because, on the one hand, it seems evident that most natural numbers are not perfect squares, so that the set of perfect squares is smaller than the set of all natural numbers; but, on the other hand, since every natural number is the square root of exactly one perfect square, it would seem that there are just as many perfect squares as natural numbers. For Galileo the upshot of this paradox was that, "we can only infer that the totality of all numbers is infinite, and that the number of squares is infinite . . . ; neither is the number of squares less than the totality of all numbers, nor the latter greater than the former; and finally, the attributes 'equal,' 'greater,' and 'less,' are not applicable to infinite, but only to finite quantities."[10]

I have quoted Galileo at some length, because it is with him that we have the first signs of the modern attitude toward the actual infinite in mathematics. If infinite sets do not behave like finite sets, this does not mean that infinity is an inconsistent notion. It means, rather, that infinite numbers obey a different "arithmetic" from finite numbers. If using the ordinary notions of "equal" and "less than" on infinite sets leads to contradictions, this is not a sign that infinite sets cannot exist, but, rather, that these notions do not apply without modification to infinite sets. Galileo himself did not see how to carry out such a modification of these notions; *this* was to be the task of Georg Cantor, some 250 years later.

One of the reasons that Galileo felt it necessary to come to some sort of terms with the actual infinite was his desire to treat space and time as continuously varying quantities. Thus, the results of an experiment on motion can be stated in the form that $x = f(t)$, that space position is a certain function of continuously changing time. But this variable t that grows continuously from, say, zero to ten is *apeiron*, both in the sense that it takes on arbitrary values, and in the sense that it takes on infinitely many values.

This view of position as a function of time introduced a problem that helped lead to the founding of the Calculus in the late 1600s. The problem was that of finding the instantaneous velocity of a moving body, whose distance x from its starting point is given as a function $f(t)$ of time.

It turns out that to calculate the velocity at some instant t_0, one has to imagine measuring the speed over an infinitely small time interval dt. The speed $f'(t_0)$ at t_0 is given by the formula $(f(t_0 + dt) - f(t_0))/dt$, as everyone who has ever survived a first-year calculus course knows.

The quantity dt is called an *infinitesimal,* and obeys many strange rules. If dt is added to a regular number, then it can be ignored, treated like zero. But, on the other hand, dt is regarded as being different enough from zero to be usable as the denominator of a fraction. So is dt zero or not? Adding finitely many infinitesimals together just gives another infinitesimal. But adding infinitely many of them together can give either an ordinary number, or an infinitely large quantity.

Bishop Berkeley found it curious that mathematicians could swallow the Newton–Leibniz theory of infinitesimals, yet balk at the peculiarities of orthodox Christian doctrine. He wrote about this in a 1734 work, the full title of which was, *The Analyst, Or A Discourse Addressed to an Infidel Mathematician. Wherein It is examined whether the Object, Principles, and Inferences of the modern Analysis are more distinctly conceived, or more evidently deduced, than Religious Mysteries and Points of Faith. "First cast out the beam out of thine own Eye; and then shalt thou see clearly to cast out the mote out of thy brother's Eye."*[11]

The use of infinitely small and infinitely large numbers in calculus was soon replaced by the limit process. But it is unlikely that the Calculus could ever have developed so rapidly if mathematicians had not been willing to think in terms of actual infinities. In the past fifteen years, Abraham Robinson's non-standard analysis has produced a technique by which infinitesimals can be used without fear of contradiction. Robinson's technique involves enlarging the real numbers to the set of *hyperreal* numbers, which will be discussed in Chapter 2.

After the introduction of the limit process, calculus was able to advance for a long time without the use of any actually infinite quantities. But as mathematicians tried to get a precise *description* of the continuum or real line, it became evident that infinities in the foundations of mathematics could only be avoided at the cost of great artificiality. Mathematicians, however, still hesitated to plunge into the world of the actually infinite, where a set could be the same size as a subset, a line could have as many points as a line half as long, and endless processes were treated as finished things.

In was Georg Cantor who, in the late 1800s, finally created a theory of the actual infinite which by its apparent consistency, demolished the Aristotelian and scholastic "proofs" that no such theory could be found. Although Cantor was a thoroughgoing scholar who later wrote some very interesting philosophical defenses of the actual infinite, his point of entry was a mathematical problem having to do with the uniqueness of the representation of a function as a trigonometric series.

To give the flavor of the type of construction Cantor was working with, let us consider the construction of the Koch curve shown in Figure 4. The Koch curve is found as the limit of an infinite sequence of approximations. The first approximation is a straight line segment (stage 0). The middle third of this segment is then replaced by two pieces, each as long as the middle third, which are joined like two sides of an equilateral triangle (stage 1). At each succeeding stage, each line segment has its middle third replaced by a spike resembling an equilateral triangle.

Now, if we take infinity as something that can, in some sense, be at-

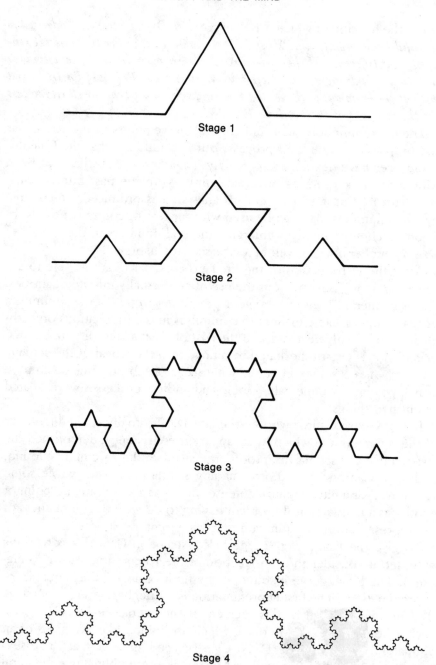

Stage 1

Stage 2

Stage 3

Stage 4

Figure 4. Adapted from Benoit Mandelbrot, *Fractals*.

tained, then we will regard the limit of this infinite process as being a curve actually existing, if not in physical space, then at least as a mathematical object. The Koch curve is discussed at length in Benoit Mandelbrot's book, *Fractals,* where he explains why there is reason to think of the Koch curve in its infinite spikiness as being a better model of a coastline than any of its finitely spiky approximations.[12]

Cantor soon obtained a number of interesting results about actually infinite sets, most notably the result that the set of points on the real line constitutes a *higher* infinity than the set of all natural numbers. That is, Cantor was able to show that infinity is not an all or nothing concept: there are degrees of infinity.

This fact runs counter to the naive concept of infinity: there is only one infinity, and this infinity is unattainable and not quite real. Cantor keeps this naive infinity, which he calls the Absolute Infinite, but he allows for many intermediate levels between the finite and the Absolute Infinite. These intermediate stages correspond to his *transfinite numbers* . . . numbers that are infinite, but none the less conceivable.

In the next section we will discuss the possibility of finding *physically* existing transfinite sets. We will then look for ways in which such actual infinities might exist *mentally.* Finally we will discuss the Absolute, or metaphysical, infinite.

This threefold division is due to Cantor, who, in the following passage, distinguishes between the Absolute Infinite, the physical infinities, and the mathematical infinities:

> The actual infinite arises in three contexts: *first* when it is realized in the most complete form, in a fully independent other-worldly being, *in Deo,* where I call it the Absolute Infinite or simply Absolute; *second* when it occurs in the contingent, created world; *third* when the mind grasps it *in abstracto* as a mathematical magnitude, number, or order type. I wish to make a sharp contrast between the Absolute and what I call the Transfinite, that is, the actual infinities of the last two sorts, which are clearly limited, subject to further increase, and thus related to the finite.[13]

PHYSICAL INFINITIES[14]

There are three ways in which our world appears to be unbounded and thus, perhaps, infinite. It seems that time cannot end. It seems that space cannot end. And it seems that any interval of space or time can be divided and subdivided endlessly. We will consider these three apparent physical infinites in three subsections.

TEMPORAL INFINITIES

Suppose that the human race was never going to die out—that any given generation would be followed by another generation. Would we not then have to admit that the number of generations of man is actually infinite?

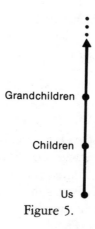

Figure 5.

Aristotle argued against this conclusion, asserting that in this situation the number of generations of man would be but potentially infinite; that is, infinite only in the sense of being inexhaustible. He maintained that at any given time there would only have been some finite number of generations, and that it was not permissible to take the entire future as a single whole containing an actual infinitude of generations.

It is my opinion that this sort of distinction rests on a view of time that has been fairly well discredited by modern relativistic physics. In order to agree with Aristotle that, although there will never be a last generation, there is no infinite set of all the generations, we must believe that the future does not exist as a stable, definite thing. For if we have the future existing in a fixed way, then we have all of the infinitely many future generations existing "at once."

But one of the chief consequences of Einstein's Special Theory of Relativity is that it is space-time that is fundamental, not isolated space which evolves as time passes. I will not argue this point in detail here, but let me repeat that on the basis of modern physical theory we have every reason to think of the passage of time as an illusion. Past, present, and future all exist together in space-time.

So the question of the infinitude of time is not one that is to be dodged by denying that time can be treated as a fixed dimension such as space. The question still remains: is time infinite? If we take the entire space-time of our universe, is the time dimension infinitely extended or not?

Fifty, or even twenty, years ago it would have been natural to assert that our universe has no beginning or end and that time is thus infinite in both directions. But recently it has become an established fact that the universe does have a beginning in time known as the Big Bang. The Big Bang took place approximately 15 billion years ago. At that time our universe was the size of a point, and it has been expanding ever since. What happened before the Big Bang? It is at least possible to answer, "Nothing." The apparent paradox of having a *first* instant in time is sometimes avoided by saying that the Big Bang did not occur *in* time . . . that time is open, rather than closed, in the past.

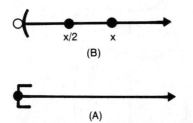

(B)

(A)

Figure 6A (bottom) and Figure 6B (top).

This is a subtle distinction, but a useful one. If we think of time as being all the points greater than or equal to zero, then there is a first instant: zero. But if we think of time as being all the points strictly greater than zero, then there is no first instant. For any instant t greater than zero, one has an earlier instant $t/2$ that is also greater than zero.

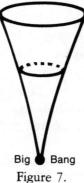

Big ● Bang

Figure 7.

But in any case, if we think of time as not existing before the Big
Bang, then there are certainly not an infinite number of years in our
past. And what about the future? There is no real consensus on this.
Many cosmologists feel that our universe will eventually stop expanding
and collapse to form a single huge black hole called the Big Stop or the
Gnab Gib; others feel that the expansion of the universe will continue
indefinitely.

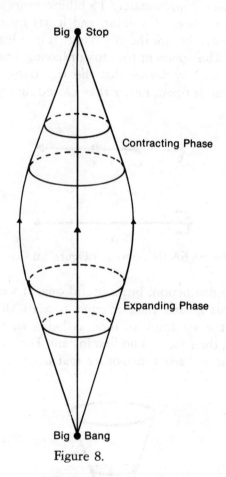

Figure 8.

If the universe really does start as a point and eventually contract
back to a point, is it really reasonable to say that there is no time except
for the interval between these points? What comes before the beginning and after the end?

One response is to view the universe as an oscillating system, which

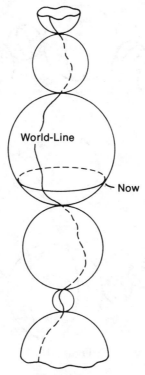

Figure 9.

repeatedly goes through expansions and contractions. This would rein-
troduce an infinite time, which could, however, be avoided.

The way in which one would avoid infinite time in an endlessly oscillating universe would be to adopt a belief in what used to be called "the
eternal return." This is the belief that every so often the universe must
repeat itself. The idea is that a finite universe must return to the same
state every so often, and that once the same state has arisen, the future
evolution of the universe will be the same as the one already under-
gone. The doctrine of eternal recurrence amounts to the assumption
that time is a vast circle. An oscillating universe with circular time is pictured in Figure 10.

There is a simpler model of an oscillating universe with circular time,
which can be called *toroidal space-time*. In toroidal space-time we have an
oscillating universe that repeats itself after every cycle. Such a model is
obtainable by identifying the two points, "Big Bang" and "Big Stop," in
Figure 11.

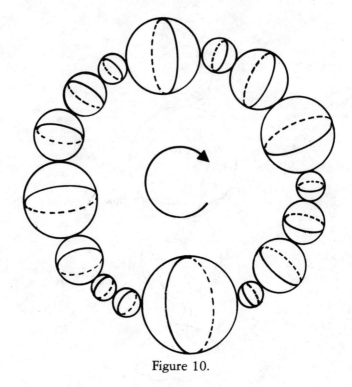

Figure 10.

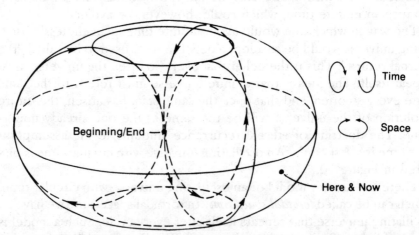

Figure 11. From R. v.B. Rucker, *Geometry, Relativity, and the Fourth Dimension.*

Note, however, that if the universe really *expands* forever, then it cannot ever repeat itself, as the average distance between galaxies is a continually increasing quantity that never returns to the same value.

SPATIAL INFINITIES

We now turn to a consideration of the possibility of spatial infinites. The potential versus actual infinity distinction is sometimes used to try to scotch this question at the outset. Immanuel Kant, for instance, argues that the world cannot be an infinite whole of coexisting things because "in order therefore to conceive the world, which fills all space, as a whole, the successive synthesis of the parts of an infinite world would have to be looked upon as completed; that is, an infinite time would have to be looked upon as elapsed, during the enumeration of all coexisting things."[15]

Kant's point is that space is in some sense not already really there— that things exist together in space only when a mind perceives them to do so. If we accept this, then it is true that an infinite space is something that no finite mind can know of after any finite amount of time. But one feels that the world does exist as a whole, in advance of any efforts on our part to see it as a unity. And if we take all of space-time, it certainly does not seem to be meaningless to ask whether the spatial extent of space-time is infinite or not.

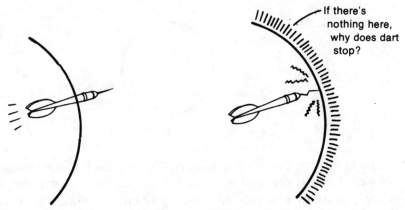

Figure 12A. Dart goes beyond "boundary."   Figure 12B. Dart stops at boundary.

In *De Rerum Natura,* Lucretius first gave the classic argument for the unboundedness of space: "Suppose for a moment that the whole of space were bounded and that someone made his way to its uttermost boundary and threw a flying dart."[16] It seems that either the dart must go past the boundary, in which case it is no boundary of space; or the dart must stop, in which case there is something just beyond the boundary that stops it, which again means that the purported boundary is not really the end of the universe.

So great was their revulsion against the *apeiron* that Parmenides, Plato, and Aristotle all held that the space of our universe is bounded and finite, having the form of a vast sphere. When faced with the question of what lies outside this sphere, Aristotle maintained that "what is limited, is not limited in reference to something that surrounds it."[17]

In modern times we have actually developed a way to make Aristotle's claim a bit more reasonable. As Lucretius realized, the weak point in the claim that space is a finite sphere is that such a space has a definite boundary. But there is a way to construct a three-dimensional space which is finite and which does *not* have boundary points: simply take the hypersurface of a hypersphere. Such a space is endless but not infinite.

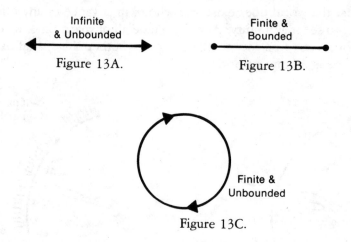

Infinite
& Unbounded

Figure 13A.

Finite &
Bounded

Figure 13B.

Finite &
Unbounded

Figure 13C.

To understand how something can be endless but not infinite, think of a circle. A fly can walk around and around the rim of a glass without ever coming to a barrier or stopping point, but none the less he will soon retrace his steps.

Again, the surface of the Earth is a two-dimensional manifold which is finite but unbounded (unbounded in the sense of having no edges). You can travel and travel on the Earth's surface without ever coming to any truly impassible barrier . . . but if you continue long enough, you will begin to recross your steps.

The reason that the two-dimensional surface of the Earth is finite but unbounded is that it is bent, in three-dimensional space, into the shape of a sphere. In the same way, it is possible to imagine the three-dimensional space of our universe as being bent, in some four-dimensional space, into the shape of a hypersphere. It was Bernhard Riemann who first realized this possibility in 1854. There is, however, a traditional belief that anticipates the hypersphere. This tradition, described in the essay, "The Fearful Sphere of Pascal," by Jorge Luis Borges, is summarized by the saying (attributed to the legendary magician Hermes Trismegistus) that "God is an intelligible sphere, whose center is everywhere and whose circumference is nowhere."[18] If the universe is indeed a hypersphere, then it would be quite accurate to regard it as a sphere whose center is everywhere and whose circumference is nowhere.

To see why this is so, consider the fact that if space is hyperspherical, then one can cover all of space by starting at any point and letting a sphere expand outwards from that point. The curious thing is that if one lets a sphere expand in hyperspherical space, there comes a time when the circumference of the sphere turns into a point and disappears. This fact can be grasped by considering the analogous situation of the sequence of circular latitude lines on the spherical surface of the earth.[19] This line of thought appears in Dante's *Paradisio* (1300).[20]

Aristotle had believed that the world was a series of nine spheres centered around the Earth. The last of these crystalline spheres was called the *Primum Mobile* and lay beyond the sphere upon which were fastened all of the stars (other than the sun, which was attached to the fourth sphere). In the *Paradisio,* Dante is led out through space by Beatrice. He passes through each of the nine spheres of the world: Moon, Mercury, Venus, Sun, Mars, Jupiter, Saturn, Fixed Stars, Primum Mobile. Beyond these nine spheres lie nine spheres of angels, corresponding to the nine spheres of the world. Beyond the nine spheres of angels lies a *point* called the Empyrean, which is the abode of God.

The puzzling thing about Dante's cosmos as it is drawn in Figure 14 is that here the Empyrean appears not to be a point, but rather to be all of space (except for the interior of the last sphere of angels). But this can be remedied if we take space to be hyperspherical! In Figure 15 I have drawn the model we obtain if we take the diagram on the last page and curve it up into a sphere with a point-sized Empyrean. In the same way, the three-dimensional model depicted by the first picture can be turned

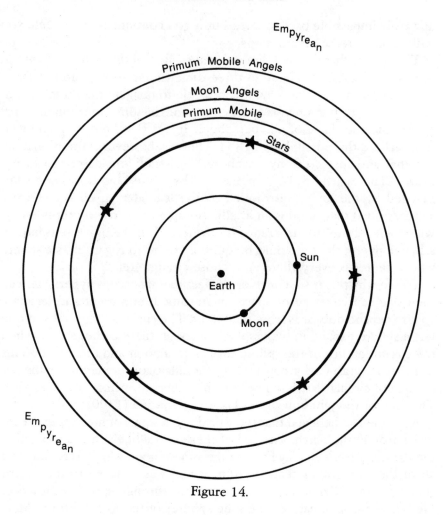

Figure 14.

into the finite unbounded space of the second picture if we bend our three-dimensional space in such a way that all of the space outside our last angelic sphere is compressed to a point.[21] Figure 16 is Doré's engraving of the Empyrean surrounded by its spheres of angels.

This whole notion of hyperspherical space was not consciously developed until the mid-nineteenth century. In the Middle Ages there was a general and uncritical acceptance of Aristotle's view of the universe—without Dante's angelic spheres.

Lucretius, of course, had insisted that space is infinite, and there were

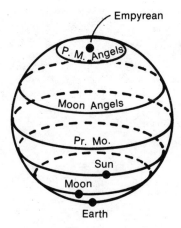

Figure 15.

many other thinkers, such as Nicolas of Cusa and Giordano Bruno, who believed in the infinitude of space. Some kept to the Aristotelian world system, but suggested that there were many such setups drifting around; others opted for a looser setup under which stars and planets are more or less randomly mixed together in infinite space.

Bruno strongly advocated such viewpoints in his writings, especially his dialogue of 1584, "On the Infinite Universe and Worlds."[22] Bruno travelled freely around Europe during his lifetime, teaching his doctrine of the infinite universe at many centers of learning. In 1591, a wealthy Venetian persuaded Bruno to come from Frankfurt to teach him "the art of memory and invention." Shortly after Bruno arrived, the trap was sprung. His host had been working closely with the ecclesiastical authorities, who considered Bruno a leading heretic or heresiarch. Bruno was turned over to the Inquisition. For nine years Bruno was interrogated, tortured, and tried, but he would not give up his beliefs; early in 1600 he was burned at the stake in the Roman *Piazza* Campo di Fiori. Bruno's example caused Galileo to express himself a good deal more cautiously on scientific questions in which the Church had an interest.

Whether or not our space is actually infinite is a question that could conceivably be resolved in the next few decades. Assuming that Einstein's theory of gravitation is correct, there are basically two types of universe: i) a hyperspherical (closed and unbounded) space that expands and then contracts back to a point; ii) an infinite space that expands forever. It is my guess that case i) will come to be most widely accepted, if only because the notion of an actually infinite space extend-

Figure 16. From Gustav Doré's *Divine Comedy* (Dover).

ing out in every direction is so unsettling. The fate of the universe in case i) is certainly more interesting, since such a universe collapses back to an infinitely dense space-time singularity that may serve as the seed for a whole new universe. In case ii), on the other hand, we simply have cooling and dying suns drifting further and further apart in an utterly

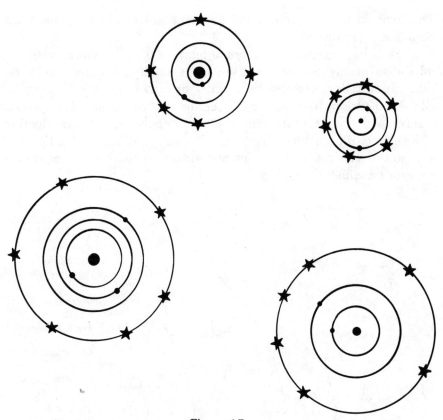

Figure 17.

empty black immensity . . . and in the end there are only ashes and cinders in an absolute and eternal night.

Even though I am basically pro-infinity, my emotions lie with the hyperspherical space. But is there any way of finding a spatial infinity here? Well, what about that four-dimensional space in which our hyperspherical universe is floating? Many would dismiss this space as a mere mathematical fiction . . . as a colorful way of expressing the finite, but unbounded, nature of our universe. This widely held position is really a more sophisticated version of Aristotle's claim that what is limited need not be limited with reference to something outside itself.

But what if one chooses to believe that the four-dimensional space in which our universe curves is real? We might imagine a higher 4-D (four-dimensional) world called, let us say, a duoverse. The duoverse would be 4-D space in which a number of hyperspheres were floating.

The hypersurface of each of the hyperspheres would be a finite, unbounded 3-D universe.

Thus, a duoverse would contain a number of 3-D universes, but no
inhabitant of any one of these universes could reach any one of the
others, unless he could somehow travel through 4-D space. By lowering
all the dimensions by one, one can see that this situation is analogous to
a universe that is a 3-D space in which a number of spheres are floating.
The surface of each sphere or planet is a finite, unbounded 2-D space;
and no one can get from one planet surface to another planet surface
without travelling through 3-D space.

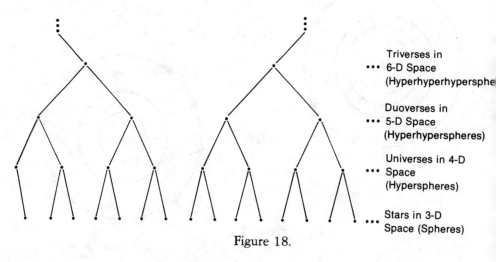

Triverses in
••• 6-D Space
(Hyperhyperhypersphe

Duoverses in
••• 5-D Space
(Hyperhyperspheres)

Universes in 4-D
••• Space
(Hyperspheres)

Stars in 3-D
••• Space (Spheres)

Figure 18.

Following the Hermetic principle, "As above, so below," one is
tempted to believe that the duoverse we are in is actually a finite and
unbounded 4-D space (the 4-D surface of a 5-D sphere in 5-D space),
and that there are a number of such duoverses drifting about in a 5-D
triverse. This could be continued indefinitely. One is reminded of those
Eastern descriptions of the world as a disk resting on the backs of ele-
phants, who stand upon a turtle, who stands upon a turtle, who stands
upon a turtle, who stands upon a turtle, etc.

Note that in that particular sort of cosmos there is only one universe,
one duoverse, one triverse, and so on. But in the kind of infinitely regressing cosmos that I have drawn in Figure 18, we have infinitely many
objects at each level. Note also that to get from star A to star B one
would have to move through 5-D space to get to a different duoverse. It
is a curious feature of such a cosmos that, although there are an infinite

ETCETERA

Figure 19.

number of stars, no one *n*-dimensional space has more than a finite
number of them.

The question we are concerned with here is whether or not space is
infinitely large. There seem to be three options: i) There is some level *n*
for which *n*-dimensional space is real and infinitely extended. The situa-

tion where our three-dimensional space is infinitely large falls under this case. ii) There is some *n* such that there is only one *n*-dimensional space. This space is to be finite and unbounded, and there is to be no reality to *n* + 1 dimensional space. The situation where our three-dimensional space is finite and unbounded, and the reality of four-dimensional space denied, falls under this case. iii) There are real spaces of every dimension, and each of these spaces is finite and unbounded. In this case we either have an infinite number of universes, duoverses, etc., or we reach a level after which there is only one *n*-verse for each *n*.

So is space infinite? It seems that we can insist that at some dimensional level it is infinite; adopt the Aristotelian stance that space is finite at some level beyond which nothing lies; or accept the view that there is an infinite sequence of dimensional levels. In this last case we already have a qualitative infinity in the dimensionality of space, and we may or may not have a quantitative infinity in terms, say, of the total volume of all the 3-D spaces involved.

INFINITIES IN THE SMALL

In this subsection I will discuss the existence of the infinity in the *small,* as opposed to the infinity in the *large,* which has just been discussed. Since a point has no length, no finite number of points could ever constitute a line segment, which *does* have length. So it seems evident that every line segment, or, for that matter, every continuous plane segment or region of space, must consist of an infinite number of points. By the same token, any interval of time should consist of an infinite number of instants; and any continuous region of space-time would consist of an infinite number of events (*event* being the technical term for a space-time location, i.e., point at an instant).

It is undeniable that a continuous region of *mathematical* space has an infinite number of *mathematical* points. Right now, however, we are concerned with *physical* space. We should not be too hasty in assuming that every property of the abstract mathematical space we use to organize our experiences is an actual property of the concrete physical space we live in. But what *is* "the space we live in"? If it is not the space of mathematical physics, is it the space of material objects? Is it the space of our perceptions? In terms of material objects or of perceptions, points do not really exist; for any material or perceptual phenomenon is spread over a certain finite region of space-time. So when we look for the infinity in the small in matter, we do not ask whether matter consists of an infinity of (unobservable) mass-points, but, rather, whether matter is infinitely divisible.

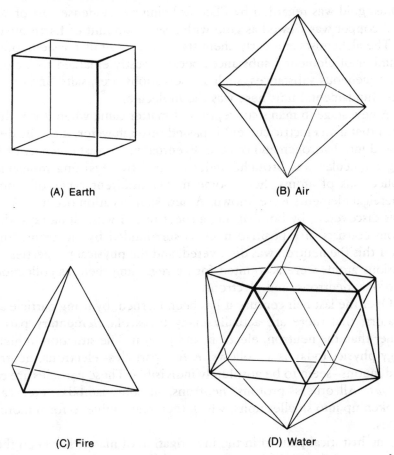

Figure 20. (A–D). From D. Hilbert and H. Cohn-Vossen, *Geometry and the Imagination.*

A commitment to avoiding the formless made it natural for Greek atomists such as Democritus to adopt a theory of matter under which the seemingly irregular bodies of the world are in fact collections of indivisible, perfectly formed atoms. (The four kinds of atoms were shaped, according to Plato, like four of the regular polyhedra. There is one other polyhedron, the twelve-sided dodecahedron, and this was thought somehow to represent the Universe with its twelve signs of the zodiac.) For the atomists, it was as if the world were an immense Lego set, with four kinds of blocks. The diverse substances of the world—oil, wood, stone, metal, flesh, wine, and so on—were regarded as being mixtures of the four elemental substances: Earth, Air, Fire, and Water. Thus, gold was regarded by Plato as being a very dense sort of Water, and copper was viewed as gold with a small amount of Earth mixed in.

The alchemists and early chemists adopted a similar system, only the number of elemental substances became vastly enlarged to include all homogeneous substances, such as the various ores, salts, and essences. The fundamental unit here was the molecule.

A new stage in man's conception of matter came when it was discovered that if an electric current is passed through water, it can be decomposed into hydrogen and oxygen. Eventually, the vast diversity of existing molecules was brought under control by regarding molecules as collections of atoms. Soon some ninety different types of atoms or chemical elements were known. A new simplification occurred when it was discovered, by bombarding a sheet of foil with alpha rays, that an atom consists of a positive nucleus surrounded by electrons. Shortly after this the neutron was discovered, and the physical properties of the various atoms were accounted for by regarding them as collections of protons, neutrons, and electrons.

Over the last half century it has been learned, by using particle accelerators, that there are actually many types of "elementary particles" other than the neutron, electron, and proton. The situation in high-energy physics today is as follows. A few particles—electrons, neutrinos, and muons—seem to be absolutely indivisible. These particles are called *leptons*. All others—protons, neutrons, mesons, lambdas, etc.—can be broken up into smaller units, which then reassemble to form more particles.

The historical pattern in the investigation of matter has been the explanation of diverse substances as combinations of a few simpler substances. Diversity of form replaces diversity of substance. So it is no surprise that it has been proposed that the great variety of divisible particles that exist can be accounted for by assuming that these particles are all built up out of *quarks*.

A second element in the historical pattern is that as more powerful tools of investigation are used, it becomes evident that there are more types of new building blocks than had been suspected initially. This is the phase that high-energy physics is currently moving into. First there were three kinds of quark: up, down, and strange. Now, the charmed quark has been admitted, and there are two new possible quarks: the top quark and the bottom quark. It seems likely that the many diverse types of quark will eventually be accounted for by assuming that each quark is a combination of a few, let us say, *darks* . . . and that there are only a very small number of possible kinds of dark. The cycle will then repeat, with more and more different sorts of dark being indirectly observed, the new diversity being accounted for by viewing each dark as a collection of a few smaller particles of which there are a limited variety, this limited variety beginning to proliferate, and so on.

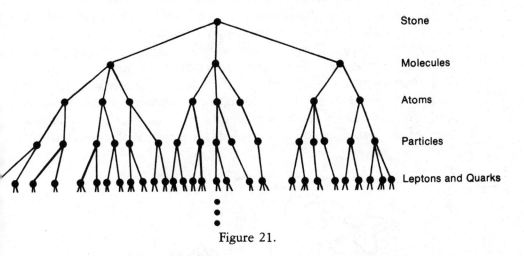

Figure 21.

If this sort of development can indeed continue indefinitely, then we are left with the fact that a stone is a collection of collections of collections of. . . . The stone thus consists of an infinite number of particles, no one of which is indivisible. There is, finally, no matter—only form. For a stone is mostly empty space with a few molecules in it, a molecule is a cloud of atoms, an atom is a few electrons circling a tiny nucleus. . . . What if any seemingly solid bit of matter proves on closer inspection to be a cloud of smaller bits of matter, which are in turn clouds, and so on? Note that the branching matter tree that I began to draw for the stone has only a finite number of forks or nodes at each level, but that since there are infinitely many levels, there are in all an infinite number of nodes or component particles.

There are various objections to this sort of physical infinity. One is the Aristotelian argument that unless one is actually smashing the stone down to the quark level, the quarks are only potentially (as opposed to actually) there. The point would be that the stone may be indefinitely divisible, but that since no one will ever carry out infinitely many divisions, there are not really infinite numbers of particles in the stone right now.

There is a more practical objection as well. This is that no quark has ever been observed in isolation; the existence of quarks is deduced only indirectly as a way of explaining the symmetries of structure that occur in tables of the elementary particles. This argument is not very strong, however. For one thing, a great number of the things we believe in can be observed only indirectly; and, more practically, if we can continue to increase the energy of our measuring tools, there is no reason to think that quarks cannot be more convincingly detected.

A more fundamental objection to the whole idea of particles, subparticles, etc., is that the underlying reality of the world may be field-like, rather than particle-like. By splitting particles indefinitely we arrived at the conclusion that there is only form, and no content; many physicists prefer to start with this viewpoint. For these physicists, the various features of the world are to be explained in terms of the geometry of

Figure 22.

space-time. To get a feeling for this viewpoint, one should look carefully at the surface of a river or small brook. There are circular ripples, flow bulges, whirlpools and eddies, bubbles that form, drops that fly up and fall back, waves that crest into foam. The *geometrodynamic* worldview regards space-time as a substance like the surface of a brook; the various fields and particles that seem to exist are explained as features of the flow.

Does the space-time of geometrodynamics allow an infinity in the small? There is really no answer to this question at present. According to one viewpoint there should be a sort of graininess to space-time, and the grain size would represent a sort of indivisible atom; a different viewpoint suggests that space-time should be as infinitely continuous as mathematical space.

What if there really is nothing smaller than electrons and quarks? Is there then any hope of an infinity in the small? One can argue that a given electron can have infinitely many locations along a given meter stick, so that our space really does have infinitely many points. It is sometimes asserted that the uncertainty principle of quantum mechanics nullifies this argument, but this is not the case.

Quantum mechanics puts no upper limit on the precision with which one can, in principle, determine the position of an electron. It is just that the more precisely the electron's position is known, the less precisely are its speed and direction of motion known. Infinite precision is basically a nonphysical notion, but any desired finite degree of precision is, in principle, obtainable. The precision with which something can be measured is thus a good example of something that is potentially infinite, but never actually infinite.

But this still gives us an actual infinity in the world. For if our electron is located somewhere between zero and one, then each member of the following infinite collection is a possible outcome of a possible measurement:

$$.2 \pm .1, .23 \pm .01, .235 \pm .001, .2356 \pm .0001, \ldots,$$
$$.235608947 \pm .000000001, \ldots$$

Although infinite precision is impossible, an electron can be found to occupy any of the infinitely many points between zero and one whose distance from zero is a terminating decimal.

There are, however, some modern physical speculations that regard "space" and "time" as being abstractions which apply to our size level, but which become utterly meaningless out past the thirtieth decimal place. What would be there instead? Our old friend the *apeiron*. But even if we cannot really speak of infinitely many space locations, we might hope to find infinitely many sorts of particle.

It is sometimes thought that quantum mechanics *proves* that there is a smallest size of particle that could exist. This is not true. Quantum me-

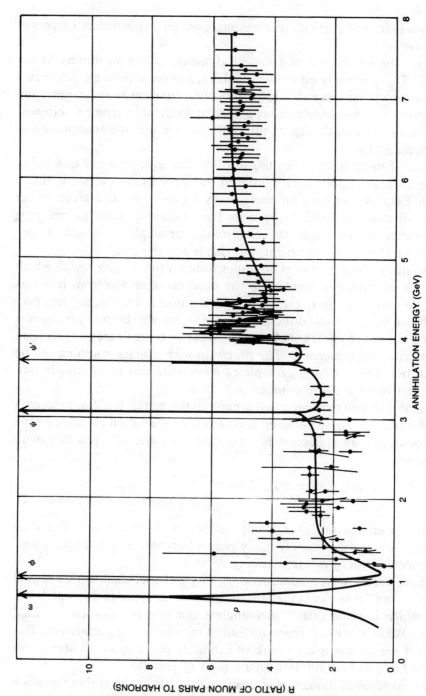

Figure 23. From *Scientific American*, Oct. 77, p. 65.

chanics insists only that in order to "see" very small particles, we must use very energetic processes to look for them.

It is illuminating, after all this, to learn how the high-energy physicists actually go about finding new particles. The process is a little like finding stations on the radio by inching the dial back and forth until you hear music instead of static. One uses a particle accelerator in which collisions (between electrons and positrons) are continually taking place. The energy of the collision processes is varied by turning the voltage on the accelerator up and down. There is number R that measures the "particleness" of the reaction taking place. R can be thought of as being a little like the information parameter that enables you to tell whether you have found a station, even though the sound of music is no louder than the sound of the static. When an energy is found at which the graph of R versus energy has a sudden peak, then it is assumed that the energy in question is characteristic of the rest-mass of a new particle. This process is called "bump-hunting." It is interesting to note that the sharper and narrower the peak, the more long-lived, and, thus, more "real" the particle is.

The question of whether or not matter is infinitely divisible may never be decided. For whenever an allegedly minimal particle is exhibited, there will be those who claim that if a high enough energy were available, the particle could be decomposed; and whenever someone wishes to claim that matter is infinitely divisible, there will be some smallest known particle which cannot be split. One is almost tempted to doubt if the question of the infinite divisibility of matter has any real meaning at all, particularly in view of the fact that such concepts as "matter" and "space" have no real meaning in the micro-world of quantum mechanics.

To return to something a little more concrete, let us consider the divisibility of our perceptual field. There is a limit to the subdivisions that this field can undergo. If two clicks happen close enough together in time, they cannot be distinguished; if a spot of ink is small enough, we can no longer see it. Hume makes much of this fact in his *Treatise of Human Nature* of 1739:

> Put a spot of ink upon paper, fix your eye upon that spot, and retire to such a distance, that at last you lose sight of it; 'tis plain, that the moment before it vanish'd the image or impression was perfectly indivisible.[23]

The best way to understand Hume's view of the world is to regard our space-time as being supplemented by an additional dimension of

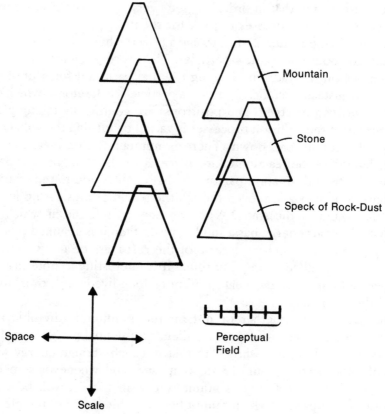

Figure 24.

scale. To represent what I have in mind, let us forget about time and drop all the space dimensions but one. In Figure 24 I have drawn the space-scale continuum for a one-dimensional world. An individual's perceptual field has a certain fixed size, as drawn; the field is made up of a certain finite number of slots or tiles—minimal perceptual units. In this model, the one-dimensional creature has two dimensions in which he can move his perceptual field. He can move to the left and right in space, and he can enlarge and contract his perceptual field. Rather than thinking of the field as enlarging and contracting, we think of the field moving up and down on the scale axis.

If the labelled objects (mountain, stone, speck of rock dust) occupy the appropriate regions of the space-scale continuum, then we can think of the ordinary perceptual level as being when the field is placed somewhere in the middle of the picture. At this perceptual level stones are

visible, but one has neither enlarged one's field of vision enough to see
the mountain as a single object, nor contracted one's attention enough
to see the specks of dust on the rock. Notice that changing the size of
one's perceptual field amounts just to moving this field about in the
space-scale continuum.

Hume takes perceptions as primary. Although he is often thought of
as an empiricist, his is actually an extremely idealistic viewpoint. The
perceptions are "out there"; one's consciousness seems to move among
them like a butterfly flitting from flower to flower.

One's perceptual field has minimal elements, yet these minimal elements can be resolved into smaller elements by altering one's field (by
paying closer attention, using a telescope, or moving closer to the object in question). The only way to reconcile these two apparently contradictory aspects of our perceptual world is to view the world as a five-
dimensional, space-time-scale continuum.

The question of the existence of an infinity in the small now becomes
the question of whether or not the space-scale continuum drawn in Figure 24 extends *downward* indefinitely; similarly, the question of the existence of infinity in the large is the question of whether or not the continuum extends *upward* indefinitely.

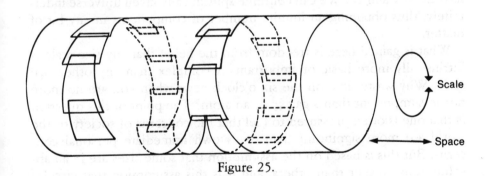

Figure 25.

I have long been interested in a curious trick that eliminates the infinity in the large and the infinity in the small without introducing any *absolute* perceptual minimum or maximum. This is simply the trick of bending the space-scale diagram into a tube, by turning the scale axis into a
circle. Here the universe could consist of many galaxies, which consist
of many star systems, which consist of many planets, which consist of
many rocks, which consist of many molecules, which consist of many
atoms, which consist of many elementary particles, which consist of

many quarks and leptons, which consist of many darks, which could consist of many universes.[24]

A problem with the circular scale model is that if our universe is broken down far enough, one gets many universes, each of which will break down into many more universes. Are all of these universes the same? Perhaps, but then it would be hard to see how there could really be more than one object in the world. Another difficulty is that if there are many universes, each of which breaks up into many more universes, how can each of the component universes be one of the starting universes?

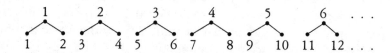

There is no problem if we have infinitely many universes. To illustrate this, I have drawn a picture of the simplest case: the case in which each universe is made up of two universes. We can see that 1 splits into 1 and 2, 2 splits into 3 and 4, 3 splits into 5 and 6, and in general n splits into $2n - 1$ and $2n$. We can continue splitting any given universe indefinitely, thus obtaining an infinite number of components in any bit of matter.

What is gained here is freedom from the belief that any size scale is intrinsically more basic or important or complex than any other size scale. Why waste time on the six o'clock news when you are no more nor less important than a galaxy or an atom? The point of this question is that one is often pressured to feel that the concerns of society or the world are more significant than one's own immediate personal concerns. But this is based on the assumption that some sizes are in an absolute sense bigger than others, and it is this assumption that circular scale undermines.

CONCLUSION

In conclusion, note that it is entirely possible that our universe is in every sense finite. A toroidal space-time of the sort mentioned in the section on temporal infinities eliminates all infinities in the large; and if circular scale is introduced as in the section on infinities in the small, then there are no discrete infinities in the small. These finitizations can be accomplished smoothly: there need be no end of time, edge of space, or smallest particle.

But it is hard to believe that there would be only *one* of these totally finite universes. First, it is difficult to see how to apply circular scale unproblematically unless there are infinitely many universes; second, the principle of sufficient reason is violated if only *this* particular finite universe exists; and third, there is the feeling that the "space" in which our space-time is curved should be real.

In the section on spatial infinities it was pointed out that if, on the one hand, one repeatedly finitizes by replacing lines with circles, and if, on the other hand, one never accepts some particular finite *n*-verse as the end of the line–if, in other words, one thinks along the lines sketched in the last two paragraphs–then one is forced to conclude that space is infinite dimensional and that there are infinitely many objects in this cosmic space.

INFINITIES IN THE MINDSCAPE

In the last section I discussed some of the ways in which an actual infinity could physically arise. But there are things that are not physical. There are minds, thoughts, ideas, and forms. In this section we will see if any of these familiar nonphysical entities are actually infinite.

In order to appreciate the section at hand, it is necessary to keep an open mind on the question of whether or not mind equals brain, for if one assumes *a priori* that a thought is nothing more than a certain biochemical configuration in a certain finite region of matter, then (unless one has infinite divisibility of matter) it seems to follow automatically that infinite thoughts are impossible.

To cast a few preliminary doubts on the hypothesis that brain equals mind, let me quickly raise a few questions. Is what you thought yesterday still part of your mind? If you own and use an encyclopedia, are the facts in that encyclopedia part of your mind? Does a dream which you never remember really exist? How can you grasp a book as a whole, even though you only read it a word at a time? Would the truths of mathematics still exist if the universe disappeared? Did the Pythagorean theorem exist before Pythagoras? If three people see the same animal, we say the animal is real; what if three people see the same idea?

I think of consciousness as a point, an "eye," that moves about in a sort of mental space. All thoughts are already there in this multi-dimensional space, which we might as well call the Mindscape. Our bodies move about in the physical space called the Universe; our consciousnesses move about in the mental space called the Mindscape.

Just as we all share the same Universe, we all share the same Mindscape. For just as you can physically occupy the same position in the Universe that anyone else does, you can, in principle, mentally occupy the same state of mind or position in the Mindscape that anyone else does. It is, of course, difficult to show someone exactly how to see things your way, but all of mankind's cultural heritage attests that this is not impossible.

Just as a rock is already in the Universe, whether or not someone is handling it, an idea is already in the Mindscape, whether or not someone is thinking it. A person who does mathematical research, writes stories, or meditates is an explorer of the Mindscape in much the same way that Armstrong, Livingstone, or Cousteau are explorers of the physical features of our Universe. The rocks on the Moon were there before the lunar module landed; and all the possible thoughts are already out there in the Mindscape.

The mind of an individual would seem to be analogous to the room or to the neighborhood in which that person lives. One is never in touch with the whole Universe through one's physical perceptions, and it is doubtful whether one's mind is ever able to fill the entire Mindscape.

One last analogy. Note that there is always a certain region of physical space that only I can ordinarily know of—barring surgery, no one but me is in a position to assess the physical conditions obtaining within my stomach. In the same way, there is a certain part of the Mindscape that only I can ordinarily know of—unless I am to be greatly favored by the Muse, the feelings that pass over me when I think of my childhood will always remain private and inexpressible. Nevertheless, these almost ineffable feelings are part of the common Mindscape—they are simply difficult for anyone else to get to.

The point of all this is that just as the finiteness of our physical bodies does not imply that every physical object is finite, the finiteness of the number of cells in our brains does not mean that every mental object is finite.

Well . . . *are* there any infinite minds, thoughts, ideas, or forms or what have you in the Mindscape?

The most familiar candidate is the set N of all natural numbers. If I try to exhibit N, all I can really do is show you something like this: $N = \{1, 2, 3, \ldots\}$. What the "..." stands for is something that is evident, yet basically inexpressible. The idea, of course, is that all of the natural numbers are to be collected together into a whole. Each of them would seem to exist individually in the Mindscape, and one would suppose that the set consisting of exactly the natural numbers would be in the Mindscape as well—one almost feels as if one can *see* it.

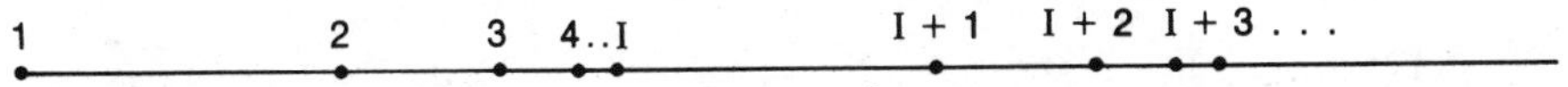

Figure 26.

We might try to avoid the use of the "..." by saying something like this: "N is the set that has the following property: one is in N, and for any number x that is in N, x plus one is in N as well." The trouble with this definition is that it does not uniquely single out one particular set. If, for instance, there were some infinitely large number I, and if N^* were the set consisting of all the numbers in N and all the numbers of the form $I + n$ for some n in N, then N^* would satisfy the property that for every x in N^*, x plus one is in N^* as well ... but N^* would be different from N.

We might try to get around this difficulty by saying that N is the *smallest set in the Mindscape* that has one in it, and that has x plus one whenever it has x. But, for reasons that I will begin to explain in the next section, the word "Mindscape" cannot be meaningfully used in a definition. The concept of "Mindscape" is too vast to be represented by any word or symbol.

If we try to avoid *this* difficulty by substituting some sort of finite description of the mental universe for the word "Mindscape," then we get the same problem as before. By the classic work of the logician Thoralf Skolem, we know that for *any* finite description of N one might come up with, there will be a different set N^* that also satisfies the description. So it is quite literally true that what is really meant by the "..." is inexpressible.

Some thinkers have taken this to mean that there is, after all, no unique N in the Mindscape. This could be true. But one need not take this to mean that there are no infinite sets in the Mindscape: if there are many, many versions of the set of natural numbers, then there are many, many infinite sets. However, it is normally more desirable to assume that there is a simple unique N in the Mindscape, just as it is simpler to assume that there is only one universe instead of a whole slew of "parallel worlds."

I might note here that if time is indeed infinite, then just as we can indicate Earth by saying, "this planet," we could indicate our N by saying, "the number of seconds left in this time." This is, in fact, what people do when they attempt to define N by saying, "N is what you get if you start with one and keep adding ones forever."

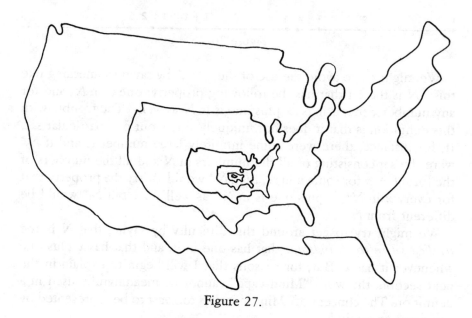

Figure 27.

If infinite forms are actually out there in the Mindscape, then maybe we can, by some strange trick of mental perspective, *see* some of these forms. The philosopher Josiah Royce maintained that a person's mental image of his own mind must be infinite.[25] His reason is that one's image of one's own mind is itself an item present in the mind. So the image includes an image that includes an image, and so on. This infinite regress can be nicely visualized by imagining a United States in which a vast and fanatically accurate scale model of the country occupies most of the Midwest. The scale model, being absolutely accurate, includes a copy of the scale model, etc. This regress is occasionally used to make a striking label for a commercial product. The old can of Pet Milk, for instance, bore a picture of a can of Pet Milk, which bore a picture of a can of Pet Milk, etc.

In a physical situation we would probably never actually be able to finish making such a label in all its infinite detail. But this is not to say that no such label or country-plus-scale-model could *exist*. There would be no problem, if matter were infinitely divisible. (If scale is indeed circular, then everything is, in a sense, already an object of this nature!)

There is certainly no reason why a nonphysical mind should not be infinite; and Royce's point is that if you believe that one of the things present in your mind is a perfect image of this mind and its contents, then your mind *is* infinite. One might try to avoid this conclusion by adopting a circular scale attitude and insisting that there is no difference between the mind and the mind's image of itself, so that the allegedly infinite set of thoughts {image of the mind, image of the image, image of the image of the image, . . . } is really the same as the set {mind, mind, mind, . . . }, which is just a set with one member: {mind}.

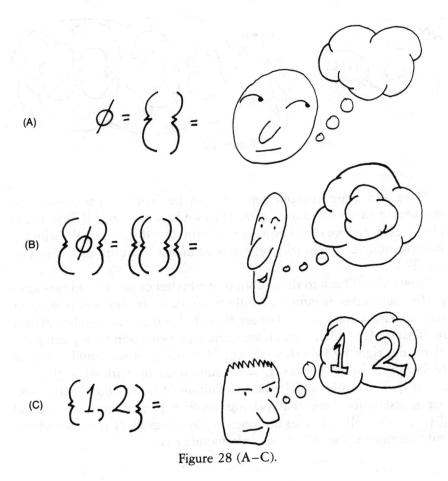

Figure 28 (A–C).

I would like to discuss this a bit more, but first let me formally introduce some of the apparatus of set theory. In Cantor's words, "A set is a Many that allows itself to be thought of as a One."[26] A set is usually given as a pair of curly brackets enclosing some description of the contents of the set. It is easiest to think of the curly brackets as a thought balloon. Thus the set $\{1, 2\}$ is the unity obtained by taking the multiplicity consisting of the numbers 1 and 2 and treating this multiplicity as a unity. That is, we can think of the set $\{1, 2\}$ as being represented by a thought balloon that has 1 in it and 2 in it.

Of particular interest in set theory is the *empty set,* $\varnothing$. $\varnothing$ is the One obtained by taking together . . . nothing. If we write out $\varnothing$ in the ordinary way we get $\{\ \}$, which I have drawn as an empty thought balloon.

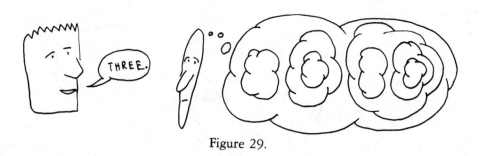

Figure 29.

More and more complicated sets can be built up using only the brackets in various arrangements. Thus we have the set $\{\{\ \}\}$ depicted in Figure 28B, and we could equally well form $\{\{\ \}, \{\{\ \}\}, \{\{\ \}, \{\ \}\}\}$ which is how the number 3 is usually represented in terms of pure sets. (See Figure 29.)

Now let's get back to the question of whether or not a mind that has a perfect self-image is infinite. Really to get down to the bare bones, say that we have a mind or label or set M such that the only member of M is M. That is, $M = \{M\}$. Now, if we change this equation by replacing the M on the right by $\{M\}$ then we get $M = \{\{M\}\}$. If we could continue replacing M by $\{M\}$ forever, we would wind up with $M = \{\{\{\{\{ . . . \ . . . \}\}\}\}\}$. This could actually be a definition of an M whose only member is itself, for note that placing another pair of brackets around $\{\{\{\{\{ . . . \ . . . \}\}\}\}\}$ changes nothing. In plain English, M is the set whose only member is the set whose only member is . . .

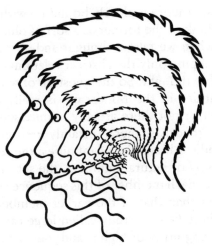

Figure 30. Based on a drawing from Robert Crumb, *Your Hytone Comix* (San Francisco: The Print Mint, 1976).

But if the only member of M is indeed M itself (rather than a copy of M), then M really only has one element. It is just that if we try to describe this element by using brackets we get an infinite description. We call thoughts like M *self-representative.* Whether or not such an M is to be regarded as infinite depends on whether you experience the M subjectively (in the way you experience your own mind), or objectively (as a feature of the Mindscape that is to be precisely described in the language of set theory).

Set theory is, indeed, the science of the Mindscape. *A set is the form of a possible thought.* Set theory enables us to put various facts about the Mindscape into one framework in the same way that the atomic theory of matter provides a framework in which the diverse physical and chemical qualities of matter can be simultaneously accommodated.

Before the atomic theory of matter, such phenomena as melting and burning, rusting and freezing were regarded as qualitatively different. Once a good atomic theory was developed, however, all of these phenomena could be thought of in more or less the same way. The notion of set was consciously introduced only at the turn of the century. Before long, it became evident that all of the objects that mathematicians discuss—functions, graphs, integrals, groups, spaces, relations, sequences—all can be represented as sets. One can go so far as to say that mathematics is the study of certain features of the universe of set theory.

The universe of set theory is closely bound up with the Mindscape—one can, perhaps, think of the former as a sort of blueprint of the latter. A set is obtained when we take a thought and abstract from it all the emotive content, keeping only the abstract relational structure. A set is the form of a possible thought. So the question of whether or not there are any infinite entities in the Mindscape is really equivalent to the question of whether or not there are any infinite sets.

According to set theorists, there certainly *are* infinite sets. Indeed, there is to be an endless hierarchy of infinities: the set of natural numbers, the set of all sets of natural numbers, the set of all sets of sets of natural numbers, etc. Each member of this sequence can be shown to be of an infinity greater than that of the earlier members. In modern set theory there is a whole field of study called large cardinals, whose specialists study a dizzying array of higher and higher infinities.

But many mathematicians and philosophers do not go along with the set theorists. The traditional *finitist* viewpoint is still with us. According to the finitists, there is nothing that is infinite, in heaven or on earth.

Those who assert that infinite sets of every size have a secure existence in the Mindscape are usually called *Platonists*. This name is a bit inapt, since Plato did not believe in infinity; but he *did* believe in the existence of ideas independent of thinkers, and it is for this aspect of his thought that the Platonists are named.

It is not likely that the finitist vs. Platonist debate will ever be concluded. On the one hand, it is probably impossible to meet the demands of a finitist who says that he will believe in infinity only if he is *shown* an infinite set right now; on the other hand, the notion of infinite sets appears to be logically consistent, so the finitist can never prove that infinite sets do not exist.

I incline towards Platonism; but if you are stubborn enough, how can I possibly convince you that infinite things are real? All I can do, after all, is to make a finite number of marks on a finite number of sheets of paper. If you are truly committed to disbelief in the infinite, then you will not be satisfied by anything less than my simultaneously exhibiting each member of some infinite set . . . and whenever I claim that I have done so, you will triumphantly point at the finiteness of the number of marks on paper which I have really shown you.

In pre-Cantorian times finitists sometimes thought that they had proved the impossibility of actually infinite sets. These proofs, however, were always fallacious. Such proofs usually deal with some particular property P of numbers that each natural number happens to enjoy.

P might be the property of being odd or even, having an immediate predecessor, being the sum of finitely many units, or being greater than any predecessor. The false proof that no infinite numbers exist then takes the form: "Every number has property *P*. If *x* is an infinite number, then *x* cannot have property *P*. Therefore no infinite numbers can exist." The fallacy in such a circular proof is that when it is asserted that "every number has property *P*," it is being quietly assumed that anything that fails to have property *P* does not exist.

But, of course, one cannot assume that the infinite sets must have certain properties before one has ever looked at them! Galileo's paradox, for example, showed that an infinite set can be put into a one-to-one correspondence with a proper subset of itself. Had we assumed in advance that no set could be put into a one-to-one correspondence with a proper subset of itself, then we would have had a proof that no infinite set can exist. But such an assumption is *totally unwarranted;* indeed, to make such an assumption is essentially to assume in advance that every set is finite . . . which does not make for a very productive debate.

But are we quite sure that the finitists will never come up with some *valid* proof that the notion of infinite sets is incoherent and fundamentally meaningless? A Platonist would answer that yes, he is sure that there is no inconsistency in the theory of infinite sets. He is sure of this because the theory in question is a description of certain features of the Mindscape that "anyone can see."

But the finitist can still hope. There is a curious proof, discovered by Kurt Gödel in 1930, that the consistency of set theory cannot be finitely proved. The time will never come when the finitist is absolutely forced to admit that it is safe to talk about infinite sets.

In mathematics no other subject has led to more polemics than the issue of the existence or nonexistence of mathematical infinities. We will return to some of these polemics in the last chapter. For now, let us reprint Cantor's opening salvo in the modern phase of this age-old debate:

> The fear of infinity is a form of myopia that destroys the possibility of seeing the actual infinite, even though it in its highest form has created and sustains us, and in its secondary transfinite forms occurs all around us and even inhabits our minds.[27]

Strong words! But what does Cantor mean when he says that the highest form of infinity created us? Read on!

THE ABSOLUTE INFINITE

There is a certain type of nonphysical entity that was not discussed in the last section. God, the Cosmos, the Mindscape, and the class V of all sets—all of these are versions of what philosophers call the Absolute. The word "Absolute" is used here in the sense of "non-relative, non-subjective." An Absolute exists by itself, and in the highest possible degree of completeness.

As I mentioned earlier, Plotinus held that the One could not be limited in any sense. As Aquinas, the quintessential theologian, says: "The notion of form is most fully realized in existence itself. And in God existence is not acquired by anything, but God is existence itself subsistent. It is clear, then, that God himself is both limitless and perfect."[28]

The limitlessness of God is expressed in a form closer to the mathematical infinite by St. Gregory: "No matter how far our mind may have progressed in the contemplation of God, it does not attain to what He is, but to what is beneath Him."[29] We have here the rudiments of the infinite dialectic process that takes place if we systematically try to build up an image of the whole Mindscape.

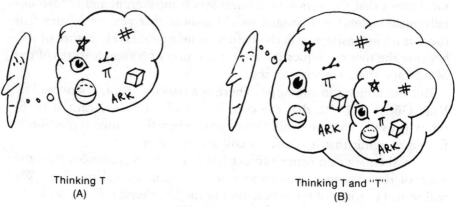

Thinking T
(A)

Thinking T and "T"
(B)

Figure 31 (A–B).

Suppose that I want to add thought after thought to my mind until my mind fills the whole Mindscape. Whenever I make an attempt at this, I am collecting together a group of thoughts into a single thought T. Now, when I become conscious of my state of mind T, I realize that this is a new thought that I had not yet accounted for . . . so I improve my image of the Mindscape by passing to the thought that includes all the elements of T plus T *itself,* viewed objectively.

This is a dialectic process in the sense that the thetic component is one's instantaneous unconscious image of the Absolute, the antithetic component is the conscious formalization of this image, and the synthetic component is the formation of a new unconscious image of the Absolute that incorporates one's earlier images and the awareness that they are inadequate.[30]

This process is most clearly understood if we start with nothing at all, as in the cartoon strip of Wheelie Willie in Figure 32. (Wheelie Willie is a character whose adventures I occasionally used to draw for the *Rutgers Daily Targum* when I was in graduate school there.) Notice that in each of the shifts, what takes place is that Wheelie Willie forms a thought that has as its members the members of the last thought plus the last thought itself. Looked at another way, the thought at each stage has all of the previous stages as components.

If we call the nth thought T_n, we can define T_n in two ways. On the one hand, we can use an inductive definition: $T_0 = \varnothing$ and $T_{n+1} = T_n \cup \{T_n\}$, where for any sets A and B, $A \cup B$ means the set of all the sets that are members of A or of B. On the other hand, we can use a different sort of inductive definition: $T_n = \{T_m : m < n\}$, which means "T_n is the set of all T_m such that m is less than n."

Some readers may have asked themselves if the thought T plus "T" really has to be different from the thought T. And the answer is, not always. In the last section we were looking at a mind, M, which has M as one of its components. Such an M is already fully self-aware, and M plus "M" is no different from M. In terms of sets, $M \cup \{M\} = M$.

It would seem, in particular, that God should be able to form a precise mental image of Himself. Insofar as the Mindscape is God's mind, what I am saying is that one of the objects in the Mindscape should be the Mindscape itself. That is, the Mindscape is an M that has M as one of its members. Now, any object in the Mindscape is, in principle, something that one can perceive through one's consciousness. So it would seem to be possible for our minds actually to attain a vision of God or of the whole Mindscape.

Now this seems to contradict St. Gregory's dictum and the general feeling that the Absolute is unknowable. But there are two *kinds* of knowing: the rational and the mystical.

If I know something rationally, then I have some thought that is built up from simpler thoughts, which are in turn built up from still simpler thoughts. This regress is not infinite, but goes only through some finite number of stages before certain simple and unanalyzable perceptions

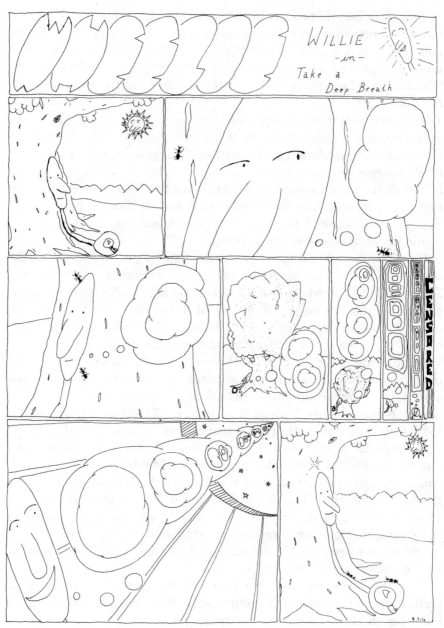

Figure 32.

and ideas are reached. My idea of "house" consists of a collection of ideas, each one of which represents a certain type of house (e.g., my house, brick house, hovel). Each idea of a type of house consists of ideas of various components and functions (doors, windows, shelter), which can in turn be explicated in terms of certain simple ideas (walking, vision, warmth).

When I communicate a rational thought, what I do first is to show what the components of my thought are, and then to show how the components fit together. If one of the components of the final thought were to be the final thought itself, then this rational communication would be blocked by an infinite regress. To explain the thought, I would first have to explain the thought. I could not finish unless I had already finished.

In terms of rational thoughts, the Absolute is unthinkable. There is no non-circular way to reach it from below. Any real knowledge of the Absolute must be mystical, if indeed such a thing as mystical knowledge is possible.

Mathematics and philosophy do not normally have a great deal to say about the mystical way of knowing things. Mystically speaking, it is possible to experience a direct vision of the whole Mindscape. This vision cannot be rationally communicated for the reasons just outlined. Of course, it is possible to communicate mystical knowledge in an indirect way, for example, by advocating that a person prepare his or her mind through carrying out some physical or spiritual exercises. But, ultimately, mystical knowledge is attained all at once or not at all. There is no gradual path by which to build up an M that has M as one of its elements.

Even if full knowledge of the Absolute is only possible through mysticism, it is still possible and worthwhile to discuss *partial* knowledge of the Absolute rationally. A significant thing about the Mindscape and the other Absolutes is that they are actually infinite. Indeed, in 1887 Cantor's friend, Richard Dedekind, published a proof that the Mindscape is infinite, where Dedekind's word for Mindscape was *Gedankenwelt,* meaning thought-world.[31]

Dedekind's argument for the infinitude of the Mindscape was that if s is a thought, then so is "s is a possible thought," so that if s is some rational non-self-representative thought, then each member of the infinite sequence $\{s, s$ is a possible thought, s is a possible thought is a possible thought, . . .$\}$ will be in the Mindscape, which must, therefore, be infinite.

A very similar argument proves that the class of all sets is infinite. The

class of all sets is normally called **V**, or Cantor's Absolute. We can use
the Wheelie Willie sequence of sets to see that there are infinitely many
different sets in **V**.

Dedekind modelled his argument after an argument that appears in
Bernard Bolzano's *Paradoxes of the Infinite* (ca. 1840):

> "The class of all true propositions is easily seen to be infinite. For if we fix
> our attention upon any truth taken at random . . . , and label it A, we
> find that the proposition conveyed by the words 'A is true' is distinct
> from the proposition A itself. . ."[32]

So we can see that the Mindscape, the class of all sets, and the class of
all true propositions are all infinite. Does this guarantee that infinite objects exist? Not really. For a case can be made for the pluralist claim that
the Mindscape, the class of all sets, and the class of all true propositions
do *not* exist as objects, as unities, as finished things.

In more familiar terms, it is not hard to prove that God is infinite . . . but what if you don't believe that God exists? It may seem
hard to doubt that the more impersonal Absolutes—such as "everything," or the Mindscape—exist, but there are those who do doubt this.
The issue under consideration is a version of the old philosophical
problem of the One and the Many. What is being asked is whether the
cosmos exists as an organic One, or merely as a Many with no essential
coherence. It is certainly true that the Mindscape, for instance, does not
exist as a single rational thought. For if the Mindscape is a One, then it
is a member of itself, and thus can only be known through a flash of
mystical vision. No rational thought is a member of itself, so no rational
thought could tie the Mindscape into a One.

Normally the word "set" is restricted by definition to apply only to
collections that are not members of themselves. Under this use of the
word, the class **V** of all sets cannot be a set, for if it were, we would have
a set **V** such that **V** is a member of itself. So **V** becomes a collection that
can never be formed into a One.

Suppose that we do not believe in circular scale and assume that any
physical thing is not a part or component of itself. Is the Cosmos, the
collection of all physical things, a thing? If it is, then it has to be a component of itself, which we do not allow. So the Cosmos is not a thing,
but only a Many that can never be a One.

There is a highly relevant passage in a letter Cantor wrote to Dedekind in 1905:

> "A multiplicity can be such that the assumption that *all* its elements 'are
> together' leads to a contradiction, so that it is impossible to conceive of

the multiplicity as a unity, as 'one finished thing.' Such multiplicities I call *absolutely infinite* or *inconsistent multiplicities*. As we can readily see, the 'totality of everything thinkable,' for example, is such a multiplicity. . ."[33]

Again, the reason that it would be a contradiction if the collection of all rational thoughts were a rational thought T is that then T would be a member of itself, violating the rationality of T (where "rational" means non-self-representative). The upshot of all this is that God, the Mindscape, the class of all sets, and the class of all true propositions all seem to be infinite, but it is at least possible to question whether any of these Absolutes exists as a single entity. Certainly they do not exist as entities that can be fully grasped by the rational mind.

CONNECTIONS

In this section I would like to explore some of the connections between the various sorts of infinities that have been discussed.[34] In his 1887 essay, "Contributions to the Study of the Transfinite," Cantor quotes a passage from Aquinas's *Summa* and states repeatedly that in this passage appear the only two really significant objections that have ever been raised against the actual infinite.[35] Let us examine this quote from Aquinas here, reproducing Cantor's italics:

> The existence of an actually infinite multitude is impossible. 1) For any set of things one considers must be a *specific* set. *And sets of things are specified by the number of things in them. Now no number is infinite,* for number results from counting through a set in units. So no set of things can actually be inherently unlimited, nor can it happen to be unlimited. 2) Again, every set of things existing in the world has been created, and anything created is *subject to some definite purpose* of its creator, *for causes never act to no purpose.* All created things must be subject therefore to definite enumeration. Thus even a number of things that happens to be unlimited cannot actually exist.[36]

It seems clear that Aquinas's first point is that an infinite set can occur only if infinite numbers exist, and he does not believe that infinite numbers exist. Cantor's theory of transfinite numbers stands as the only adequate response to this objection. For many years, it was believed that the notion of actually infinite numbers was fundamentally incoherent. It was only with the birth of Cantor's theory in the late 1800s that a consistent and reasonable theory of infinite, or transfinite, numbers was developed. As Cantor remarks in his discussion of Aquinas's objection, this objection against the existence of actually infinite collections is to be met *positively* by exhibiting a theory of infinite numbers.

It is not so obvious what Acquinas's second point might be. It might be taken to be simply a variation on the first point. Under this reading, the first point says that any set must have a number of cardinality, but all numbers are finite; and the second point says that any set must have a purpose or significance, but any definite purpose is finite. If this is indeed Aquinas's meaning, then we can say that once again the Cantorian theory of infinite sets provides a positive rebuttal.

Aquinas's whole view of the infinite is not really tenable, for he held that God is infinite, but that no created thing is infinite. This contradicts a widely accepted principle known as the *Reflection Principle*. The Reflection Principle as formulated in set theory goes as follows: every conceivable property that is enjoyed by V is also enjoyed by some set. (Recall here that V is Cantor's Absolute, the class of all sets.) Philosophically it would run: every conceivable property of the Absolute is shared by some lesser entity; or, every conceivable property of the Mindscape is also a property of some possible thought.

The motivation behind the Reflection Principle is that the Absolute should be totally inconceivable. Now, if there is some conceivable property P such that the Absolute is the only thing having property P, then I can conceive of the Absolute as "the only thing with property P." The Reflection Principle prevents this from happening by asserting that whenever I conceive of some very powerful property P, then the first thing I come up with that satisfies P will *not* be the Absolute, but will instead be some smallish rational thought that just happens to reflect the facet of the Absolute that is expressed by saying it has property P.

Let me give an example of a Reflection Principle argument. *For every thought S in the Mindscape, the thought "S is a possible thought" is also a thought in the Mindscape.* By Reflection there must, therefore, be some thought W such that For *every thought S in W, the thought "S is a possible thought" is also in W.* This W reflects, or shares, the italicized property of the Mindscape. But note now that this W must be infinite. So an infinite thought exists.

Again, it is true that each of the Wheelie Willie sets T_n is a member of V. By the Reflection Principle there must, therefore, be some set N such that each of the Wheelie Willie sets T_n is a member of N. Therefore an infinite set N exists.

The point I wish to make is that if one accepts the existence of any of the various infinite Absolutes, then one is fairly well committed to accepting the existence of infinite thoughts and sets. For to deny the Reflection Principle is practically to assert that the Absolute can be finitely described, which is most unreasonable.

The passage from St. Augustine that I referred to earlier contains a kind of Reflection Principle argument for the reality of the set N of all natural numbers. In that passage Augustine argues that God must already know each and every natural number and that he even knows "infiniteness" in the form of all the natural numbers taken at once—for otherwise the set of natural numbers would exhaust his abilities. God, according to Augustine, must lie *beyond* the set of natural numbers.

To summarize the points in this chapter:

1. The infinite normally inspires such feelings of helplessness, fatility, and despair that the natural human impulse is to reject it out of hand.
2. There are, however, no conclusive proofs that everything is finite; and the question of whether or not anything infinite exists remains as an open, almost empirical problem.
3. There are various sorts of physical infinites that could actually exist: infinite time, infinitely large space, infinite dimensional space, infinitely continuous space, and infinitely divisible matter. Each of these infinites is, in principle, avoidable; whether or not our Cosmos actually does avoid infinities remains to be seen.
4. In Cantor's set theory we have a great number of infinite sets. This simple and coherent theory of the infinite provides a logical framework in which to discuss infinities. Moreover, if we feel that the things that mathematicians discuss are real, then we can conclude that actually infinite things exist.
5. Attempts to analyze the phenomenon of consciousness and self-awareness rationally appear to lead to infinite regresses. This seems to indicate that consciousness is essentially infinite.
6. The Absolute is certainly infinite. So one must either deny the reality of the Absolute or accept the existence of at least one infinity.
7. According to the Reflection Principle, once one has an infinite Absolute, one must also have many conceivable infinities as well.

PUZZLES AND PARADOXES
(*Answers on Page 295*)

1. It is sometimes said that if infinitely many planets existed, then every possible planet would have to exist, including, for instance, a planet exactly like Earth, except with unicorns. Is this necessarily true?

2. Consider a very durable ceiling lamp that has an on-off pull string. Say that the string is to be pulled at noon every day, for the rest of time. If the lamp starts out off, will it be on or off after an infinite number of days have passed?

3. For each observer O, there is some fixed upper bound N_O to the number of stars that O can physically see. Therefore, for each observer the universe is finite. Does this imply that the universe is finite?

4. "I have five fingers on my left hand," means the same thing as, "When I count up all the fingers on my left hand, the last number I say is *five*." What might "I have ∞ fingers on my left hand" mean?

5. Suppose that we find an infinite number I that is the largest possible number. But now, what about $I + 1$?

6. In the little-known field of "enumerative geometry," it is said that there are ∞ points on a line and ∞^2 points in a plane. There are said to be ∞^2 *lines* in the plane as well: "To get the correct number ∞^2 of straight lines in the plane, we must divide the number ∞^4 of pairs of points in the plane by the number ∞^2 of pairs of points on a straight line."[37] How many circles should there be in the plane? How many ellipses?

7. Can you prove, without circularity, that seven is a finite number?

8. The universe has lasted about 10^{10} years since the Big Bang. There are about 3×10^7 seconds in a year. According to quantum mechanics, the usual conception of continuous time does not extend to intervals shorter than 5×10^{-44} seconds, so we might think of this unit as being a kind of "instant," faster than which nothing can happen. How many "instants" of time does that come to so far? Is it reasonable to argue that larger numbers, such as 10^{100}, do not yet exist?

9. Say that the space we live in is infinitely large. Consider an infinite line L contained in our space. L is infinity yards long, and L is infinity feet long. But since each yard is three feet, L is also three-times-infinity feet long. How can infinity equal three times infinity?[38]

10. Here is an example of an infinite regress. Suppose that some person wishes to prepare a text in which every appearance of the letters "man" is replaced by the letters "woman." If this is rigidly adhered to, then "man and woman" becomes "woman and wowoman," then "wowoman and wowowoman," and so on. what do you reach in the limit?

CHAPTER TWO
ALL THE NUMBERS

In this chapter we will begin by tracing the development of the familiar real number system with its infinity of irrational numbers. Once one has accepted irrational numbers, there is really no reason not to accept infinitely large or transfinite numbers. So the second section of the chapter will be devoted to the transfinite ordinals and cardinals. The ordinals form a gappy number sequence somewhat like the natural numbers. It is a natural move to fill in these gaps as densely as possible, just as one fills in the space between, for instance, two and three with rational and real numbers. If we fill in as much as possible we end up with what might be called an *absolutely continuous* ordering. In the section on infinitesmals and surreal numbers I will present some examples of such orderings, any one of which can be viewed as comprising "all the numbers" (including the infinitesmals). In the final section of this chapter I will return, once again, to the question of whether the infinitely big and infinitely small numbers have any real existence, physically or otherwise.

FROM PYTHAGOREANISM TO CANTORISM

Pythagoras lived in Greece and in Italy in the sixth century B.C. He is an extremely shadowy and ambivalent figure. On the one hand, he was a wizard, the shamanistic leader of a religious sect. On the other, he has frequently been credited with bringing about the birth of modern mathematics and mathematical physics.

The sect of Pythagoreans is best known for their belief in *metempsychosis,* or reincarnation. They believed that there is one cosmic mind or soul, that you are alive because a small piece of this soul is imprisoned in your body, and that the bit of soul that animates you will animate many other bodies before returning to full unity with the one big soul. The Pythagoreans adhered to a great number of rules and taboos (never look back when crossing a border, always put the right shoe on first, never pick up food that drops from the table), apparently in an effort to bring themselves into a closer harmony with the cosmos. Presumably it was hoped that if in the course of your lifetime you could bring yourself into a close enough relationship with the One, then when your body died, the soul that vivified it might be able to return to the source instead of being forced into another body.

Pythagoras was said to be able to remember several of his previous lives, and he was believed to have many other supernatural powers as well. There is a whole series of ancient miracle tales about Pythagoras, such as stories that he was once seen in widely separated places at the same time, and that once when he was crossing a river it hailed him in an audible voice saying, "Greetings, Pythagoras."

Part and parcel of the Pythagorean religious beliefs were a number of numerological notions. There was a feeling that the essential nature of the cosmos was somehow numerical, with certain numbers seeming to embody particular abstract concepts. The Pythagoreans made the following identifications: 1 was mind (the One); 2 was opinion (the first moving away from unity); 3 represented wholeness (beginning, middle, and end); 4 was justice (a "square" deal); 5 stood for marriage (since 5 = 2 + 3, and even numbers were regarded as female, odd as male). Under a later system the numbers one through four were identified with the point, line, plane, and solid, respectively.

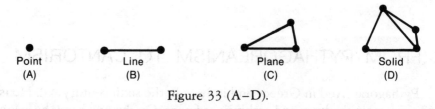

Point   Line   Plane   Solid
(A)   (B)   (C)   (D)

Figure 33 (A–D).

The number ten was singled out for special attention and was said to symbolize perfection. One reason for this is obvious: people have ten

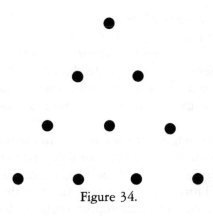

Figure 34.

fingers, and most of our systems of numeration are based on the number ten. But a more important reason for the importance of the number ten is that $10 = 1 + 2 + 3 + 4$, and the numbers one through four and their interrelations were regarded as primary. This fact about 10 was represented by the Pythagorean *tetractys* depicted in Figure 36. A Pythagorean would feel right at home in a bowling alley, ritually building and destroying the *tetractys* with a sphere punctuated by a triad of holes, and recording his progress with a series of numbers inscribed in squares.

The Pythagoreans assumed that since ten was so important, there should be ten heavenly bodies. At the time there were only nine known celestial objects (not counting the stars), so the Pythagoreans postulated the existence of a counter-Earth that is never seen, because it is always on the opposite side of the sun.

It is interesting to realize that this *type* of argument is the stock in trade of modern mathematical physics. For example, a three-dimensional chart of all the known elementary particles is drawn up. The chart looks, let us say, like a regular dodecahedron with one corner missing. It would look prettier, more symmetric, if there were an additional particle with such and such characteristics to fill the missing corner, so the physicists postulate the existence of such a particle. The surprising thing is that, often as not, such an argument turns out to be correct: a particle with exactly the predicted properties is discovered.

The fact is that *a priori* mathematical considerations can lead to empirically determined physical truths. The structure of the physical universe is deeply related to the structure of the mathematical universe. The Pythagoreans were aware of examples of this relationship, having observed, for instance, that if the lengths of two stretched strings are in a simple numerical ratio (such as 2 : 1 or 3 : 2 or 4 : 1), then the notes produced by plucking the strings are consonant.

The conclusion that the Pythagoreans drew was, according to Aristotle, "that the elements of numbers are the elements of things, and that the whole heaven is a harmony and a number." Again, Aristotle states that the Pythagoreans "considered number as the substance of all things."[1]

This sort of viewpoint is not uncongenial to the modern scientist, for whom any phenomenon can be expressed in terms of numbers, vectors, functions, operators, groups, and the like. If one believes that the universe is basically all form and no content and that the forms that arise in nature all admit of mathematical representation, then one can reasonably conclude that anything that exists is ultimately a mathematical object.

Take my right shoe, for example. I can, of course, state the size, count the number of eyelets, or determine the weight in grams. But even independently of my efforts, the shape exists mathematically as the set of coordinates of points that happen to lie within the substance of the shoe; and the color of the shoe is precisely specified by a function giving the wave lengths of light reflected at each point of the surface of the shoe. As for the actual particles that make up the shoe, they may very well be nothing more than small irregularities in the curvature of space-time. So it is not really so odd to believe with the Pythagoreans that ultimate reality is precise mathemetical form.

LIMITED	UNLIMITED
ONE	MANY
REST	MOTION
STRAIGHT	CROOKED
GOOD	BAD

So far, so good. But the story gets more interesting. The Pythagoreans did not believe in infinite forms. They are credited with the creation of a "table of opposites," which I have partially reprinted. Looking at this table, it is evident that the Pythagoreans would not have been big fans of infinity. In the original Greek they have *apeiron* for "unlimited."

Now, if i) everything is a mathematical form and ii) nothing infinite exists, then everything is basically either a natural number or some relationship between a few natural numbers. Note that you have to give up either i) or ii) if it can be proved that there is some feature of the world that cannot, even in principle, be fully represented by a finite number of natural numbers.

Imagine the coast of Southern Italy. You're out on the water
. . . brilliant, ultramarine water and dry, rugged rocks. Pythagoras
himself is aboard the boat on this outing of the brotherhood. There's
been a lot of hassles with the locals, and everything is finally mellow,
kind of *merged* out here on the water. Pythagoras is sitting on the deck
talking with Hippasus, a guy in his thirties who laughs a lot. Hippasus is
scratching lines on the smooth deck with a knife, showing Pythagoras
and some of the others a construction he's been fooling with. It goes
something like this.

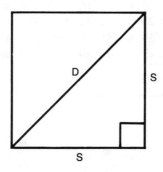

Figure 35.

Once a square and the diagonal of the square have been drawn, one
can ask about the ratio $d:s$ of the length of the diagonal to the length of
the side. If all things are expressible in terms of natural numbers, then
one would expect that there are two natural numbers m and n such that
$d:s::m:n$. *But it can be conclusively proved that no such natural numbers m
and n exist.* The ratio $d:s$ is irrational, nameless, *apeiron*.

In the mental movie we were just watching, Hippasus is letting Pythagoras in on this. One version of the movie's ending is that when the
Pythagoreans returned from their sail, it turned out that Hippasus had
"drowned at sea"!

Given the Pythagorean theorem, which states that the square of the
hypotenuse of a right triangle is equal to the sum of the squares of the
sides, we can see that $d^2 = s^2 + s^2$, so $d^2 = 2s^2$, so $d^2/s^2 = 2$, so the ratio
$d/s = \sqrt{2}$. In modern terms, what the Pythagoreans learned was that
$\sqrt{2}$ is an *irrational* number. But as far as they were concerned, they had
discovered that there was a physical relationship, the ratio $d:s$, which
was not representable in terms of numbers. Since they did not recognize the existence of ratios other than the natural number ratios, the
ratio $d:s$ was called *alogos,* meaning "inexpressible." It was also called
arratos, meaning "not having a ratio."

It is interesting to see how one might go about trying to find a representation of $\sqrt{2}$ as a fraction, or ratio, of natural numbers. This amounts to the problem of finding an n such that for some m, $m^2/n^2 = 2n^2/n^2$. In the table below, I have sketched the beginning of a search for such an n. The curious thing is that we can say with certainty that this search must remain forever fruitless. The proof is covered in almost every survey course of mathematics.

$(2/2)^2 = 4/4 < 8/4 < 9/4 = (3/2)^2,$ so $2/2 < \sqrt{2} < 3/2$
$(4/3)^2 = 16/9 < 18/9 < 25/9 = (5/3)^2,$ so $4/3 < \sqrt{2} < 5/3$
$(5/4)^2 = 25/16 < 32/16 < 36/16 = (6/4)^2,$ so $5/4 < \sqrt{2} < 6/4$
$(7/5)^2 = 49/25 < 50/25 < 64/25 = (8/5)^2$ so $7/5 < \sqrt{2} < 8/5$
$(8/6)^2 = 64/36 < 72/36 < 81/36 = (9/6)^2,$ so $8/6 < \sqrt{2} < 9/6$
$(9/7)^2 = 81/49 < \underbrace{98/49} < 100/49 = (10/7)^2,$ so $9/7 < \sqrt{2} < 10/7$

Continue forever with all
fractions equal to two in this
column.

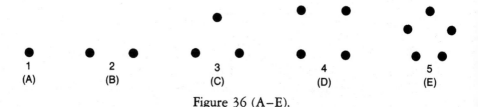

Figure 36 (A–E).

For the Greeks there were two kinds of magnitudes: discrete and continuous. Discrete magnitudes could be counted, set into correspondence with natural numbers that were sometimes visualized as patterns of dots. But continuous magnitudes simply did not correspond to any *number* at all. Just as we can add and multiply numbers, we can manipulate continuous magnitudes by means of the techniques called geometrical algebra. The Greeks developed these techniques to the point where they could, in effect, solve most quadratic equations involving continuous magnitudes.

Consider, for example, the geometric technique for finding the mean proportional between two lengths a and b. That is, given line segments a and b, we wish to find a line segment m such that $a:m::m:b$. The construction of m is as follows.

1. Put a and b end to end, forming AC.
2. Construct a semicircle having AC as diameter.

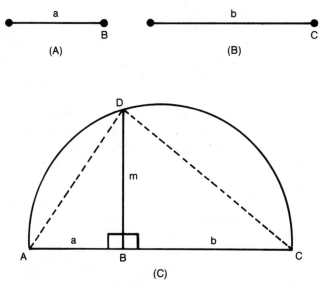

Figure 37 (A–C).

3. Erect a perpendicular to AC at B, meeting the semicircle at D, let AD have length m.

4. $a:m::m:b$ because triangle ABD is similar to triangle DBC.

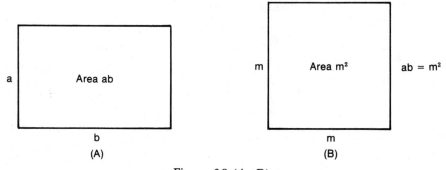

Figure 38 (A–B).

In modern terms we would say that m is a solution of the equation $a/x = x/b$, $a \cdot b = x^2$, or $x = \sqrt{a \cdot b}$. The fact that we can solve the second to last equation is expressed geometrically in Proposition 14 of Book II of Euclid's *Elements*, which says in part that, "It is possible to construct a square equal [in area] to any given rectangle."[2]

Notice how differently the problem of finding the mean proportional is treated today. Since i) we think of there being a real number corresponding to every length, and since ii) we have extended all of the usual operations, such as multiplication and square root, to the real numbers, we are able to assert that i) any given line segments have some real number lengths a and b and that ii) there is a real number $m = \sqrt{ab}$.

What are these protean real numbers of ours? In general, a non-negative real number has the form $n.r_1r_2r_3r_4r_5 \ldots$, where n is a natural number, and each of the r_i is one of the digits 0 through 9. The interesting thing about these "real" numbers is that they are, in point of fact, very *ideal* objects. The string of digits to the right of the decimal place is infinite. Strictly speaking, a real number can never be completely written out.

Of course, some real numbers, such as 25.000 . . . or 3.123123123 . . . , eventually begin repeating themselves. For convenience we write these numbers as follows: $25.\overline{0}$ and $3.\overline{123}$, where it is understood that the string of digits under the bar is repeated over and over. There is an interesting little theorem about repeating decimals. To state this theorem, we must keep in mind that a real number is *rational* if it is equal to some fraction, such as $^7/_{18}$.

Theorem: A real number r is rational if and only if it has a repeating decimal expansion.

$$
\begin{array}{r}
.28571428 \ldots = \overline{.285714} \\
7)\,②.00000000 \ldots \\
\underline{1\ \ 4} \\
⑥0 \\
\underline{56} \\
④0 \\
\underline{35} \\
⑤0 \\
\underline{49} \\
①0 \\
\underline{7} \\
③0 \\
\underline{28} \\
②0 \\
\underline{14} \\
⑥0 \\
\underline{56} \\
4 \ldots
\end{array}
$$

Instead of giving a formal proof of the theorem, let's just see an example of how each direction works. First, imagine that you have a real number r that is equal to the fraction $^2/_7$. In order to get the decimal expansion of r, we begin dividing 7 into 2. I have circled the successive remainders that occur. Notice that i) when you are dividing by 7 the remainder is always one of the natural numbers 0 through 6, and ii) if the same remainder occurs twice, then the decimal starts repeating, since the same sequence of actions will follow.

Second, imagine that there is a real number r that has the repeating decimal expansion .123. Now, r has an infinite number of blocks of "123" to the right of the decimal, so if we move one of these blocks to the left of the decimal, by taking 1000 r, then there will still be an infinite number of "123" blocks to the right of the decimal.

$$
\begin{aligned}
1000\ r &= 123.123123123... \\
-\qquad r &= \quad\ \ .123123123... \\
\hline
999\ r &= 123 \\
r &= {}^{123}/_{999} \\
r &= {}^{41}/_{333}
\end{aligned}
$$

So, in the indicated calculation, if we subtract r from 1000 r, then there is nothing remaining on the right of the decimal.

It is satisfying to see that the two finite ways of describing a real number—as a fraction, or as a repeating decimal—coincide. Given that we have proved that $\sqrt{2}$ is irrational, we can be sure that the decimal expansion of $\sqrt{2}$ *never repeats*. That is, when we write $\sqrt{2} = 1.4159$ . . . , we do not have any simple way of describing the pattern that the ". . ." stands for.

We can also use the theorem in the reverse way. That is, we can artificially construct a non-repeating decimal and be sure that this represents an irrational real number. There is, for instance, the artificial number of Liouville, .01001000100001000001000001 . . . , where the building principle of steadily increasing the number of zeros between the ones guarantees that the number never repeats itself. A different sort of non-repeating decimal can be obtained by sticking together all of the natural numbers to get the "number number," .1234567891011121314151617181920212223. . . .

Figure 39.

But how can we be sure that such artificial decimal expansions are really numbers? What exactly is *meant* by .12345. . . ? The understanding is that .12345 . . . stands for the infinite series or sum, $^1/_{10}$ + $^2/_{100}$ + $^3/_{1000}$ + $^4/_{10000}$ + $^5/_{100000}$ + ... It is easy to visualize a geometric interpretation.

One is tempted just to say that a given real number, such as .12345 . . . , is really to be thought of as a point on the idealized real number line. The problem with this approach is that one has not really explained where the "real number line" comes from. The real number line is basically something to be found with certainty only in the Mindscape; there is no reason to assume in advance that our physical space is filled with copies of the real number line.

This problem was finally dealt with only about one hundred years ago, by our old friends Cantor and Dedekind. Cantor basically defined a real number simply as an infinite sequence of digits, just as was done above. The original element of his approach was that one does not act as if the limit or sum of the infinite series expressed by a real number is anything *other than* or *external to* the series itself. Thus, the sum of the series $^2/_{10}$ + $^5/_{1000}$ + $^7/_{10000}$ + $^9/_{100000}$ + . . . is nothing other than the series itself, also known as .20579. . . . By using various weird definitions one can learn to add and multiply such series with each other without having to pretend that one is really working with finitely given limits. The point is that Cantor gave up the pretense that the real numbers are primarily finite lengths. He treated them rather as arbitrary infinite series of the form $\pm n.r_1r_2r_3r_4$. . . .

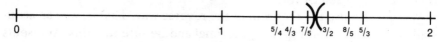

Figure 40.

Dedekind also defined real numbers in terms of infinite sets. His approach was to characterize a real number as a *cut* $[L, R]$ of the rational numbers. The idea is that every rational is either in L or in R and every member of L is less than every member of R. For instance, the square root of two would be represented by the cut $[\{a/b: a^2/b^2 < 2\}, \{a/b: a^2/b^2 > 2\}]$. The crucial thing about Dedekind's definition of real number is that, again, the real number itself is an *infinite* set. To be more precise, a Dedekind real number is a *pair* $[L, R]$ of infinite sets.

It is a curious fact in the history of mathematics that Dedekind's definition of real numbers is taken over almost unaltered from the Eudoxian theory of proportion given in Euclid's *Elements*, Book V. The problem Eudoxus had been concerned with was how we can compare and manipulate ratios (such as the $d:s$ ratio mentioned above) that are not equal to the ratio of any two natural numbers. His solution was, essentially, to regard an irrational ratio $X:Y$ as a cut of the form $[\{m:n \mid mY < nX\}, \{m:n \mid mY > nX\}]$. One can see that this makes sense if one realizes that $m/n < X/Y$ if $mY < nX$, and likewise for $>$.

The difference between what Eudoxus and Dedekind did is that Eudoxus thought of the ratio between two magnitudes as the fundamental thing, with the description in terms of infinite sets arising only in a practical and *potentially infinite* way (since one would not, in practice, ever need *all* of the members of each side of the cut). Unless someone had constructed two specific magnitudes to be compared, the equivalent cut had no meaning . . . as it was an infinite, and thus unreal, thing.

Dedekind, on the other hand, accepted the *actually infinite* sets of the cut as fundamental. Whether or not one has a particular trick for constructing a length that drops a point down into the cut's gap is immaterial. All the different actually infinite cut-sets exist in the Mindscape, and all the real numbers are already there, whether or not they can be finitely named or constructed.

The point is that the only way to get a stable mathematical representation of the notion "arbitrary real number" is to represent real numbers by actually infinite sets. There is no other way to get an absolute foundation of the real number system in terms of discrete mathematical objects.

Once it was realized that real numbers can be represented in terms of infinite sets, the dam broke. Ten years after Cantor's death it was already a commonplace that every mathematical object can be represented by a set. If you have ever picked up a mathematics text in any field, be it analysis, algebra, or topology, you will have noticed that the book begins with a short chapter or section on set theory. This is because everything the book mentions can be best represented as a set.

For the Pythagoreans everything was a natural number. Their belief became untenable when it was realized that certain things are in their inmost essence infinite. The modern mathematical credo called Cantorism asserts that everything (at least everything mathematical) is a set.

Just as the existence of the actual infinite forces a revision of the Pythagorean position, the existence of the Absolute infinite forces a revision of the Cantorian position. If there are, indeed, Absolutes of the kind discussed in the earlier section on the Absolute infinite, then there are things that are not sets. Set theorists are still not quite certain what to do about this. But let us not worry about Absolute infinity before discussing the transfinite.

Once we realize that the irrational numbers are fundamentally infinite, in that they can be fully grounded only on a theory of infinite sets, then it is natural to start looking at infinitely large, or *transfinite*, numbers. In Cantor's words, "One can without qualification say that the transfinite numbers *stand or fall* with the infinite irrationals; their inmost essence is the same, for these are definitely laid out instances or modifications of the actual infinite."[3]

A last remark on Cantorism. Just as chemistry was unified and simplified when it was realized that every chemical compound is made of atoms, mathematics was dramatically unified when it was realized that every object of mathematics can be taken to be the same *kind* of thing. There are now other ways than set theory to unify mathematics, but before set theory there was no such unifying concept. Indeed, in the Renaissance, mathematicians hesitated to add x^2 to x^3, since the one was an area and the other a volume. Since the advent of set theory, one can correctly say that all mathematicians are exploring the same mental universe.

TRANSFINITE NUMBERS

In my novel *White Light*, I describe a mountain that is higher than infinity.[4] This mountain, called Mount On, consists of alternating cliffs and meadows. The curious thing about it is that even after one has climbed ten cliffs, a thousand cliffs, infinitely many cliffs . . . there are always more cliffs. The climbers of Mount On are able to make some progress because they are able to execute a procedure called a "speed-up." By using speed-ups they are able, for instance, to zip past the first infinity of cliffs in two hours.

How is this done? The idea is to climb the first cliff in one hour, the next cliff in half an hour, the one after that in a quarter of an hour, and, in general, the nth cliff in $1/2^n$ hours. Since $1 + 1/2 + 1/4 + 1/8 + . . .$ sums to 2, we see that after two hours our climbers have passed infinitely many cliffs. But there are more, many more.

In this section we will climb up through the transfinite numbers, which are usually called *ordinal numbers*, or just *ordinals*. Typically, one describes some ordinal a by giving an example of an ordered set M such that *if* one could count M in the correct order, *then* one would count up to a. a is then viewed as the *abstract order type of M*, called $\overline{M}$ for short. The ordinal $\overline{M}$ is gotten from the ordered set M by ignoring the actual appearance of the individual members of M and instead concentrating on the arrangement, or order, of these members.

FROM OMEGA TO EPSILON-ZERO

The transfinite ordinal numbers can be thought of as arising through counting. There are two principles for generating ordinal numbers: I) if you have the ordinal number a, then you can find a *next* ordinal, called $a + 1$; II) if you have some definite sequence of increasing ordinals a, then you can find a last ordinal which is greater than all the a's, called $\lim(a)$.

We also need a first ordinal to start with, called 0. (Strictly speaking, the second principle for generating ordinals gives us 0, since zero is the first ordinal after the empty sequence.) In any case, once we have zero, the first principle can be repeatedly applied to get the ordinal numbers $0, 1, 2, \ldots$. Now, to get past the infinite sequence of finite ordinals we use principle II to get $\lim(n)$, usually called ω, pronounced "omega." (Omega is also sometimes called "alef-null.")

Omega is the last letter in the Greek alphabet (they put zeta somewhere in the middle), which may be why Cantor chose to use it as the number after all the finite numbers. The word "omega" is somewhat familiar, as it appears in the Book of Revelations, where God is quoted as saying, "I am the Alpha and the Omega," meaning, "I am the beginning and the end."

Now we have $0, 1, 2, \ldots \omega$. Using principle I repeatedly we get the sequence $0, 1, 2, \ldots \omega, \omega + 1, \omega + 2, \omega + 3, \ldots$. To go further, we use principle II to form $\lim(\omega + n)$, which is usually called $\omega + \omega$ or $\omega \cdot 2$.

You might wonder why $\lim(\omega + n)$, $\omega + \omega$, and $\omega \cdot 2$ should all be the same. It turns out that there is a definite way to define addition and multiplication of infinite ordinals so that everything works out. Let me briefly explain. The ordinal number $a + b$ is obtained by counting out to a and then counting b steps further. The ordinal number $a \cdot b$ is obtained by counting up to a, b times in a row. That is, $a \cdot b$ is obtained by sticking together b copies of a, treating this as an ordered set M, and then abstracting to get the ordinal number $\overline{M} = a \cdot b$.

As long as one sticks to finite ordinals, these operations are the same as the ordinary plus and times, and they are commutative. However, once we start working with infinite ordinals, commutativity no longer holds. Thus,

$$\underbrace{\blacktriangle}_{1} + \underbrace{\blacktriangle\blacktriangle\blacktriangle\blacktriangle\cdots}_{\omega} = \underbrace{\blacktriangle\blacktriangle\blacktriangle\blacktriangle\cdots}_{\omega}$$

$$\underbrace{\text{XXXX}\cdots}_{\omega} + \underbrace{\text{X}}_{1} = \underbrace{\text{XXXX}\cdots\text{X}}_{\omega + 1}$$

$1 + \omega$ is just the same as ω, but $\omega + 1$ is the next number after ω. Again,

$$2 \cdot \omega = \underbrace{2}_{\text{☆☆}} + \underbrace{2}_{\text{☆☆}} + \underbrace{2}_{\text{☆☆}} + \ldots = \underbrace{}_{\omega}\text{☆☆☆☆} \ldots$$

$$\omega \cdot 2 = \underbrace{}_{\omega}\square\square\square\square\ldots + \underbrace{}_{\omega}\square\square\square\square\ldots = \underbrace{}_{\omega + \omega}\square\square\square\square\ldots\square\square\square\square\ldots$$

$\omega \cdot 2$ is two omegas placed next to each other, which gives an ordinal $\omega + \omega$, but $2 \cdot \omega$ is omega twos placed next to each other, which makes an ordered set with ordinal number ω.

Moving on with the ordinals, by using principle I repeatedly, we now have 0, 1, 2, . . . ω, $\omega + 1$, $\omega + 2$, . . . $\omega \cdot 2$, $\omega \cdot 2 + 1$, $\omega \cdot 2 + 2$, ... It is evident that $\lim(\omega \cdot 2 + n)$ should be the ordinal $\omega \cdot 2 + \omega$ also known as $\omega \cdot 3$. Continuing in this vein, we can arrive at $\omega \cdot n$ for each finite n, and using principle II we form $\lim(\omega \cdot n)$, which should be omega copies of omega, that is, $\omega \cdot \omega$, also known as ω^2.

In order to see how to pass from ω to ω^2 to ω^3 and finally on to ω^ω, it will be useful to look at the pictures in Figure 41. I will now describe four sets of points, M_1, M_2, M_3, and M_ω. In each case, the set of points can be thought of as lying between zero and one on the real number line, and each M_i will be such that if we view it as an ordered set and abstract to form the ordinal number M_i, we arrive at ω^i. One may wonder how to fit an ordered set with ordinal number ω^2 in between zero and one on the real line, since the space available seems to be finite. The trick is based on Zeno's paradox: if you start out going from zero to one, but do it by first going halfway, and then going half the remaining distance, and then half the remaining distance, and so on . . . always going just half of what's left . . . then it will take you omega steps to get there. This is basically what the first picture shows.

The second picture is obtained by plugging a copy of the first picture into each of the spaces between points in the first picture. The third picture is obtained by plugging a copy of the second picture into each of the spaces between points in the first picture . . . . at least, this is what we would say if we think of ω^3 as being $\omega^2 \cdot \omega$. If, on the other hand, we think of ω^3 as being $\omega \cdot \omega^2$, then we would say that the third picture is obtained by plugging a copy of the first picture into each of the spaces between points in the second picture, thus obtaining ω^2 copies of ω (as opposed to ω copies of ω^2.) Both come to the same thing, since al-

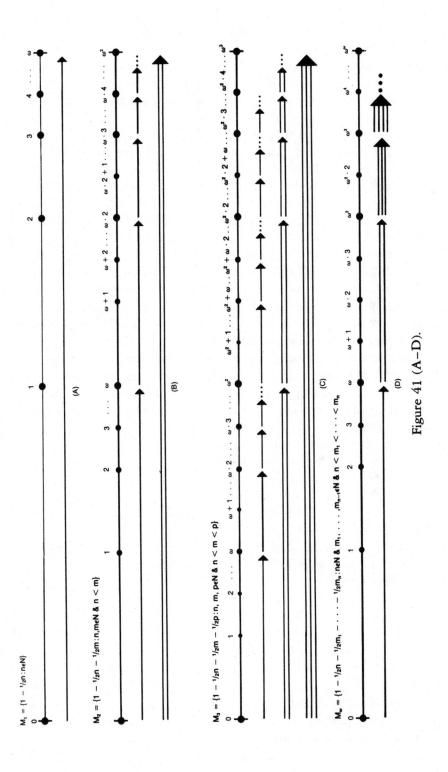

Figure 41 (A–D).

though ordinal multiplication is not commutative, it *is* associative. The fourth picture is obtained by first continuing the process started in the first three pictures endlessly . . . and then putting a copy of each of the pictures together.

I have written next to each picture the real number coordinates of the points involved if we think of the interval as the unit interval on the real line. These set definitions are not particularly important for us, but what *is* important is to realize that a transfinite arrangement of points can be fitted into a finite space. By using Zenonian squeezing we can sort of see an ordering of type ω^2 all at once! Transfinite ordinals are not really so inconceivable after all.

It turns out that one can actually fit *any* countable ordinal into a picture of this nature. (In the next subsection we will look at the *un*countable ordinals.) But the illustration of ω^3 is already something of a mess, and had I not used the arrow symbolism, the picture of ω^ω would be so lacking in detail as to be quite uninformative. We will look at a different technique for picturing ordinals shortly.

But first let me describe just how far we want to go in this section. One way of characterizing ω^2 is that it is an ordinal a such that $\omega + a = a$. This can be seen by thinking of ω^2 as being $\omega + \omega + \omega + \omega + \ldots$. Clearly, putting an "$\omega +$" in front of this symbol changes nothing. In point of fact, ω^2 is the *first* such ordinal.

What about the first ordinal a such that $\omega \cdot a = a$? If we take ω^ω to be $\omega \cdot \omega \cdot \omega \cdot \omega \ldots$, we can see that placing an "$\omega \cdot$" in front of ω^ω changes nothing. Or, assuming that the familiar laws of exponents hold for ordinals (and they do), we can reason that $\omega \cdot \omega^\omega = \omega^1 \cdot \omega^\omega = \omega^{1+\omega} = \omega^\omega$, since $1 + \omega = \omega$. As before, it is possible to prove that ω^ω is the *first* ordinal a such that $\omega \cdot a = a$.

The first ordinal a such that $\omega^a = a$ is called ϵ_0, pronounced "epsilon zero." Simply by manipulating symbols, we would expect that ϵ_0 has the following form:

$$
\vdots \\
\omega \\
\;\;\omega \\
\;\;\;\omega \\
\;\;\;\;\omega
$$

Evidently, putting such a symbol in the exponent position over an omega does not change anything, since a stack of omegas $1 + \omega$ high is the same as a stack ω high.

But we can describe epsilon zero a little better. Suppose that $^a b$ is defined to mean "an exponentiated stack of a many b's." $^a b$ is pronounced, "b tetrated to the a." The name tetration is used since *tetra* is the Latin root for *four,* and tetration occurs in fourth place in the logical progression: addition, multiplication, exponentiation, tetration. You don't ordinarily hear much about tetration because it is so powerful an operation that tetrating even very small numbers with each other produces inordinately large numbers. A tetration is worked out below.

$$
\begin{aligned}
^4 2 &= 2^{(2^{(2^2)})} \\
&= 2^{(2^4)} \\
&= 2^{16} \\
&= 64{,}536
\end{aligned}
$$

Note that we must be careful to associate from the top down, rather than from the bottom up, if we want to get the largest possible numbers.

As further examples, $^3 3$ is $3^{(3^3)} = 3^{27}$, which is just under eight trillion. $^3 10$ is $10^{(10^{10})} = 10^{\text{billion}}$, which is the number that is written by putting a one and then *ten billion* zeros (as opposed to, say, a million, which is written by putting a one and then *six* zeros. Now, $^2 \omega$ is just ω^ω. And $^3 \omega = \omega^{(\omega^\omega)}$, which is kind of hard to get at. One way of visualizing this number is to go back to the picture of ω^ω, and to imagine replacing each of the dots on the line by the symbol "$\omega \cdot$" to get the product of ω^ω omegas.

In any case, the point of all this is that ϵ_0 is $^\omega \omega$. And the countable ordinals don't stop there. If, for instance, one could make some sense of the following symbol one would have an even larger ordinal.

$$
\begin{matrix}
& & & \cdot \\
& & \cdot & \\
& \cdot & & \\
\omega & & & \\
& \omega & & \\
& & \omega & \\
& & & \omega
\end{matrix}
$$

In trying to think of bigger and bigger ordinals, one sinks into a kind of endless morass. Any procedure you come up with for naming larger ordinals eventually peters out, and the ordinals keep on coming. Finally, your mind snaps, and maybe you get a momentary glimpse of what the

Absolute infinite is all about. Then you try to formalize your glimpse, and you end up with a new system for naming ordinals . . . which eventually peters out. . . .[5]

But this ω^ω is just the beginning. Let us look at a different type of picture of ω^ω. Suppose that we let *PN* be the set of all polynomials in x with natural number coefficients. Examples of members of *PN* would be $x, x + 3, 5x^2 + 2x + 4, 3x^8 + 6x^3 + 163$. Now, suppose that given two polynomials $p(x)$ and $q(x)$ from *PN*, we define the following order relation: $p(x) <_{bep} q(x)$ if and only if the graph of the polynomial q eventually manages to get above, and stay above, the graph of the poly-nomial p. (The letters "bep" stand for "by-end-pieces.")

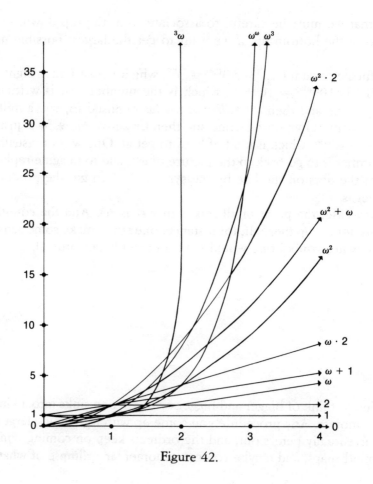

Figure 42.

The reason I have brought all this up is that if we take PN with the $<_{bep}$ ordering, then the ordinal number PN of this ordering is ω^ω. The correspondence is simplicity itself: the polynomial $p(x)$ represents the transfinite number $p(\omega)$ (with the one stipulation that the coefficients of the polynomial must be moved to the right in the translation process that is, $3x^2 + 2x$ is to correspond to $\omega^2 \cdot 3 + \omega \cdot 2$).

In Figure 42 I have illustrated the fact that $0 <_{bep} 1 <_{bep} 2 <_{bep} x + 1 <_{bep} 2x <_{bep} x^2 <_{bep} x^2 + x <_{bep} 2x^2 <_{bep} x^3 <_{bep} x^x <_{bep} {}^3x$, where ${}^3x$ means x tetrated to the three, $x^{(x^x)}$.

Strictly speaking $<_{bep}$ has only been defined for polynomials, but it is evident how we can extend it to arbitrary expressions or functions in x. If we allow tetration as a standard operation and let PPN be the set of all pseudo-polynomials formed by using natural number coefficients and tetration, as well as exponentiation, then it is not difficult to see that PPN is ϵ_0 when PPN is ordered by $<_{bep}$. An example of a pseudopolynomial from PPN would be $({}^5x) + 2({}^3x)^4 + 7({}^3x) + ({}^2x)^8 + 13({}^2x)^2 + 11({}^2x) + 3x^7 + 9x^3 + 2x + 78$. Note that ${}^xx$ represents ϵ_0.

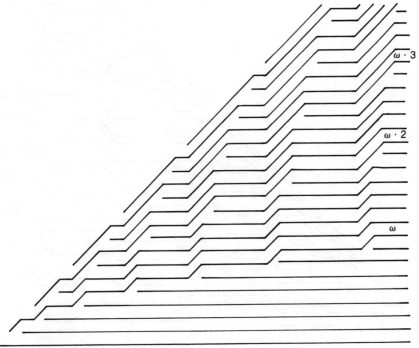

Figure 43. Ziggurat to ω^2.

The by-end-pieces ordering was first studied by DuBois-Reymond, whose work is interestingly presented in G. H. Hardy's book, *Orders of Infinity*.[6] Felix Hausdorff improved upon Reymond's techniques in his monumental series of papers, *"Untersuchen über Ordnungstypen,"* in the early 1900s. The term "by-end-pieces ordering" is taken from Kurt Gödel, whose research revived interest in this ordering in the 1960s.[7]

We could really restrict our attention to functions from the natural numbers into the natural numbers. It is amusing to represent ordinal numbers as stacks of such functions, filling in the lines between graph points in the natural way. In order to make the pictures look nice, you leave out parts of the lines to avoid crossings. I call these pictures "zig-

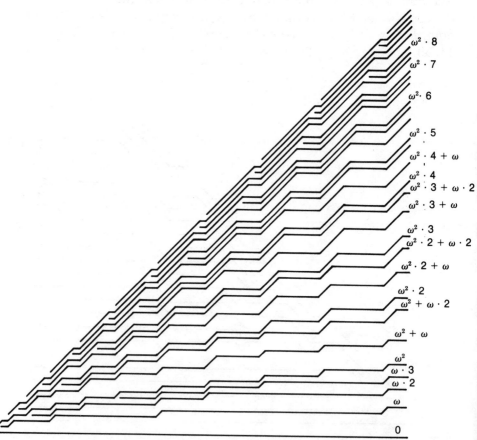

Figure 44. Ziggurat to ω^3.

gurats" because they look like Babylonian towers or Aztec pyramids. Figure 43 is a ziggurat of height ω^2, and Figure 44 shows one of height ω^3. In the latter I have not drawn in all the lines, so that it is not so confusing. In every case, the missing line is a line that would hug the line right underneath itself as closely as possible.

THE ALEFS

The famous mathematician David Hilbert used to illustrate his popular lectures with stories about a hotel with infinitely many rooms.[8] This mythical hotel, usually called Hilbert's Hotel, is supposed to have omega rooms: Room 0, Room 1, Room 2, . . . , Room n, and so on. As in the last section, it is convenient to start counting with 0.

To fix the ideas, I have drawn a picture of Hilbert's Hotel in Figure 45. In order to fit it on the page, I have assumed that each floor is equipped with a science-fictional space condenser, a device that makes each succeeding story two-thirds as high as the one before. The shrinking field also affects the guests. Thus, although the ceilings on Floor 3 are only two or three feet high, the space condenser on that story shrinks the guests to one or two feet, and they are perfectly comfortable. I will leave it as an exercise for the reader to check that if the first floor is ten feet high, and each successive floor is two-thirds as high as the one before, then the total height of the hotel's ω stories is thirty feet.

One of the most paradoxical things about Hilbert's Hotel is that even after it fills up, more and more people can be squeezed in, without making anyone share a room! Say, for instance, that ω guests have arrived, and every room is occupied with a guest n in each Room n. Now, say that one more guest arrives: Guest ω. How to fit him in?

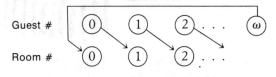

ω + 1 Guests in ω Rooms

Easy! We put Guest ω in Room 0, which is emptied by moving Guest 0 to Room 1, which is emptied by moving Guest 1 to Room 2, which is emptied by . . .

Fine! But what if there had been an *infinite* number of new guests? Even such a procession of ω + ω guests could be lodged in Hilbert's

Individual Guest Number: 0, 1, 2, . . . , $\underbrace{\omega}$

Total Number of Guests: $\underbrace{\hspace{2em}}_{\omega}$ $+1$

Figure 45.

Hotel. We simply put the first ω guests in the even-numbered rooms, and the next ω guests in the odd-numbered rooms.

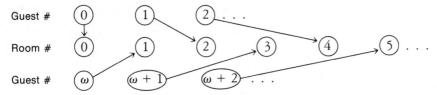

Now it is in fact possible, by suitable rearrangements, to fit ω^2, ω^ω, or even ϵ_0 guests into Hilbert's Hotel. But eventually we do reach a limit to this wonderful hotel's powers of absorption: *alef-one,* also known as $\aleph_1$. Alef-one is a hard number to describe. One way of putting it is that alef-one is the first ordinal number a such that no possible rearrangement can fit a set of a guests into ω rooms. Alef-one represents an order of infinity that is *essentially greater than* ω in a way that $\omega + \omega$ is not.

To get a better idea of alef-one, let's go back to the idea of people climbing a mountain as high as all the ordinals. How hard will it be for them to get to alef-one?

We will assume that, in the marvelous land where Mount On rises, the climbers can attain any desired finite speed. Then, as we mentioned at the beginning of the "Transfinite Numbers" section, they can scale the infinity of cliffs leading up to ω in a finite time. If they keep accelerating so that the nth cliff only takes $1/2^n$ hours, then they reach ω after two hours. By repeating this they could reach $\omega + \omega$ in four hours, or, if they did everything four times as fast, they could actually get to $\omega + \omega$ in one hour. Turning to Figure 46, we can see how our climbers could even reach ω^2 in one hour. The idea is to devote to each of the ω^2 cliffs the alloted finite interval of time. (Thus, for instance, the stretch between $\omega \cdot 2$ and $\omega \cdot 2 + 1$ is to be covered during the time interval between $3/4$ hour and $13/16$ hour.)

Now, the point I want to make is that these climbers could never reach alef-one. There is no way that we can fold together various finite bursts of speed and cover alef-one cliffs in a finite amount of time. The only way to get out to alef-one is by actually going ahead and travelling alef-one miles per hour.

One last picture of alef-one. Go back to Figure 42. In this picture we saw how various ordinals can be represented as sequences of functions ordered according to steepness (the $<_{bep}$ ordering). How long a sequence of functions can be found with each function steeper than all the ones before? At least alef-one. That is, if S is a set of functions so that

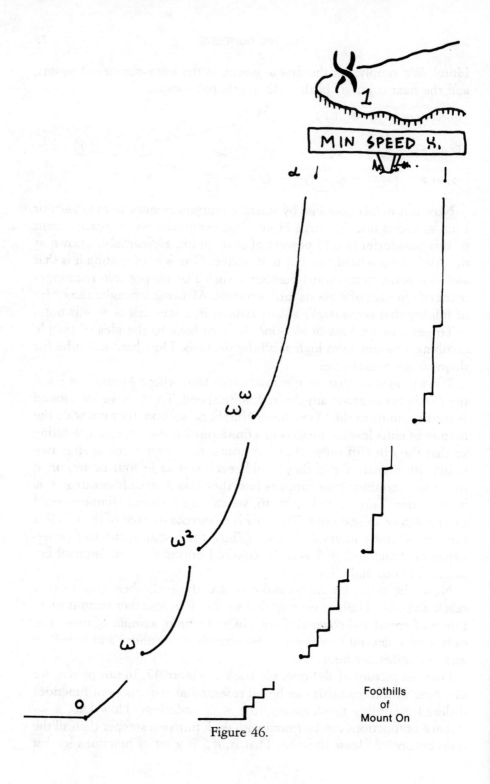

Figure 46.

for every g (no matter how steep), there will be an f in S that is steeper, then S must have at least alef-one members.

It is now time to give a formal definition of alef-one. This definition hinges upon the notion of cardinality. Given two ordinals A and B, we say that A *has the same cardinality as* B if there exists a one-to-one map from A onto B. (When I speak of a map from A onto B, I really mean a map from the set of ordinals less than A onto the set of ordinals less than B.)

In our discussion of Hilbert's Hotel we learned, for instance, that $\omega + \omega$ has the same cardinality as ω. For there is a way of matching up, in a one-to-one fashion, the members of $\{0, 1, 2, \ldots \omega, \omega + 1, \omega + 2, \ldots\}$ with the members of $\{0, 1, 2, \ldots\}$. That is, there is a one-to-one map from $\omega + \omega$ onto ω. Now, alef-one is defined to be the first ordinal with cardinality greater than ω. Alef-one is the first ordinal that cannot be mapped one-to-one onto ω.

In general, we say that the ordinal number A is a *cardinal number* if and only if A does not have the same cardinality as any B less than A. All of the natural numbers are cardinals. There is, for instance, no way to find a *one-to-one* map from three onto two. (Keep in mind that we commonly identify the number N with the set $\{0, 1, \ldots, N - 1\}$ of N numbers less than N.)

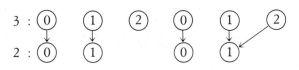

Not from all of 3.   Not one-to-one.

ω is, of course, a cardinal number. The infinite number ω cannot be mapped one-to-one into any of the finite numbers before it. No finite hotel, no matter how large, can sleep infinitely many guests one to a room.

The infinite cardinals are also called *alefs*. In general, alef-a, or $\aleph_a$, means the ath infinite cardinal. Thus, alef-null, or $\aleph_0$, is just ω, the first (0th) infinite cardinal. It turns out that just as we can always find more ordinals, we can always find more cardinals. After $\aleph_0$ come $\aleph_1, \aleph_2, \aleph_3, \ldots \aleph_\omega, \aleph_{\omega+1}, \ldots \aleph_{\omega^\omega}, \ldots \aleph_{\aleph_1}, \ldots \aleph_{\aleph_\omega}, \ldots$, and so on. After a while we can even find a number θ such that $\theta = \aleph_\theta$. One way of getting such a θ is indicated below.

$$\theta = \aleph$$
$$\aleph$$
$$\aleph$$
$$\aleph$$
$$\cdot$$
$$\cdot$$
$$\cdot$$

Is this the end? It's never the end, when you're discovering transfinite numbers. After θ come $\aleph_{\theta+1}$, $\aleph_{\theta+\omega}$, $\aleph_\theta + \aleph_\omega$, and so on and on, world without end.

According to the Reflection Principle, it is impossible ever to conceive of an end to the ordinals. We do have a symbol Ω, the big Omega, which we use to stand for the Absolute Infinity that lies beyond all the ordinals. But Ω is inconceivable. The Reflection Principle makes this precise by saying that any description D of Ω that we might come up with will apply to some ordinal a short of Ω.

Ω is called Absolute Infinity because it is not a relative notion. The line of ordinals leading out to Ω contains *all* the ordinals, all the possible stages of counting. It is because every possible ordinal occurs before Ω that Ω is not really a definite ordinal number. Confusing talk! If you would like to read more about Ω and the transfinite numbers plan to take Excursion I.

INFINITESIMALS AND SURREAL NUMBERS

The best-loved of Zeno's paradoxes is the one that states that you can never leave the room you are in. For, Zeno reasons, in order to reach the door you must first traverse half the distance there. But then you are still in the room, and to reach the door you must traverse half the remaining distance. But then. . . .

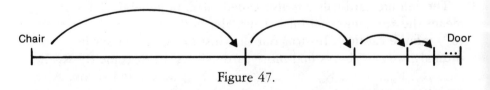

Figure 47.

What exactly is paradoxical here? The problem seems to be that we can analyze the passage from chair to door in two ways. First, in the normal way, as a unit, a single undivided action, as a 1. Second, as the limit of an infinite sequence of actions, as the sum of a series ($1/2 + 1/4 + 1/8 + \ldots$). The modern tendency is to resolve Zeno's paradox by insisting that the sum of the infinite series $1/2 + 1/4 + 1/8 + \ldots$ *is equal to* 1.[9] One then goes on to add that since each of the successive steps is done in half as much time, the actual time taken to complete the infinite series is no different from the real time taken to leave the room.

But still . . . there is some residue of dissatisfaction. The feeling is that if you just keep halving the distances you never really *get* to the door. You may get arbitrarily close, but you never quite reach the limit.

The paradox can be expressed in a different way. In our ordinary real number system, we say that the number K with decimal expansion .99999 . . . is the same as 1. An informal argument for this is sketched below.

$$
\begin{aligned}
10K &= 9.999 \quad \ldots \\
- \quad K &= .9999 \ldots \\
\hline
9K &= 9 \\
K &= 1
\end{aligned}
$$

But maybe this argument is misleading. What if there is some number, call it $1 - 1/\omega$, that is greater than any finite string .9 . . . 9 of nines, yet less than 1? If K were actually equal to $1 - 1/\omega$, the informal argument used in the last paragraph would not work, for this argument overlooks the fact that the difference between $10K$ and 10 is ten times as great as the difference between K and 1. There is a residual infinitesimal quantity below that does not get canceled out.

$$
\begin{aligned}
10K &= 10 - 10/\omega \\
- \quad K &= 1 - 1/\omega \\
\hline
9K &= 9 - 9/\omega \\
K &= 1 - 1/\omega
\end{aligned}
$$

Intuitively, nothing could be more natural than to go ahead and talk about $1/\omega$, $1/\aleph_1$, and so on. Just as we move from the natural numbers to the fractions and then on to the reals, should we not be able to move from the whole ordinal numbers to some richer number field?

Curiously, Cantor himself was very much opposed to this step. When a fellow mathematician attempted to use Cantor's transfinite numbers to develop a theory of infinitely small quantities, Cantor accused him of trying to "infect mathematics with the Cholera-Bacillus of infinitesimals."[10] Cantor even constructed a proof that no number can be infinitesimal. This proof, however, is just as circular and worthless as finitist attempts to prove that no number can be infinite. In both cases, the desired conclusion is smuggled in as part of the definition of "number."

Why was Cantor so vehemently opposed to infinitesimals? In his valuable essay, "The Metaphysics of the Calculus," Abraham Robinson suggests that Cantor already had enough problems trying to defend transfinite numbers.[11] It seems likely that, consciously or otherwise, Cantor deemed it politically wise to go along with orthodox mathematicians on the question of infinitesimals. Cantor's stance might be compared to that of a pro-marijuana Congressional candidate who advocates harsh penalties for the sale or use of heroin. Yet, as we shall see, there is almost as much justification for infinitesimals as there is for Cantor's transfinite ordinals.

Formally speaking, it is as consistent to say that there is a number between all of .9, .99, .999, . . . and 1 as it is to say that there is a number greater than all of 1, 2, 3, ... And just as we go on to find more and more ordinals piled atop one another, we can go on to find more and more infinitesimals squeezed beneath each other.

Part of the great attractiveness of Cantor's theory stems from the fact that all of his transfinite numbers can be seen as steps towards the single Absolute Infinity that lies beyond them all. Indeed, as I have mentioned before, modern formalizations of set theory often proceed by introducing a symbol Ω for Absolute Infinity and by then assuming the Reflection Principle: every conceivable property of Ω is shared by some ordinal less than Ω. Thus, for instance, since we know that Ω is

$$0, 1, 2, \ldots \qquad\qquad\qquad \Omega.$$
$$0, 1, 2, \ldots \qquad \overset{\diagdown|\diagup}{\underset{\diagup|\diagdown}{=\omega=}} \qquad \Omega \Bigg\} \text{ Reflection}$$

greater than all the finite numbers n, we know by Reflection that there must be some existing ordinal, call it ω, that is also greater than all the finite n.

The Reflection Principle is really a different way of saying "Ω is inconceivable." For "Ω is inconceivable" is the same as "there is no conceivable property P that uniquely characterizes Ω," and this is the same as "whenever P is a conceivable property of Ω, then there must be other ordinals also enjoying property P." Ω can never be the *first whatever*, just as it can never be the *only whatever* . . . whatever *whatever* may be.

To sum up, we justify Cantor's transfinites from the two assumptions: i) there is an Absolute Infinity, Ω; ii) Ω is inconceivable.

A skeptical reader could, quite rightly, demand to know how it is possible to discourse rationally about an inconceivable object like Ω. I would respond that Ω is a *given,* an object of our immediate pre-rational experience. And to use the tools of symbolic logic to investigate an empirically existing phenomenon is not to commit a category mistake, any more than it is a categorty mistake to look at living cells through the inanimate lenses of a microscope.

We have a primitive concept of infinity. This concept is inspired, I suspect, by the same deep substrate of mind that conditions religious thought. Set theory could even be viewed as a form of exact theology. By means of the set-theoretic analysis of Absolute Infinity, we attain knowledge of many lower infinities—the transfinite ordinals and cardinals.

I wish now to call the reader's attention to a different sort of absolute: *Absolute Continuity.* Perfectly continuous space.

The basic intuition about an Absolutely Continuous line is that such a line cannot be conceived of as a set of points. Zeno expresses this intuition in his paradox of the arrow.[12] The paradox of the arrow seems to constitute a proof that space is not made of points. For, Zeno argues, consider an arrow that flies from the bow to the target. If space is made up of points, then the flight of the arrow can be decomposed into an infinite set of frozen movements, movements where the tip of the arrow successively occupies each of the points between bow and target. The problem is that while the arrow is at any *one* fixed point, say the halfway point, the arrow is motionless. How can the flight of the arrow be a sequence of motionless stills? Where did the motion go?

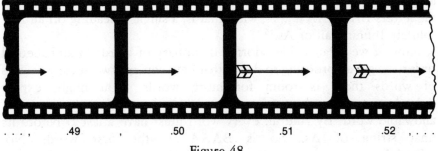

. . . ,   .49   ,   .50   ,   .51   ,   .52 , . . .

Figure 48.

A movie of an arrow's flight is, of course, a sequence of motionless stills. But this does not disturb us, as we realize that the arrow moves in between the pictures. The problem Zeno raises is that if space is made of points, and if a still is taken at *each point,* then there is no possibility of "moving between the pictures" . . . because there is *nothing between the pictures.*

Zeno's way out of the paradox is to deny that space is really made up of points. As a Parmenidean monist, Zeno viewed space as an undivided whole that cannot really be broken down into parts. We can find scattered locations in space, but space is always more than the sum of these isolated points. One can pick out higher and higher infinities of points from an Absolutely Continuous tract of space, but there will always be a residue of leftover space, of continuous little pieces, infinitesimal intervals over which the actual motion takes place.

This view of space has been held by several philosophers since Zeno, notably C. S. Peirce and, perhaps, Kurt Gödel. Gödel distinguishes between the set of points described in set theoretic analysis and the continuous line of space intuition: "According to this intuitive concept, summing up all the points, we still do not get the line; rather the points form some kind of scaffold on the line."[13]

Peirce goes further than this. According to him, a truly continuous line is so richly packed with points that no conceivable set, no matter how large, can exhaust the line. There should not just be *one* point between all of $^1/_2$, $^2/_3$, $^3/_4$, $^4/_5$, $^5/_6$, . . . and 1. There should be ω points, $\aleph_1$ points, Absolutely Infinitely many![14]

The early set theorist Felix Hausdorff demonstrated the logical possibility of such an Absolutely Continuous ordering. Hausdorff's construction goes as follows. We are to imagine a superdictionary of a very special sort: i) All the words in Hausdorff's dictionary will be spelled using only As and Bs. ii) Each word will be of Absolutely Infinite length, with one letter for each ordinal. iii) Each word must end with the letters BAAAAA . . . , with, that is, a single B leading off an Absolutely Infinite tail of As.[15]

Now, if we arrange Hausdorff's dictionary of words in alphabetical order, we get an ordering so dense that between *any* two successive sets of words there is room for more words. You might expect AAAA . . . to be the first word in the dictionary, but this is not an allowable word. By rule iii, every word must have a last B before its final string of As. So is BAAA . . . the first word? No. ABAAA . . . comes first. And between these two comes, of course, ABBAAA. . . . In order to describe some of the longer words I will write A^ω and B^ω to mean strings of ω As and Bs, respectively.

$A^\omega A^\omega BAAA \ldots$	$1/(\omega + \omega)$
$A^\omega BAAA \ldots$	$1/\omega$
$AABAAA \ldots$	$1/4$
$ABAAA \ldots$	$1/2$
$ABBAAA \ldots$	$3/4$
$AB^\omega BAAA \ldots$	$1 - 1/\omega$
$BAAA \ldots$	1
$BBAAA \ldots$	2
$B^\omega A^\omega BAAA \ldots$	$\omega/2$
$B^\omega ABAAA \ldots$	$\omega - 1$
$B^\omega BAAA \ldots$	ω
$B^\omega BBAAA \ldots$	$\omega + 1$
$B^\omega B^\omega BAAA \ldots$	$\omega + \omega$

Exhibited above are some of the words in Hausdorff's dictionary, along with more customary number names that are appropriate for where these words appear. These number names are expressive, but basically a bit unjustified, as we do not have in our hands a definition for how to add and multiply these words like numbers.

But quite recently, the English mathematician and puzzlist John Horton Conway has discovered an Absolutely Continuous class of numbers that do have a simply defined addition and multiplication. Conway's new numbers are called *the class of surreal numbers,* or simply *No.*[16]

Conway speaks of his numbers as being "born" on an endless succession of days, one day of creation for each ordinal. In general, on the ath day, new numbers are placed in all gaps between successive sets of surreal numbers born on earlier days. In Conway's ingenious system one can write down and manipulate virtually any wild number one might think of:

$$\aleph_1/\aleph_2, \ \sqrt{\omega + \pi}, \ \aleph_3 + \frac{\sqrt{2}}{\aleph_5}$$

or what have you. He even gets a definition of the traditional symbol ∞, for potential infinity. ∞ is defined as the gap between the finitely large and the infinitely large surreal numbers, and Conway derives the weird equation, $\infty = \sqrt[n]{\omega}$, which almost magically ties together potential infinity ∞, the simplest actual infinity ω, and the Absolute Infinite Ω.

Although Conway's surreal numbers are aesthetically pleasing and philosophically significant, they have not attained any wide usage among more practically minded mathematicians. One problem could be

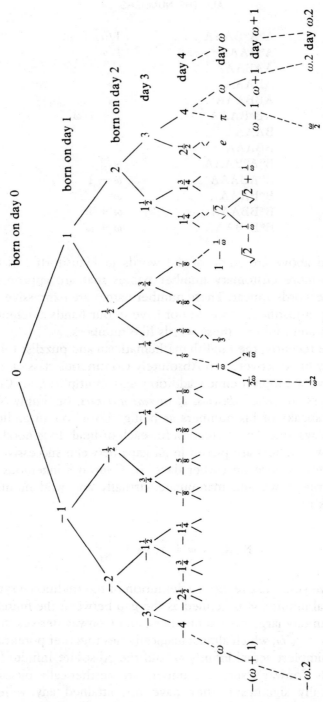

Figure 49. From J. H. Conway, *On Numbers and Games* (Academic Press, 1976), p. 11.

that it is hard to define the higher-order operations (such as exponentiation and tetration) correctly on the number field No.

Instead of using the surreal numbers, those mathematicians who need infinitesimals use a smaller, and somewhat different, extension of the reals, the so-called *hyperreal numbers*. These numbers, also known as non-standard numbers, were systematically introduced by Abraham Robinson in the 1960s.[17]

Conway's numbers are introduced as "gaps" between pairs of sets. Thus, $\sqrt{\omega}$ is $(0, 1, 2, \ldots | \ldots \omega/4, \omega/2, \omega)$ and, more obviously, $1/\omega$ is $(0 | \ldots 1/4, 1/2, 1)$. Robinson's hyperreal numbers are best thought of as sequences of reals. Thus, a hyperreal is a function f from the set N^+ of positive natural numbers ($N^+ = \{1, 2, 3, \ldots\}$) into the set R of real numbers.

This is convenient, since all of the operations on the ordinary reals can be carried over in a "pointwise" fashion to the functions that we call hyperreal numbers. Thus, $f + g$ is the sequence such that $f + g(n) = f(n) + g(n)$, and f^g is defined by $f^g(n) = f(n)^{g(n)}$.

As was already mentioned in the "From Omega to Epsilon-zero" subsection, we can order such functions by-end-pieces: $f <_{\text{bep}} g$ means that the graph of f eventually gets under, and stays under, the graph of g.

Figure 50 is a drawing of some of the functions representing hyperreal numbers. Notice that "$1/\omega$" is given by the function $I(n) = 1/n$. Any standard real number r will be represented by a constant function $c_r(n) = r$. It is easy to see that for any positive real r we will have $c_0 <_{\text{bep}} I <_{\text{bep}} c_r$, which is to say that I represents an infinitesimal.

As is well-known, Newton and Leibniz used both infinities and infinitesimals to develop the differential and integral calculus. They used the symbol "∞" for infinity, and the symbol "dx" for an infinitesimal number. To put a better face on things, Leibniz assured his correspondents that his ∞ was merely a potential, or "syncategorematic," infinity. Newton called his infinitesimal dx a "fluxion."

Bishop Berkeley derided Newton's fluxions as "the ghosts of departed quantities," but the progress of infinitesimal analysis continued unimpeded throughout the eighteenth century. No one was quite sure it was all right to use infinities, but the process gave the right answers. As Jean Le Rond d'Alembert said, *"Allez en avant, la foi vous viendra."* Finally, in the mid-nineteenth century, Weierstrass and others found a way to avoid the use of actual infinities in calculus.[18] Let us briefly describe the technique that was adopted.

In calculus one discusses different sorts of calculation processes

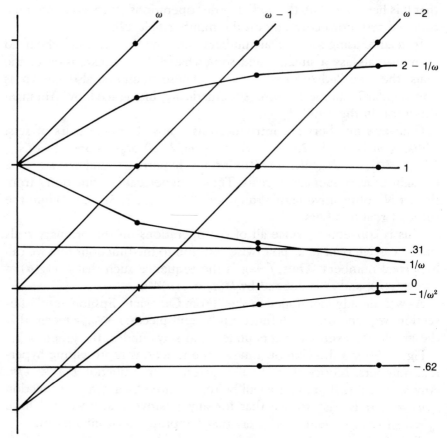

Figure 50.

$C(\ ,\ ,\ )$ whose outcome depends on what numbers are fed in. A typical use of infinitely large and infinitely small numbers is to say that for certain real numbers a and b, the outcome of $C(\infty, dx, a)$ is b. Both the derivative and the integral are defined in this way: as calculations that accept a real number (or numbers) and some infinitely large and/or small numbers as input, and that then give some definite real number as output.

Weierstrass's technique was to replace a statement such as "$C(\infty, dx, a) = b$" by "if we take I to be a *large enough* real number, and i to be a *small enough* real number, then the value of $C(I, i, a)$ will be *very close* to b." By letting I grow and i shrink, one is supposed to be able to bring $C(I, i, a)$ to within any desired degree of closeness to b. This is called the *limit process*.

Of course, if one believes in infinitely large and infinitely small numbers, it is much simpler just to ask if $C(\infty, dx, a)$ equals b. But so great is the average person's fear of infinity that to this day calculus all over the world is being taught as a study of *limit processes* instead of what it really is: *infinitesimal analysis*.

As someone who has spent a good portion of his adult life teaching calculus courses for a living, I can tell you how weary one gets of trying to explain the complex and fiddling theory of limits to wave after wave of uncomprehending freshmen.

I often think of C. H. Hinton's words from a similar context:

"How pleasant it would be to let pass away some of that verbiage I learnt at school—learnt because teachers must live, I suppose. The apeing and prolonged caw called grammar, the cackling of the human hen over the egg of language—I should like to unlearn grammar."[19]

But there is hope for a brighter future. Robinson's investigations of the hyperreal numbers have put infinitesimals on a logically unimpeachable basis, and here and there calculus texts based on infinitesimals have appeared.[20]

HIGHER PHYSICAL INFINITIES

Suppose that our universe were, in fact, more than ω miles in diameter. Could you ever travel a transfinite number of miles from Earth?

This *would be possible*, assuming you had the ability to accelerate up to any speed short of the speed of light (some .7 billion miles per hour). The technique hinges on Einstein's Special Theory of Relativity. As you travel closer and closer to the speed of light, your own time runs slower and slower relative to the rest of the universe. By a suitable acceleration, you could travel, say, ω billion miles in what feels like four hours.

The first billion miles (done at about .4 the speed of light) would seem to take two hours. The next billion miles (done at about .6 the speed of light) would take only an hour of your time. The third billion (done at .7 the speed of light) would take you only half an hour. And so on. After four hours you would be past milestone ω.

In my novel, *White Light,* I describe such a journey from the viewpoint of a massless soul and spirit who is able to accelerate himself endlessly by some obscure force of will. Let me reprint the description of his journey, purportedly made with a fellow spirit named Kathy:

The first part of the trip was dull. Although we were accelerating steadily, it still took an hour to get out of the solar system. And then we had an
hour and a half of vacuum till the next star.

About three hours into our trip it began to get interesting. Objectively
we were doing about .7 the speed of light. Because of our distorted time
and length standards, it felt like we were doing three times that. Weird
relativistic effects began setting in.

It seemed like we were looking out of a cave. All behind us and on
both sides of us there was the dead absolute nothing called 'Elsewhere' in
relativity theory. The stars had somehow all scooted their images around
to in front of us. We accelerated harder.

The thousand light-year trip across the galaxy only seemed to take half
an hour. But what a half-hour. I would be looking out our speed-cone at
the vast disk of stars that lay ahead of me . . . most of them clinging to
the edge. Slowly one of the stars would detach itself from the clustered
edge and accelerate along a hyperbolic path towards the center, then,
ZOW, it would whip past us and go arcing back out to the edge of our
visual field.

There was a pattern to the flicker of passing stars, and I began to get
into it. It was like listening to the clicking of train wheels. Everything but
the swooping pulses of light faded from my attention. I pushed to make
the flickering come faster.

There were patterns to the flicker . . . star clusters . . . and as we
accelerated more I began to see second and third order patterns. Suddenly the stars stopped. We were out of the galaxy.

Our visual field had contracted so much that I felt myself to be looking
out of a porthole. There was dark on all sides and I knew fear. My back
was a knot of pain, but I drove myself to accelerate more and more, to
make the porthole smaller.

A few squashed disks of light tumbled out from infinity and whizzed
back. Then more and more came twisting past. Galaxies. I felt like a gnat
in a snowstorm. We flew through some of the galaxies. Inside was a
happy blur. We were going much too fast to see the individual stars hur-
tle past.

We pushed harder, harder. We hit a galaxy every few seconds now,
and as before I began to detect higher-order patterns in the stroboscopic
flicker.

From then on that was all I could see . . . a flicker which would build
and build to an almost constant flash, abruptly drop in frequency, and
then build again. At the end of each cycle we reached a higher level of
clustering and the light became brighter.

I was on the ragged edge of exhaustion. The strobing was building cas-
tled landscapes in my mind. My lucidity was fading fast as I stared into
the more and more involuted blur of light before me. I tried to make it
come faster.

There was still a certain depth to the pattern of light ahead of us, but I noticed that the harder I pushed the acceleration, the shallower and more two-dimensional the scene in front of me became. I concentrated on flattening it out.

The energy to push no longer seemed to emanate from me or from Kathy. It was as if I were somehow ram-jetting the incoming light right through us . . . applying only a certain shift of perspective to move us even faster.

With a final effort we turned the universe into a single blinding point of light.[21]

I have reprinted such a long excerpt so as to make the idea of travelling out past infinity feel a bit more natural. Of course, actually to do such a trip in a real, and massive, spaceship would require an infinite amount of energy. But this could perhaps be taken care of by scooping in star after passing star for fuel!

A side effect of using relativistic time dilation to travel ω miles is that one would finish the trip not only ω miles away from earth, but also ω years in the future. The concept of a transfinite cosmic time scale has appeared in some modern astrophysical discussions. In particular, it has been predicted that if one were to fly at a rotating black hole, one would bounce off and land ω years in the future—in, effectively, a different universe.[22]

A different method of achieving very high physical infinities in the large is to postulate the existence of many parallel universes. Indeed, if our universe is determined by $\aleph_0$ bits of information, and if every possible universe exists, then there must be at least $\aleph_1$ other universes. Hugh Everett's notorious Many-Worlds Interpretation for quantum mechanics postulates a situation like this. One difficulty for the doubting Kantian is that it is, in principle, impossible to detect the other parallel worlds.

The idea of one universe that lasts $\aleph_1$ years is about the same as the idea of $\aleph_1$ universes that last ω years each. In either case, one has to do something very extraordinary to break out of the particular space-time one starts in. Whether one calls this extraordinary action "travelling infinitely far into the future" or "jumping to a parallel time-stream" is, perhaps, only a matter of taste.[23]

What about higher infinities in the small? As is discussed in Excursion I, Cantor proved in 1873 that mathematical space contains at least $\aleph_1$ points. So, if our physical space is actually rich enough so that every real number (such as 3.14159 . . .) represents a distinct point in space, then there are at least $\aleph_1$ points in space. Whether mathematical space might contain more than $\aleph_1$ points is still an open problem: the Continuum Problem.

Assuming that our physical space is, in fact, the same as mathematical space, Cantor was led to a very strange speculation: that matter-objects and aether-objects are made up of, respectively, $\aleph_0$ and $\aleph_1$ "atoms" each. Let us see how he arrived at this position.[24]

Cantor believed that since any tiny ball-type atom can be cut in half, the ultimate constituents of matter must be point-sized objects, which he called *mass-monads*. (The word "monad" is taken from Leibniz, who in his *Monadology* describes the universe as a collection of simple and indivisible monads acting in harmony.)[25] Cantor held that any piece of matter is made up of $\aleph_0$ mass-monads, packed densely together like the rational numbers on the number line.

In Cantor's time it was widely believed that there is another fundamental substance besides matter: *aether*. Cantor did not regard the aether as an all-pervading substance filling all of space. He believed, rather, that the aether is clumpy, and that such phenomena as light, heat, electricity, and magnetism could be explained in terms of tendrils and globs of aether.

In so far as aether should be a finer and subtler substance than matter, Cantor proposed that each chunk of aether is made up of $\aleph_1$ *aether-monads*. Thus, mass is thought of as something similar to a heap of sand, a congeries of bright dots, while aether is like water, a continuous substance that can trickle through the infinitesimal interstices in matter.

Can anything be made of Cantor's theory? Present-day science has no need of aether, or of any other higher types of matter. But something may yet come of Cantor's theory. Peirce, for instance, has suggested that the mind-body problem might be solved by regarding the human mind as being able to act on Cantorian aether-objects. Perhaps we have ectoplasmic souls, each made of $\aleph_1$ aether-monads!

Infinities much larger than $\aleph_1$ could occur in the small, if space were to be Absolutely Continuous in the sense discussed in the previous section. It is hard to imagine ever observing such nested infinitesimals, but some useful theory, with observable consequences, may find its theoretical foundations in Absolutely Continuous space. Or it could be that particles can be viewed as singularities of varying transfinite order.

PUZZLES AND PARODOXES
(Answers on page 297)

1. Describe how you would organize your time so as to travel $\omega + \omega$ miles in one hour.

2. Let a be some real number between -1 and 1. Consider the sum S of the geometric series given by the equation $S = 1 + a + a^2 + a^3 + \ldots$. Prove that S is $1/(1-a)$ by multiplying both sides of the defining equation by a, and by then subtracting the new equation from the old. Test the formula out on $a = 1/2, -1/2, 1/3, 2/3$, and .1. What happens for $a = 1, -1$, or 2?

3. How exactly would you fit $\omega \cdot \omega$ guests into the ω rooms of Hilbert's Hotel?

4. Consider the guest register at Hilbert's Hotel. Each page holds only finitely many names, and new arrivals must always sign in at the next available space. How many pages must the register have so that as long as guests can still be squeezed into the hotel (possibly by rearranging them), there is still room for the signatures of the new arrivals (without rearranging)?

5. In his short story, "The Book of Sand," Jorge-Luis Borges describes an infinite book that i) has no first or last page.[26] The book is written in a foreign alphabet, but includes illustrations on some of the pages. The narrator in the story observes that ii) each page of the Book of Sand has a different, apparently random, natural number in the corner. He also reports that iii) "The small illustrations, I verified, came two thousand pages apart." What type of ordering would the pages of this book resemble?

6. What is the difference between $\aleph_1 \cdot \omega$ and $\omega \cdot \aleph_1$?

7. We say the ordinal number a is even, if for some ordinal b, $2 \cdot b = a$. Is ω even? Is $\omega + 4$ even?

8. The ordinal number a is said to be *regular,* if a cannot be written as the sum of less than a numbers, all of which are less than a. ω is regular since it is not the sum of finitely many finite numbers. $\omega + \omega$ is not regular since it is the sum of two (which is less than $\omega + \omega$) numbers (ω and ω) less than $\omega + \omega$. Ten is not regular since it is the sum of less than ten numbers less than ten, (say $3 + 5 + 2$, or $9 + 1$). Is $\aleph_1$ regular? Only three of the finite natural numbers are regular. Which are these numbers?

9. Points in the usual Cartesian plane are given as pairs (x, y) of real numbers. Let the points in the *Dehn plane* be given as pairs (x, y) of finitely large surreal numbers.[27] (Infinitely small numbers allowed, but infinitely large numbers excluded.) Show that Euclid's Fifth Postulate fails in the Dehn plane–that is, show that there are *many* straight lines through $(0, 1)$

that do not intersect the x-axis at any finite point—and conclude that in the Dehn plane a line can have *more* than one "parallel" through a given point.

10. A recent science-fiction novel states that the following is a law of nature: "Every string which has one end also has another end."[28] Is this necessarily true?

CHAPTER THREE
THE UNNAMEABLE

This chapter discusses three famous logical paradoxes: the Berry paradox, Richard's paradox, and the paradox of the Liar. It is my contention that each of these paradoxes points to the existence of mental concepts that defy any exact formalization. Insofar as the human mind understands these paradoxes, it is in some sense infinite.

The first section deals with the Berry paradox—a paradox having to do with the impossibility of ever explaining exactly how we use language. This paradox has certain implications about the limitations of man-made "thinking machines."

The second section discusses a number of topics relating to Richard's paradox. In particular, we ask if there is, in heaven or on Earth, anything truly random—random in the sense of having no finite description, random in the sense of being infinitely complex.

The "What is Truth?" section delves into the rich variety of problems coming out of the simple sentence: THIS SENTENCE IS NOT TRUE.

THE BERRY PARADOX[1]

What is the biggest natural number you can think of? If your mind is infinite (as it may very well be), then you might insist that you can think of *all* the natural numbers, so I'd better change the question to this: What is the biggest natural number that you can describe? On the one hand, it seems that there can be no greatest number G that you can describe . . . for if you can talk about G, then why shouldn't you be able to talk about $G + 1$, $G + G$, $G \cdot G$, and G^G? On the other hand, it seems that there must be some limit to the sizes of the natural numbers that you can describe . . . for whether or not your mind is really infinite, you certainly will not specifically mention each of the natural numbers in your lifetime.

This last fact seems at first to provide a way out of the dilemma posed by the question of what is the largest number I personally can describe. The fact is that I might once discuss some number G, but then die before ever getting around to mentioning $G + 1$, so it is not, after all, true that if one can talk about G, then one can invariably talk about $G + 1$ as well. But what about talk like this: "Let u_0 be the smallest natural number that I will never name," or "Let W be the smallest natural number that is greater than every number that I will ever name." Can it be that these phrases are not really names? But who can deny that the numbers u_0 and W actually exist, even though their precise values will not be clear until I die?

There is a realm of natural numbers that can be humanly described, and beyond a somewhat variable transition point, there lies a whole other realm of natural numbers that cannot be singled out by any description that is short enough to be humanly comprehensible. It is these unnameable natural numbers that are of interest here.

There is, of course, a sense in which any natural number has a name —the base ten positional representation we are so familiar with. But for you or me, something like 543 . . . 784 cannot really serve as a description of a number if the " . . . " stands for a random string of digits stretching from here to the other side of the galaxy.

To make our original question less subject to individual variations, we will ask it about a general person rather than about you specifically; to make the question more precise, we will require our descriptions to be less than one billion words long. Thus, the question now becomes: What is the biggest natural number that a person can be told of in less than a billion words? Actually, it is more interesting to phrase the question this way: What is the first natural number that cannot be described to a person in under one billion words?

The reason I pick the limit of a billion words here is that I am about thirty-five, and I estimate that so far I have read some 300 million words and have heard some 200 million words, making a total of half a billion words. So a reasonably generous estimate of the number of words a person might absorb in a lifetime seems to be one billion. The point of my question is: What is the smallest natural number that cannot be de-scribed to a person in words?

Perhaps you have noticed that there is something fishy here. Assume there are indeed numbers that cannot be described to a person in words in the space of a lifetime; and assume that there is indeed a *least* such number, which we may as well call u_0. Now, it looks as if *I have just described a particular natural number called u_0. But u_0 is supposed to be the first number that cannot be described in words.*

Paradoxes are compact energy sources, talismans. Contemplating one of Zeno's paradoxes is like running your hand over some callipygous marble Aphrodite, or staring into a mandala. A good paradox can never be finally disposed of. The paradox touched upon in the paragraph above is a version of the problem of how we can talk about things that we cannot talk about. It was cast in the present form by an obscure li-brarian named G. G. Berry, who told it to Bertrand Russell, who put it this way: " 'The least integer not nameable in fewer than nineteen sylla-bles' is itself a name consisting of eighteen syllables; hence the least in-teger not nameable in fewer than nineteen syllables can be named in eighteen syllables, which is a contradiction."[2] The very existence of a paradox such as this can be used to derive some interesting facts about the relationship between the mind and the universe. No one has made such a derivation as boldly as Borges:

> "We (the undivided divinity operating within us) have dreamt the world. We have dreamt it as firm, mysterious, visible, ubiquitous in space and durable in time; but in its architecture we have allowed tenuous and ex-ternal crevices of unreason which tell us it is false."[3]

Actually, since the paradoxes inhere in the very nature of rational thought, I don't think that "we" *could* have chosen to dream a world free of paradoxes. Rather than saying that the paradoxes indicate that the rational world is "false," I would say that they indicate that it is *incom-plete*—that there is more to reality than meets the eye.

In order to reach a richer appreciation of what is involved in the para-dox of the first unnameable number, it is necessary to think a bit about how one goes about naming natural numbers.

NAMING NUMBERS

Imagine. Pursued by a tiny girl in a yellow dress, white ducks move deliberately across a green sward sloping down to stagnant brown water. How many ducks are there?

It is surprisingly difficult to freeze and count the items in a mental image. As soon as you muster the determination and mental clarity to focus on one part of the flock of ducks, the rest of them become hazy and variable.

In Rene Daumal's *Mount Analogue,* the character Father Sogol (who is modelled on G. I. Gurdjieff) claims that we cannot really carry more than four things in the mind at once, and gives the following example:

> "1. I get dressed to go out; 2. I go out to catch a train; 3. I catch the train to go to work; 4. I go to work to earn a living . . . ; now try to add a fifth step, and I am sure that at least one of the first three will vanish from your mind."[4]

The word "number" can be construed as meaning "that which makes numb." The point is that the assigning of numbers to things in the world replaces vibrant confusion with hard and immutable fact. For an idealist, the surprising thing is that it is possible even to *begin* to thus "numb" the world. That is, if one believes that the whole world is a dream, an illusion, an image in some Mind . . . if one believes this, then it is hard to account for the identity between the numbers that different people extract from the world. If I go into the woods and count the branches on a particular dead oak that looks like a dinosaur, and if you do the same tomorrow, then our numbers will certainly agree. You can leave your home for fourteen years, and when you go back, your house is *still* the seventh on the right. The only way to account for the numerical identities among the worlds you and I dream up must be that in some sense you and I are really the same person.

But whether or not we are in some sense the same person, I still have to write the rest of this section, so I can read it and find out about the Berry paradox! The point I was just making is that we get some sort of hold on the world by assigning numbers to things. Now, the way in which we assign numbers to finite sets of things is by *counting.* In order to learn to count we memorize, during the first few years of life, a certain sequence of sounds. To count a small collection you say the successive number sounds as you point once at each member of the collection. The last sound you make is used as the name of the number, or the size or cardinality of the collection.

Any well-remembered sequence could be used as a number list. The Greeks used the letters of their alphabet. If you liked you could use the names in your phone book or the verses of the Bible. In his story "Funes the Memorious," Borges describes a youth with a perfect memory who invents a unique name for each of the first 24,000 numbers:

In place of seven thousand thirteen, he would say (for example) *Maximo Perez;* in place of seven thousand fourteen, *The Railroad;* other numbers were *Luis Melian Lafinur, Olimar, sulpher, the reins, the whale, the gas, the caldron, Napoleon, Agustin de Vedia.* In place of five hundred, he would say *nine.* . . . . I tried to explain to him that this rhapsody of incoherent terms was precisely the opposite of a system of numbers. I told him that saying 365 meant saying three hundreds, six tens, five ones, an analysis which is not found in the 'numbers' *The Negro Timoteo* or *meat blanket.* Funes did not understand me or refused to understad me.[5]

As Borges points out, the drawback with unusual counting sequences is not only that they are hard to learn, but also that they are not based on a system which produces an unending supply of new names.

The simplest system for naming numbers is the tally system. This system names the number n by a sequence of n strokes. Thus, the name of the number we normally call five would be /////, or stroke stroke stroke stroke stroke. The first number not nameable by the tally system in less than one billion words is just stroke stroke stroke . . . stroke stroke stroke, where I have left out 999,999,994 strokes so that there will be room for something else in my life.

Our familiar number system is based on memorized names for the numbers / through /////////; an ingenious system based on powers of ten is used for larger numbers. There are, in principle, definite names for every number. In the United States the name of the number we write as a 1 followed by $3(n + 1)$ 0's is called a (Latin for n)-illion. Thus, if we have a 1 followed by 12 0's, then we have a trillion, since 12 = $3(3 + 1)$ and Latin for three is tri. The number consisting of a one followed by 100 0's is frequently called a *googol,* but it could also be called ten duotrentillion, since $100 = 1 + 3(32 + 1)$ and Latin for thirty-two is something like duo-trentum. Actually, one hardly ever uses the higher "-illion" names, and numbers which are more than thirty digits long are normally read just by listing their digits with the understanding that these digits are to be interpreted in terms of the powers of ten system.

For large numbers it is convenient to use exponential notation. Here googol is written as 10^{100}, and one can readily pass on to *googolplex,* defined as 10^{googol} or $10^{(10^{100})}$. Notice that googolplex is *not* nameable in less than a billion words if we only use the ordinary "-illion" number notation for it. Clearly there will be some numbers near googolplex which are so irregular that they cannot be named in any way much shorter than reading off their digits. These numbers are really unnameable for a human being, since a number that was googol digits long would, if written out on sheets of paper, easily fill space out to the most distant visible star—since I estimate that if we packed ten billion cubic light years with books containing digit after digit, then there would only be room for some 10^{62} digits in these books.

Making estimates of such ridiculously large numbers is an ancient and honorable pastime. There is a well-known treatise by Archimedes called "The Sand Reckoner," where he estimates that it would take under 10^{63} grains of sand to fill a sphere with a radius equal to the distance from the Earth to the Sun.[6] The really interesting thing about his essay is that the Greeks did not have the notion of exponentiation. All they had was the notion of multiplying two numbers, and the largest number that they had a name for was a *myriad* ($= 10,000 = 10^4$).

So how did Archimedes manage to get out to 10^{63}? He took myriad-myriad as his basic building block M ($= 10^8$). What he did then was, without using exponents directly, to find names for all of the numbers out to $M^{(M^2)}$ ($= 10^{(8 \cdot 10^{16})}$). His trick was, for each j and k less than or equal to M, to introduce numbers of the kth order of the jth period. If we call the greatest number of the kth order of the jth period $A(k, j)$; then Archimedes' construction can be summarized by four rules: i) $A(1, 1) = M$; ii) $A(1, j + 1) = M \cdot A(k, j)$; iii) $A(k + 1, 1) = A(k, M)$; iv) $A(k + 1, j + 1) = M \cdot A(k + 1, j)$. Actually, Archimedes could have done better. He should have used a stronger rule than iv)—call it iv*) $A(k + 1, j + 1) = A(k + 1, 1) \cdot A(k + 1, j)$. The largest number he reached was $A(M, M)$, which he called "myriad-myriad units of the myriad-myriad-th order of the myriad-myriad-th period," and which is $(M^M)^M = M^{(M^2)}$, the product of M^2 copies of M. If Archimedes had used iv*) instead of iv), he would have gotten $A^*(M, M) = M^{(M^M)}$, the product of M^M copies of M.

$$\underbrace{M \cdot \ldots \cdot M \cdot}_{\text{1st period} \to M^M} \quad \ldots \quad \underbrace{\cdot M \cdot \ldots \cdot M}_{M\text{th period} \to M^{M^2}}$$

What Archimedes Did

One wonders whether there is any limit to the numbers that could be described on the basis of M, multiplication, and nested iterations of length M. There is a sort of double iteration process that defines what is called the *Ackermann generalized exponential* $G(n, k, j)$ as follows i) $G(1, k, j) = j \cdot k$; ii) $G(n + 1, 1, j) = j$; iii) $G(n + 1, k + 1, j) = G(n, G(n + 1, k, j), j)$. This looks harmless enough, but it turns out that

$M\cdot\ \ldots\ \cdot M\ \ldots\ M\cdot\ \ldots\ \cdot M\ \ldots\ M\cdot\ \ldots\ \cdot M\ \ldots\ M\cdot\ \ldots\ \cdot M\ \ldots$

1st period* $\to M^M$

2nd period* $\to M^{M^2}$

3rd period* $\to M^{M^3}$

$\vdots$

Mth period* $\to M^M$

How He Might Have Continued

$G(2, k, j)$ is the number gotten by multiplying k many j's (also known as j^k); $G(3, k, j)$ is the number gotten by exponentiating a stack of k many j's (also known as $^k j$ or j tetrated to the k); $G(4, k, j)$ is the number gotten by tetrating a stack of k many j's (sometimes known as j pentated to the k); and so on. Clearly $G(M, M, M)$ is going to be a fabulously large natural number.

There is a certain formal criterion that regards the definition of A as singly nested and the definition of G as doubly nested. It turns out that the function taking x into $G(x, x, x)$ is by-end-pieces greater than any function that can be defined using single nesting (recall the by-end-pieces ordering from "Transfinite Numbers.") By nesting our definitions more than two times, we can get more rapidly growing functions and names for larger numbers. It would seem that the limit of what can be done here might be a number P that is greater than any $H(M, \ldots, M)$, where H is a function of M arguments defined by at most M nestings. The idea would be that one cannot systematically reach beyond P without using a systematic procedure that in some dimension is bigger than M.

This does not, however, rule out the possibility of finding a short, but non-systematic, description of some number much greater than P. In 1742, Christian Goldbach made the conjecture that every even number greater than two is the sum of two primes. It is not known whether or not Goldbach's conjecture is true. It *is* true for all the even numbers that anyone has ever looked at: $4 = 2 + 2$ (recall that 2 is the first prime number, as it has no divisors other than 1 and itself), $6 = 3 + 3, 8 = 3 + 5, 20 = 7 + 13, 84 = 23 + 61$, and so on. But maybe Goldbach's conjecture fails for some fantastically large even number. In this case "the least even number after 2 that is not the sum of two primes" could be the name of some number much greater than P.

In itself, the string of symbols $G(M, M, M)$ is not really a name. A name should be self-sufficient, containing definitions of all the symbols and words that it uses. Of course, these definitions involve more words, but the assumption is that ultimately any description should be reducible to extremely simple statements about, let us say, writing down strokes. It is not hard to imagine how one would complete the name $G(M, M, M)$ by adding a description of M, of multiplication, of the Ackerman generalized exponential, and so on. We denote this expanded name by $[G(M, M, M)]$. The idea is that $[G(M, M, M)]$ might be something like the last few pages, along with a few more pages of further amplification. Given this full description, anyone would be able to figure out what number is being named, to the point of being able to come up with a list of $G(M, M, M)$ strokes, if there were no limitation on time.

Names like [the googolth prime number] or $[G(M, M, M)]$ are what we might call constructive names. There is another sort of name, [the least even number greater than 2 that is not the sum of two primes], [the millionth perfect number], or [the first n such that there is a string of 20 sevens that ends at the nth place of the decimal expansion of π]. It is not presently known whether any of these names actually names a number, since we don't know if there are any even numbers greater than 2 that are not the sum of two primes, or if there are actually a million perfect numbers (recall that a number is perfect if it is the sum of its divisors other than itself), or if there is a string of 20 sevens anywhere in the decimal expansion of π. Trying to find the number named by each of these names involves searching through all the numbers until the right sort of number is found. But in each case, the search could be fruitless. If, however, you didn't know this in advance, you would simply continue searching forever.

UNDERSTANDING NAMES

Now that we have discussed various sorts of names of numbers, let's go back to the paradox that started all this. u_0 is supposed to be the first number that cannot be described in under one billion words. It seems that the only way to avoid paradox is to assume that u_0 is not the name or description of any number. This could be because i) every natural number *can* be described in under one billion words, or because ii) there is no way to extend our description of u_0 to a really complete description $[u_0]$ that can be understood by everyone.

Possibility i) is a bit odd. The idea here is that even though there are only a finite number of names that have a length under one billion words, there are infinitely many ways of *taking* these names. Given the open-ended nature of mind and language, a single name can be interpreted in many ways. This is actually less reasonable than it sounds. What possibility i) boils down to is that if u_0 is a name of n, then it must be the name of some $m > n$ as well, so that u_0 itself is actually the name of infinitely many different numbers. The justification for this would be that each of the infinitely many times you say, "The first number not nameable in under one billion words," you will mean something a little more comprehensive by *nameable*. In other words, you start out saying, "u_0" and then you say, "but that wasn't the *real* u_0 since I laid that name 'u_0' on it . . . what I am thinking of now is the *real* u_0," and you get a bigger number; and then you keep doing the same shift over and over forever. The point is that if you think of the concept of *naming* in endlessly more sophisticated ways, then the single name u_0 can actually serve to name each of the natural numbers!

This way of weaseling out the Berry paradox is sometimes explicitly ruled out by requiring that the names for numbers be interpreted in one and only one definite way. In this case, we are forced to accept possibility ii). Possibility ii) says that there is no way to explicate in under one billion words what we mean by "nameable in under one billion words." Where exactly does the difficulty lie? The problem is not in coming up with a list of all the possible combinations of under one billion words. In principle, that can be done quite mechanically. A sufficiently large machine could mindlessly print out all of the possible combinations of under one billion words without any difficulty at all. Assuming that we restrict ourselves to the million or so words that make up the English language, then there will be something like $10^{(6\,\text{billion})}$ sequences of under one billion words.

Let me repeat that the problem in explicating the phrase "nameable in under one billion words" is *not* the problem of actually producing each of the $10^{(6\,\text{billion})}$ possible combinations of words that might constitute a person's lifetime word intake. The problem is, rather, this: there is no way to describe in (under a billion) words a general procedure that will translate any string of (under a billion) words into the number, if any, named by that string of words. Put differently, *there is no way for a person to describe exhaustively how he goes about transforming words into thoughts.*

For suppose that you could come up with some final description of how it is that you turn words into thoughts, names into numbers. This description, call it TRANS, would be so precise that given TRANS and any sequence of words one could, without having to think at all, apply TRANS to the sequence of words and come up with the number, if any, described by the sequence in question. Now, consider this description of u_0: [Mechanically generate one after another of the possible strings of under one billion English words. Apply TRANS to each string in turn, making a list of the numbers obtained in this way. u_0 is the first number not in this list.].

This bracketed description, let us call it $[u_0\text{-TRANS}]$, has approximately the same length as TRANS; so if TRANS could be described in under one billion words, then $[u_0\text{-TRANS}]$ would also weigh in at under one billion words, leading to the unacceptable conclusion that u_0 can be described in under one billion words. So we must conclude that any TRANS that is less than one billion words cannot code up an exhaustive or programmatic description of how to understand every English phrase shorter than one billion words.[7]

The bound of one billion is perhaps too big to be meaningful. Let's drop our word limit down to 200,000, the length of a good-sized book. What we have discovered is that there can be no book-length program that will enable a computer to understand every book. One could, of course, write a "crib sheet" program that provided explicit answers to every question that could ever be asked about every book—but this program would overflow the bounds of the galaxy. The interesting fact that we have learned from the Berry paradox is that there can never be a tricky sort of program that will give a *reasonably short* description of how to understand language.

It is perhaps not too much a caricature of the later Wittgenstein's "ordinary language" philosophy to say he argued that when people talk to each other, they are really just playing a game with noises. You make such a noise and I do this and that. But if it is *even in principle* impossible for anyone ever to formulate a complete set of rules describing how he uses language, then how can anyone assert that learning to use language is just a process of learning to play a certain game according to certain rules? To express this thought in the style of Wittgenstein: If this is just a game following rules, then why can't anyone tell me what the rules are?

Wittgenstein's earlier view of language seems more tenable.[8] According to this "picture theory of language," there are certain relationships among concepts and objects that we perceive in the physical and mental universe around us. In order to point these features out to each other, we use words to set up language structures that somehow mirror or model them. I think such a view of language (which is dependent on an external, objective, but undefinable, concept of "truth") seems much more tenable than any view of language as the working out of a formalizable logical system.

It is worthwhile to give a more precise description of what the Berry paradox tells us about digital computers. The information-theoretic interpretation of the Berry paradox that I am about to describe is due to Gregory Chaitin.[9] Consider a fixed machine M. For the sake of definiteness, let us say that M is a large digital computer with an interactive APL terminal, which has the form of an IBM typewriter with a thick cable leading out the back. We can think of any string of symbols that is typed on the terminal keyboard as being a program P. We say that P is an M-name for the number n if typing the program P causes the machine to type out the number n in the customary base ten notation and then stop.

The length of a program P is defined to be the number of key strokes necessary to input P, and we define the *information-theoretic complexity of n on M* to be the number $I_M(n)$ equal to the length of the shortest program that is an M-name for n. In other words, $I_M(n)$ is the smallest number k such that k properly chosen hits on the keyboard will cause M to print out the number n in base ten notation.

It is not hard to see that $I_M(n)$ will never be longer than the actual number of digits in the number n. For if we simply type the number n on our terminal, then we will not have told M to *do* anything, so M will be stopped, and the number n will be sitting there typed out on the paper in front of us. If one types the "program" "2525252525252525252525252525," then one is immediately in the situation of having 2525252525252525252525252525 typed out and M stopped. However, we could achieve the same end result more economically by typing the instruction: "Print '25' sixteen times and stop." The latter instruction is a bit more economical since it uses 27 keystrokes instead of the 32 keystrokes of the original program. For a more extreme example, consider the case of 10^{1000}. To type out this number in base ten notation requires 1001 keystrokes. But, assuming our M uses the symbol "*" to stand for exponentiation, we can simply type "evaluate 10*1000" on the terminal, which will then busily type out all 1001 digits of 10^{1000} and stop.

We now want to consider the program R, which has a form something like, "Print the first number that has no M-name as short as this program." Now, we do not think of M as automatically endowed with self-knowledge, so the phrase "M-name" must be explicitly definied. This can be done by giving a list of the rules by which M actually operates—in effect, one gives the computer instructions that enable it internally to simulate its own behavior. Having done this, we can find some simply described number r (such as 10*10) such that the program R will have a length much less than r, where R is, "Simulate a machine M that operates as follows: [Description of M goes here]. Now print out the first number that has no M-name shorter than r."

Now, since this program has length much less than r, it must fail to be an M-name for any number, otherwise a contradiction would arise. But how can a definite program such as R fail to name a number? Anyone who has ever tried to program a computer is familiar with the phenomena of *looping* and *endless search*. Certain programs cause the machine to go into an endless loop, never outputting anything; other programs will cause the machine to output an endless string of numbers. When a machine enters either one of these situations, then it will run forever unless some external agent turns it off.

#1: Go to #2	If $n = n$, then print "n."
#2: Go to #1	If $n \neq n$, then stop.
LOOPING	ENDLESS SEARCH

If program R is fed into M, then M will run forever and never come up with an output. Why is this? When M tries to execute the program R, M must go through a process like this: i) list all possible programs of length r; ii) for each such program P, simulate the action of M on P; iii) if M acting on P outputs n, then place n in the set S; iv) let $u_0(M)$ be the first number not in the set S. We know *a priori* that this process can never terminate, so there must be some sort of loop or endless search hidden in it. Where? The loop is in the translation step ii). For one of the programs of length less than r is going to be the program R, and when M tries to simulate the action of M on R, then M must simulate M trying to work through steps i) − iv) of R . . . which eventually leads to simulating the simulation of program R . . . which leads to the simulation of the simulation of the simulation of program R . . .

It is almost as if the poor machine is trying to break through to full self-consciousness with this endless regress of self-simulation. Another way of looking at this problem is to say that M cannot prove that there are any numbers n with complexity $I_M(n) > r$. Now, r is approximately equal to the complexity of a description of M, so one might say that M cannot prove the existence of any numbers with information-theoretic complexity much greater than that of M.

I used $u_0(M)$ above to stand for the first number with complexity greater than r. Suppose that we use $W(M)$ to stand for the first number greater than every number with complexity less than r. That is, $W(M)$ is greater than any number that M can output on the basis of less than r strokes, where r, again, is a number chosen to be greater than the complexity of M. If we think of the output of M as being the number of seconds t that it runs before stopping, instead of being the number n that it prints, then we have an interesting situation. Assuming that M is as complex a thing as can be conceived of, then any instruction fed into M will be less complex than M. Therefore, the output of M with such an input must be less than $W(M)$. If we think in terms of runtime as output, then we can conclude that any instruction less complex than M will cause M to do one of two things: a) run for some time shorter than $W(M)$, or b) run forever.

So now, dropping all the terminology, we can look at a very curious fact. Assuming there is some upper limit to the complexity of the machines we can build, then there is some number W such that any machine which we set into operation must either a) turn itself off in less than W seconds, or b) run forever. It is as if there were certain regions of the future that are entirely inaccessible to us.

To make this problem more colorful, imagine that time will continue indefinitely and that you are going to build a time machine which you can set to take you D days into the future for any particular D that you program into the machine. There would be some future day D_W such that any attempt to get the machine to take you out past D_W would end with the machine running forever and taking you infinitely far into the future. The problem would be that you could in no way even *conceive* of any finite number of days large enough to reach past D_W.

Or, if you don't like time machines, think of this in terms of time bombs. There is a certain future date after which it can be said with assurance that no time bomb set up by us before the year 2000 can still go off (assuming that the timer on the bomb is some sort of digital device that does not malfunction).[10]

Let's go back to the Berry paradox for human beings, instead of for digital computers. One apparent difference between the two is that humans have a certain sort of self-knowledge, which is perhaps not enjoyed by machines. But are we really any better off with out so-called self-knowledge? Don't we run into precisely the same regress as M did when we try to understand what is meant by u_0, the first number that we can't name? For to find the first number we can't name, we have to find out what each name applies to, and take the first number that is left out —but to figure out what each name applies to, we have to figure out what the name u_0 applies to, don't we?

One way out of this regress is to regard a name like u_0 as being a *second-order name,* so that there will be all the numbers nameable (in under a million words, say) without using the second-order concepts such as "nameable"; then there will be the second-order nameable numbers, such as u_0 or $G(u_0, u_0, u_0)$; beyond all these would be u_1, the first number not nameable using second-order concepts, etc. But this type-theoretic way out of the Berry paradox is not really satisfying, for one wants to say, "look, by 'u_0' I mean the first number that can't be named in less than a million words by any means whatsoever, and that includes every order of language that you can invent in under one million words."

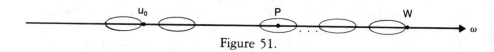

Figure 51.

This absolutist view can be visualized as in Figure 51. All the circled numbers are nameable, and u_0 is the first unnameable number. There will probably be runs of nameable numbers further up the line. For instance, "the largest perfect number" might name some extremely large number $P > u_0$, and then we would have $P + n$ and $P - n$ nameable as well, for numbers n with names not too close to the maximum word limit. Eventually, all the names will peter out and one gets a number W that is greater than *every* number nameable in less than the prescribed number of words.

To avoid paradox, one has to accept the fact that the names u_0 and W are really *not* names. So the concept of "nameability" is itself really unnameable. The symbols n-a-m-e-a-b-l-e *point* to the concept, but they do not really reach it. In somewhat the same sense, the symbol Ω, which is used to stand for the Absolute Infinite, *points* towards the Absolute Infinite, but cannot really denote it. Just as the Absolute lies beyond any possible description, the notion of *nameable in a lifetime* lies beyond any rational human description.

Is this a final solution to the Berry paradox? Not really. For we are still left with the basic problem of how a sentence like "nameability is an unnameable concept" is meaningful, even though the subject of the sentence is a word that cannot denote any single graspable concept. It is curious how interesting it can be to talk about things that we supposedly can't talk about!

One conclusion we might draw is that there are two distinct modes of consciousness: the finite and the infinite. As long as I identify with my body and my rational mind, I cannot conceive of my u_0; but it is not hard to envision my u_0 if I identify with the Absolute. This does not lead to the usual type-theoretic regress, because someone who is merged with the Absolute is in a position to "name" each and every natural number at once.

RANDOM REALS

In principle, every real number r codes up a countably infinite amount of information: the infinite sequence of digits in r's decimal expansion. In practice, all of the real numbers we ordinarily deal with are actually specified by a finite amount of information. Names such as $2/7$, $\sqrt{13}$, π^2, cos 3, $\log_{10}387$ are, in fact, compact and stylized sets of instructions for generating the endless decimal expansions in question.

We will say that a real number is *random* if it has an irreducibly infinite amount of information. That is, a sequence of digits is random if there is no finite way of describing it—no finitely given procedure that can be used to generate the sequence digit by digit. Actually, the word "random" is usually applied only to real numbers that obey certain further conditions having to do with the notion that each nameable subsequence of a random sequence should be random as well.[11] But for our purposes, it will be sufficient to equate randomness with unnameability.

The subsection, "Constructing Reals," contains a historical development of the various sorts of names for real numbers that are used, with particular emphasis on the Greek methods of constructing reals. The subsection, "The Library of Babel," presents and analyzes the idea of a Total Library containing all possible books. In the somewhat technical subsection, "Richard's Paradox," a construction discovered by Jules Richard is examined. This construction seems to produce a random real by diagonalizing over the set of reals having names in the Total Library; it raises some of the same issues as did the Berry paradox. In the final subsection, "Coding the World," the question of the physical existence of random reals is investigated.

CONSTRUCTING REALS

In this section it will be useful to think of real numbers as points on an idealized continuous line. The points called 0 and 1 must be marked arbitrarily, but from then on the correlation between real numbers and points on the marked line is more or less automatic.

The ordinary representation of a real number as an endless decimal expansion can be viewed as a description of an *infinite* procedure for locating a particular point (or infinitesimal neighborhood) on the marked line. If we allow ourselves the use of various standard curves and surfaces, then there are many real numbers for which there is a *finite* procedure for locating the corresponding point on the marked line.

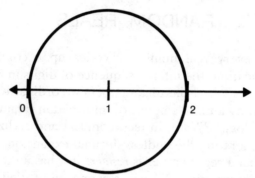

Figure 52.

Assuming that we have a compass, the point called 2 can be found in a finite amount of time by drawing the circle with its center at 1 and radius equal to the interval between 1 and 0. It should, perhaps, be pointed out that, in practice, the perfect drawing of this circle would actually be an infinite process. For even if our lines had no thickness and our compass's drawer and stabber were points, there would still be the difficulty of managing to put the stabber exactly through the point 1 and contriving to set the drawer down exactly on the point 0. In practice, matching up two points is an infinite process of repeatedly reducing one's error—a feedback loop. To avoid this objection to the finiteness of the construction of 2, one may as well assume that perfect circles with specified centers and radii will spring into existence on command, as will perfectly straight lines through specified pairs of points. This, after all, is the content of Euclid's third and first postulates, which do not actually mention "rulers and compasses."

The upshot is that given 0 and 1 we can, without idealizing too much, find the point corresponding to the real number 2 with infinite precision in a finite amount of time. The same is true of any real number whose decimal expansion terminates or repeats, for such a real number is rational, and every rational point on the marked line can be found with ruler-compass constructions.

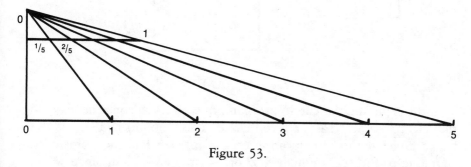

Figure 53.

As an example, I show how one might find the point corresponding to the real number 2.40000. . . . This is the same as $2^2/_5$, so the core of the construction is to find a line segment of length $^2/_5$. This is done in Figure 53 by means of a trick that uses similar triangles to divide a given interval (in this case a copy of the interval between 0 and 1) into five equal pieces.

Using ruler and compass we can not only construct any rational number, but also construct any number obtained from the rationals by combining the operations of addition, subtraction, multiplication, division, and the taking of square roots. (Recall that in the section "From Pythagoreanism to Cantorism" I showed how to take square roots with ruler and compass.) So a point on the marked line corresponding to,

say, $\dfrac{\sqrt{3 + \sqrt{21}}}{308}$ or $\sqrt{3\sqrt{\sqrt{2}}}$, can be found with infinite precision in a

finite number of time if we are allowed to call circles and lines into existence as desired.

The three famous problems of antiquity have to do with the question of which ruler-compass constructions are possible. I refer here to the problems of Duplicating the Cube, Trisecting the Angle, and Squaring the Circle. The first and third of these problems are of special interest here.

In the Duplicating the Cube problem, one wants to construct a cube that has twice the volume of a given cube. It is not hard to see that this is

tantamount to finding a finite method for locating on the marked line
the point corresponding to the real number $\sqrt[3]{2}$.

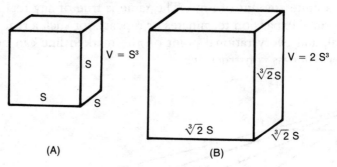

(A)   (B)

Figure 54 (A–B).

In the Squaring the Circle problem, one wants to construct a square
that has the same area as a given circle. This amounts to being able to
find a finite method for locating on the marked line the point corresponding to the real number $\sqrt{\pi}$ or, what is equivalent since we can
square and take square roots with ruler and compass, the real number
π.

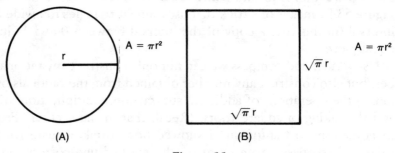

(A)   (B)

Figure 55.

Someone brought up on the modern theory of real numbers is prone
to feel that simply writing down and understanding the symbols $\sqrt[3]{2}$
solves the problem of Duplicating the Cube. For, after all, once you understand that $\sqrt[3]{2}$ means "the real number whose cube is 2," you can set
out to determine digit after digit of the decimal expansion of $\sqrt[3]{2}$.

$$1^3 < 2^3, \qquad \text{so } \sqrt[3]{2} \text{ begins with 1}$$
$$1.2^3 < 2 < 1.3^3, \quad \text{so } \sqrt[3]{2} \text{ begins with 1.2}$$
$$1.26^3 < 2 < 1.27^3, \text{ so } \sqrt[3]{2} \text{ begins with 1.26}$$

.

.

.

If you like, you can locate the points corresponding to each of the initial pieces of the endless expansion of $\sqrt[3]{2}$, and after an infinite amount of time you will reach the limit point that represents $\sqrt[3]{2}$.

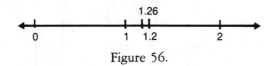

Figure 56.

But this is nothing the Greeks did not know about. They were fully aware that one can find points that better and better approximate the location of $\sqrt[3]{2}$ on the marked line, or that given a marble cube one can find larger marble cubes that come closer and closer to weighing twice as much as the original cube.

By the process of trial and error we can always come closer and closer to finding a continuous magnitude with some specified property. The *axiom of continuity* (first explicitly introduced by Dedekind) asserts that there is a single right magnitude that exists as the limit of any such process. In modern times we have found it convenient actually to identify continuous magnitudes with such processes—a particular continuous magnitude is represented by the specific process of trial and error coded up in the decimal expansion of the corresponding real number.

But when the Greeks posed the problem of duplicating the cube, they wanted a *finite* method of constructing a line segment with length *exactly* equal to $\sqrt[3]{2}$. They were suspicious of infinite processes, holding that no such process could legitimately be regarded as completed, and in the absence of a finite construction of the location $\sqrt[3]{2}$ they might even have questioned whether any such perfectly right location really exists at all.

The first person to come up with a finite method for constructing $\sqrt[3]{2}$ was Archytas, a member of the Pythagorean sect. Archytas was a friend of Plato, and is said to have once prevented Dionysius from killing Plato (Dionysius being Plato's school master). Archytas is also credited by Aristotle with having invented a model bird that flew and a special sort of baby rattle—all of this in the fourth century B.C.

Archytas's method for finding the cube root of two involves considering the intersection point of a torus, a cylinder, and a cone. One starts with the circle with diameter OC equal to 2 (Figure 57A). On this circle, find a point B such that the chord OB has length 1.

We generate the torus by rotating a copy of this circle about a line through 0 and perpendicular to OC. We generate the cylinder by mov-

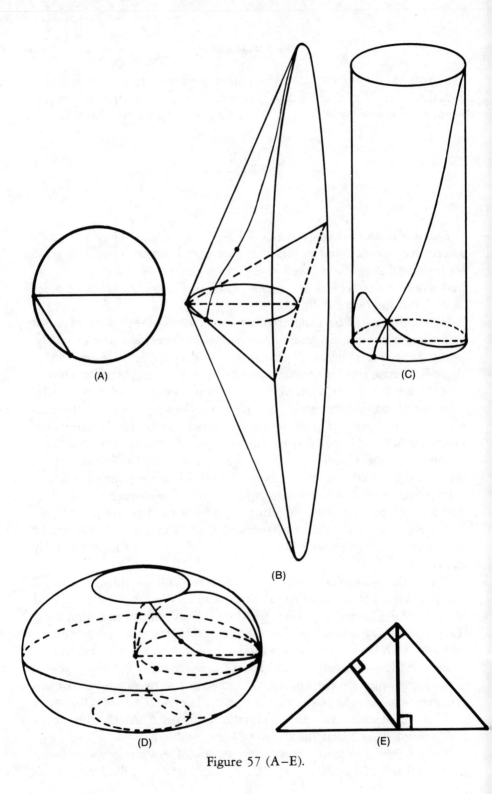

Figure 57 (A–E).

ing the circle perpendicular to itself. We generate the cone by extending the chord OB until it intersects a line passing through C perpendicular to OC, and by then rotating the triangle formed about this line.

If these three surfaces are placed so that the original circles match, then there will be a point P where all three surfaces intersect. Directly beneath P will be a point X on the original circle, and the distance OX equals $\sqrt[3]{2}$. In order to vizualize where P is, I have drawn on the cylinder the curves where the cylinder would intersect the cone and torus. P lies at the intersection of these two curves.

The reason that OX is $\sqrt[3]{2}$ has to do with the fact that this particular construction leads to a right triangle of the kind drawn in Figure 57E. Because all of the triangles in this figure are similar, we get the continued proportion that leads to the solution of the Duplicating the Cube problem.

The Duplicating the Cube problem arose when the oracle at Delos told the Delians that a plague they were currently suffering would be lifted only if they would double the size of the temple altar. The Delians were clever enough to realize that the volume of *any* three-dimensional object is doubled if each of its dimensions is increased by a factor of $\sqrt[3]{2}$, and thus the problem was born.

The Greeks seem to have realized quite soon that no ruler-compass construction would suffice to construct cube roots. There are a number of recorded solutions to the Duplicating the Cube problem, all of them involving some sort of higher-order curve or surface. A thorough catalogue of these solutions appears in T. L. Heath's classic work, *A History of Greek Mathematics,* written during the First World War.

In his preface, Heath quotes a remark of Plato's on the Duplicating the Cube Problem that must have seemed particularly apt during those war years: "It must be supposed, not that the god specially wished this problem solved, but that he would have the Greeks desist from war and wickedness and cultivate the Muses, so that, their passions being assuaged by philosophy and mathematics, they might live in innocent and mutually helpful intercourse with one another."[12]

It was not conclusively proved that $\sqrt[3]{2}$ cannot be constructed using only lines and circles until early in the 1800s, but, as I have mentioned, the Greeks suspected this from the very beginning, and took the Duplicating the Cube problem as an impetus to move on to more complicated, *but still finite,* methods of construction.

It should be noted that Archytas's method is a fairly natural extension of ruler-compass methods. Ruler-compass methods allow one to: i) form the trail of a point moved in any fixed direction; ii) form the trail of a point rotated about any other point. To get the cylinder, torus, and cone of Archytas we need only be allowed to: i) form the trail of any plane curve moved in any fixed direction; and ii) form the trail of any curve (ie., plane figure) rotated about any line.[13]

As we know since Ferdinand Lindemann's proof of 1882, the number pi is transcendental, which means that it can never be constructed by taking intersections of simple (algebraic) curves and surfaces a finite number of times. One might be inclined to say that pi can be constructed simply by rolling a circle of diameter one through one revolution, but this seems to involve the notions of motion and of time, which are perhaps extra-geometric. Actually the notion of rolling the circle

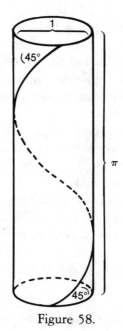

Figure 58.

can be statically represented by the helix in Figure 58. This is a helix that moves up just as fast as it moves around the cylinder, making a 45 degree angle with the cylinder's generators at each point. Evidently, the vertical change produced during one complete revolution will be equal to the circumference of the circle.

A somewhat more constructive process for getting pi in a finite amount of time is due to Archimedes, and is illustrated in Figure 59.

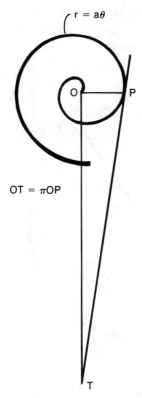

Figure 59.

The idea is that one takes an Archimedean spiral, obtained by rotating a line at a constant rate while a point moves out along this line at some other (not necessarily related) constant rate. If the spiral starts at O and completes its first turn at P, then one can draw the line PT tangent to the spiral at P and the line OT perpendicular to OP at O, and the point T where these two lines cross will determine a distance OT equal to pi times OP.

The Greeks also used a much more contemporary method for finding approximations of pi: the method of exhaustion. Originated by Antiphon and perfected by Archimedes, this method consists simply of inscribing and circumscribing polygons with more and more sides upon a given circle. One can, in general, compute the perimeter of any polygon, so that by finding the perimeter of, for instance, a 96-sided regular polygon that just fits inside a circle with diameter one, it is possible to get a good approximation of pi. By precisely this method Archimedes showed that $3^{10}/71 < \pi < 3^1/7$, and there is, in principle, no limit to the accuracy obtainable by this method (although, of course, infinite precision will never be obtained in a finite amount of time in this manner).

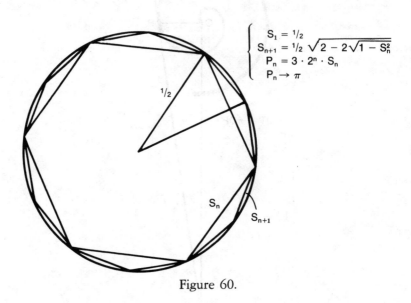

$$\begin{cases} S_1 = 1/2 \\ S_{n+1} = 1/2 \sqrt{2 - 2\sqrt{1 - S_n^2}} \\ P_n = 3 \cdot 2^n \cdot S_n \\ P_n \to \pi \end{cases}$$

Figure 60.

I have illustrated Archimedes's method of exhaustion in Figure 60. The idea is that one starts with a regular inscribed hexagon at stage 1, and that by repeatedly bisecting the subtended arcs of the circle one gets to a $3 \cdot 2^n$-gon at stage n. The length S_n of the side of a $3 \cdot 2^n$-gon inscribed in a circle of radius $1/2$ is given by the recursive formula displayed in Figure 60. To see how the formula is obtained one can think of the figure as illustrating the general transition from S_n to S_{n+1}, and by using the Pythagorean theorem twice the formula can be obtained. Given the side S_n of the regular $3 \cdot 2^n$-gon inscribed in a circle of radius $1/2$, we can approximate π by the perimeter $P_n = 3 \cdot 2^n \cdot S_n$ of this polygon.

In a sense, the three formulas in Figure 60 give a finite description of pi, and in a sense they do not. On the one hand, the three formulas can be used to compute P_n for arbitrarily large values of n, and for large enough n, P_n will be arbitrarily close to pi. On the other hand, these three formulas will never give the exact point pi on the number line after any finite length of time.

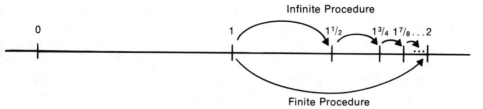

Figure 61.

Zeno's bisection paradox brings this distinction into sharp relief. If you are at 1 on the number line, there are two ways to get to 2. You can use the finite procedure of moving one unit all at once; or you can use the infinite procedure of moving $1/2$ unit, then $1/4$, then $1/8$, etc. If we ignore the possibility of there being distinct points an infinitesimal distance apart from each other, then it is evident that these two procedures lead to the same point. This fact is usually represented by the equation $1 + 1 = 1 + 1/2 + 1/4 + 1/8 + \ldots$. Zeno viewed this as paradoxical. He assumed *a priori* that no actual infinity could exist, so that no infinite process could be regarded as completed; therefore the equivalence between a finite and an infinite process seemed impossible.

As I have mentioned before, the ordinary decimal expansion of a real number actually represents an infinite process for finding a point corresponding to the number in question. Although infinite, this process is quite definite, and we feel comfortable with it. It should be realized, however, that for many of these real numbers there is no alternate finite process at all for finding the length described.

Take, for instance, the number $L = 10^{-1} + 10^{-2!} + 10^{-3!} + 10^{-4!} + 10^{-5!} + \ldots = .11000100000000000000000010000000000000000000000000000000000000000000000000000000000000000000000000000000000010\ldots$. (Recall that $n!$ means n factorial, the product of the numbers one through n). In 1844 Joseph Liouville proved that this number is transcendental, meaning that L is not the root of any polynomial with rational coefficients. There are many transcendental numbers, but L was the first number for which anyone was able to *prove* that it is transcendental. In fact, L is artificially tailor-made to make just such a proof possible.[14]

Since L is transcendental, it can never be found by a method such as Archytas's technique of intersecting algebraic curves and surfaces. Since L is highly artificial, it is most unlikely that a segment of length L can ever be found by any other finite method . . . such as the spiral and helix constructions of π. So L is a number for which the corresponding length can only be found by the sort of infinite process questioned by Zeno.

One could, of course, idealize a bit more than previously and imagine a device—call it a Summer—that will construct lengths corresponding to numbers such as L. The Summer geometrically sums up any infinite series in one second. Thus, given a decmal expansion $.r_1r_1r_3$ . . . , the Summer moves a pointer to $.r_1$ in the first $^1/_2$ second, finds $.r_1r_2$ in the next $^1/_4$ second, fnds $.r_1r_2r_3$ in the next $^1/_8$ second, and at the end of one second, the Summer's pointer indicates the point corresponding to the decimal series $.r_1r_2r_3$. . . .

This is really not *so* great an idealization—recall that we already had to idealize in order to talk about ruler-compass constructions. In a sense, when you move a pointer from 1 to 2 on the number line in one second, you are acting as a Summer on the series $1 + ^1/_2 + ^1/_4 + ^1/_8 + . . .$! This, incidentally, was part of Aristotle's answer to Zeno's bisection paradox: Since we do indeed act as Summers when we move in time, there is nothing fundamentally contradictory about them.

One must be careful, however, not to feed a divergent series into the Summer, or it will break. If, for instance one gives the Summer the Grandi series to work on, then it will shake itself to pieces. The Grandi series is the series $1 - 1 + 1 - 1 + 1 - 1 + . . .$ Looked at one way this is $(1 - 1) + (1 - 1) + . . . = 0$; but looked at another way, we have $1 + (-1 + 1) + (-1 + 1) + . . . = 1$!

Grandi discovered this series in 1703; he claimed that God must have used a technique based on this series in order to create Something from Nothing, and thus get the cosmos going.[15] This is not really as insane as it sounds—a more sophisticated way of saying something similar would be to claim that the cosmos is a moiré, an interference pattern produced by a wave function out of phase with itself.

But in any case, if one feeds the Grandi series into the Summer, the pointer is going to twitch back and forth between 0 and 1 infinitely many times at a speed that reaches infinity as the second runs out. Unless some special provision (e.g., "point to 1 when in doubt") has been built into the Summer, there is simply no reason why it should point to any one point instead of any other point at the end of a second . . . so I would prefer to say that it breaks.

The question of what a Summer does with divergent infinite series has been thoroughly discussed by philosophers of science as the prob-

Figure 62. The Summer at work on the Frandi Series.

lem of the Thompson Lamp.[16] There is no problem with the action of a Summer on a convergent series such as the decimal expansion of a real. The increments of motion become smaller and smaller, and at the end of the second the pointer has smoothly come to rest at a particular point. When you hold a hard pencil between your two fingers and let the tip bounce on a table, you see a motion something like that of the pointer of a Summer acting on a convergent series. In principle, the pencil point bounces an infinite number of times, but the total distance travelled can be computed to be finite, and the total time elapsed is finite as well. In particular, if the pencil point starts out at a height of one inch above the table, and each time it bounces back to $^9/_{10}$ the height from which it has dropped, then it will fall a total distance of $1 + ^9/_{10} + (^9/_{10})^2 + (^9/_{10})^3 + \ldots = 10$, and the time elapsed can similarly be seen to be finite.

$$
\begin{aligned}
D &= 1 + ^9/_{10} + (^9/_{10})^2 + (^9/_{10})^3 + \ldots \\
- \quad ^9/_{10}\,D &= ^9/_{10} + (^9/_{10})^2 + (^9/_{10})^3 + \ldots \\
\hline
^1/_{10}\,D &= 1 \\
D &= 10
\end{aligned}
$$

From now on, we will say that a real number r has been finitely described if there is some finite description of how to generate the decimal series expansion $R.r_1r_2r_3 \ldots$ of r. In general, such a finite description will consist of some general set of instructions that, when applied to any natural number n, will institute a process terminating with the evaluation of the digit r_n. Any such description of a general function giving an R and an r_n for each n can be fed into the Summer to locate a particular point on the real line.

There is, of course, a little vagueness in the words, "finite description" and, as we will see in the "Richard's Paradox" subsection, this vagueness leads to complications like those that came out of the Berry paradox concerning the words "describable in less than a billion words."

THE LIBRARY OF BABEL

Eventually we want to try to find a random real number—that is, a real number that has no finite description whatsoever. But before going on with this project, we need to get a clearer notion of the set of all possible finite descriptions.

It is initially easier not to be concerned with questions of meaningfulness, and to view *any* string of symbols whatsoever as a "possible description." The set of all finitely long strings of typographic symbols is, in any case, an intrinsically interesting set.

Related sets have been discussed in "The Universal Library" by Kurd Lasswitz[17] and in "The Library of Babel" by Jorge Luis Borges.[18] The Borges story is a first person account narrated by an inhabitant of a fantastic library consisting of an apparently endless number of hexagonal rooms lined with identically bound volumes. Each volume has 410 pages consisting of forty lines of eighty symbols each. Most of the books seem to be meaningless jumbles of letters.

The narrator and his fellows spend their whole lives wandering in this library, ceaselessly speculating on what it all means. Some believe that every single book has meaning—if not in Spanish, then in English or Hungarian; if not in any known language, then in a language of the future or in code. But a book consisting, say, of 410 pages of *b*'s seems to be meaningless under *any* interpretation; and the narrator draws the conclusion that the library actually includes every possible book-length string of symbols.

He feels that the library includes everything, and one of those marvellous Borgesian lists ensues:

> Everything: the minutely detailed history of the future, the archangels' autobiographies, the faithful catalogue of the Library, thousands and thousands of false catalogues, the demonstration of the fallacy of those catalogues, the demonstration of the fallacy of the true catalogue, the Gnostic gospel of Basilides, the commentary on that gospel, the commentary on the commentary on that gospel, the true story of your death, the translation of every book in all languages, the interpolations of every book in all books.[19]

In a sense a library like this is useless. Randomly selecting a book from the Library of Babel is equivalent to sitting down and randomly typing 410 pages. Even if you were, by some miracle, to find a book in the Library that seemed to provide a solution to Cantor's Continuum Problem, you would have to check very carefully to make sure that you had not obtained one of the thousands of false versions of this book . . . and even if your book seemed to be without error, it might be possible to find another error-free book that provided a very different solution to Cantor's Continuum Problem. Looking at the titles of the books would be of no help, for a book called *The Continuum Problem* might turn out to be about, say, astral travel.

The narrator of "The Library of Babel" actually claims too much for his library in the quote above. The detailed history of the future would (unless the world is to end quite soon) probably not fit into any 410-page volume; and the catalogue of the immense Library of Babel would certainly not fit into any 410 pages. To be really certain of having everything in one's library, one should allow books to have any finite length, no matter how long. But then one's Library becomes infinite.

In a belated urge to economize, Borges limits the symbols used in the books in his Library to twenty-five: the lower-case letters of the alphabet (excluding h, k, w, and x), the period, the comma, and the space. But it is too late for economy. Each book in his Library has 1,312,000 slots ($= 410 \times 40 \times 80$), so there are $25^{1,312,000} \approx 10^{2,000,000}$ books in the Library of Babel . . . barring repetitions.

Kurd Lasswitz arrives at the same figure for the number of volumes in his Universal Library, and to point out the largeness of this number he remarks that if his books were placed side by side it would make a shelf about $10^{1,999,982}$ light-years long. The fact is that there are so many books in this Library that the number of light-years of books is not substantially smaller than the number of books.

As I mentioned above, if we allow our books to be arbitrarily long, then there are infinitely many books in what could be called the Total Library. We can see that the Total Library is infinite, since for each natural number n, there must be a book consisting of n repetitions of the word "yam." Thus all of these are books in the Total Library: yam, yamyam, yamyamyam, ad inf.

Given that the Total Library is already going to be infinite, there is not much point in trying to cut corners on the number of symbols used in the books. A book such as *Infinity and the Mind* uses something like three hundred typographical symbols, but to keep the rest of the discussion manageable, we'll limit ourselves to the seventy-five most basic symbols: the space, the lower and upper case Roman alphabets, the digits 0 through 9, the apostrophe, comma, dash, semicolon, colon, period, exclamation point and question mark, the left and right quotation marks, and the left and right parentheses.

Is there any danger that the Total Library is *uncountably* infinite? If you are careless, you might think that it is, reasoning as follows: I can create an arbitrarily long finite book using the seventy-five basic symbols by choosing one of these symbols ω times. This can be done in $75 \times 75 \times 75 \times \ldots = 75^{\omega} = c$ ways; therefore, the Total Library has the uncountable cardinality of the continuum.[20]

The flaw in this argument is that what has really been calculated is the number of books of length ω, rather than the number of books having length less than ω. The Total Library has cardinality $\aleph_0$, and we prove this by constructing a one-to-one map, called CODE, from the set of all finite books into the set of natural numbers. (This proves that the cardinality of the Total Library is $\leq \aleph_0$, and since the "yam" argument shows that this cardinality is $\geq \aleph_0$, the rules of transfinite arithmetic imply that the cardinality in question is *equal* to $\aleph_0$.)

To set up the CODE map we start by assigning a ditigal code to each of the seventy-five basic symbols.

	-1	y	-28	X	-56
a	-2	z	-29	Y	-57
b	-3	A	-31	Z	-58
c	-4	B	-32	0	-59
d	-5	C	-33	1	-61
e	-6	D	-34	2	-62
f	-7	E	-35	3	-63
g	-8	F	-36	4	-64
h	-9	G	-37	5	-65
i	-11	H	-38	6	-66
j	-12	I	-39	7	-67
k	-13	J	-41	8	-68
1	-14	K	-42	9	-69
m	-15	L	-43	'	-71
n	-16	M	-44	,	-72
o	-17	N	-45	-	-73
p	-18	O	-46	;	-74
q	-19	P	-47	:	-75
r	-21	Q	-48	.	-76
s	-22	R	-49	!	-77
t	-23	S	-51	?	-78
u	-24	T	-52	"	-79
v	-25	U	-53	"	-81
w	-26	V	-54	(	-82
x	-27	W	-55	)	-83

We are careful to use only code numbers with no zeros in them, reserving zero for a different purpose. To code up a given string of symbols, we replace each symbol by its code number, put zeros in between the code numbers so we can tell them apart, and then stick everything together to get a big natural number.

Yes?
Y   e   s   ?
57   6   22   78
5706022078
5,706,022,078

Using the zeros as spacers makes it possible to decode any code number.

45,017,077
45017077
45   17   77
N   o   !
No!

Of course, most numbers don't code up anything at all under this system. Thus, 235,794 simply doesn't code up anything. But the point is that every book in the Total Library *is* coded up by some finite natural number.

The reader may enjoy checking that the number CODE (*Moby Dick*), which codes up the text of *Moby Dick,* starts out 330201401401015-060103902209015020601407 60 . . .

It will be convenient later on to think of the $\aleph_0$ books in the Total Library as being listed $B_0, B_1, B_2, \ldots$ in order of the size of their code numbers. In effect, this arrangement lists all the books one symbol long, then all the books that are two symbols long, and then the three-symbol-long books, and so on.

Coming back to the coding process for a minute, notice that there is, in principle, no reason why a child could not be taught from the beginning to read book codes instead of books. You'd teach him that 2, 3, 4, . . . is the alphabet (pronounced hay, bee, sea, . . .), teach him to use 76 at the end of a sentence, teach him always to separate his symbols by zeros, and so on.

For a person taught to read in this manner, the Total Library would simply be an infinite subset of the set of natural numbers. "Have you read 3,702,102,025,011,023,028,071,022,010,490,201,101,603,017,-026?" "Yes, I liked it even more than his 54." The point is that there is nothing sacred about our particular letter symbols. What is essential in a book is the overall pattern these symbols form.

A curious thought arises here. Suppose that we gave the code numbers for all the books in the Library of Congress to some blob from outer space. To the extent that the meaning of any word can be explained, the explanation of all the recurring symbol patterns will be provided by other symbol patterns in this welter of data. Would the alien ever be able to figure out what the books were about? Even if he couldn't understand them in the usual sense, would he be able to appreciate the abstract patterns of symbols formed by the classics?

Note that these questions would equally well arise if the alien were given the books in English. The point of considering the *codes* of the books in this example is just to bring out the fact that given a text of ours, an alien would have only the text's abstract structure to work with.

With this in mind, the scientists interested in extraterrestrial communication have devised certain very simply patterned messages to be beamed into outer space. A recent spacecraft also carried some more complex information patterns for the aliens, notably a recording of Chuck Berry's "Johnny B. Goode." Reading off the digital coding of this song, the aliens may notice certain regularities of pattern, certain mathematical progressions.

A song is a curious sort of information pattern in that it has no real content—it is appreciated simply for its form. One might be tempted to think that we can somehow teach aliens what our words mean, perhaps by means of chemical formulae, etc. Still, there seems to be a large portion of our language experience that can be taught only by direct demonstration. "There, feel that, drink . . . that's *water,* Helen, *water.*"

It may be that the aliens will enjoy our messages only sensuously, in the way we enjoy music and abstract art. Then again, they may read meanings peculiar to their own world-views into our messages.

The discussion of aliens looking at our books is quite relevant to the question of when a string of symbols *names* a real number. For some, the two symbol string "pi" names a definite real number. Someone else might prefer the longer name, "pi is the ratio of the circumference of a Euclidean circle to its diameter." An individual unfamiliar with mathematics might require this last definition to be amplified to a complete treatise on plane geometry.

Presumably, a complete enough treatise would enable any type of thinking creature to derive and use the formulae for pi set out in Figure 60, even if this creature had no idea of the kind of visual and tactile experiences that humans associate with "circles."

A different approach to naming pi would have been simply to start with these formulae and some sort of an explanation of how to use them. Alternatively, one could use the description of pi that is usually proved towards the end of a second-year calculus course: "Pi is the limit of the infinite series $4 - \frac{4}{3} + \frac{4}{5} - \frac{4}{7} + \frac{4}{9} - \ldots$."

In a sense, digital computers are the only aliens that we are able to talk to at all, so they are often taken as a standard of nameability. Every large general-purpose computer is fundamentally the same (they are all "universal Turning machines"), so we can talk about a general computer C without having to be too specific. One can formally say that the book B_i *names* the real number $r = K.e_1e_2e_3e_4 \ldots$ provided that feeding B_i into C throws C into such a state that if zero is fed into C, C prints out K, and if any n greater than zero is fed into C, then C prints out the nth digit e_n of the decimal expansion of r.

In the next subsection we will use the phrase "B names the real number r" in several ways. If the specific way just given is meant, then we will sometmes say "B is a C-name for the number r" to emphasize this.

RICHARD'S PARADOX[21]

Let M be some type of being: a computer, a human, the human race as a whole, a thinking galaxy, or God Himself. We say that the finite string of symbols B is an M-name for the real number s exactly when M is able to give the number s on the basis of the information in B. As was discussed in "Constructing Reals," we may know the value of a real number, even though we cannot give the full infinite decimal expansion all at once. We will say that M is able to give us the real number s, provided that M is in a position to give the nth digit of the decimal expansion of s for any desired n. Thus, I say that I am able to give the real number pi, not because I have the entire decimal expansion in my mind at once, but, rather, because for whatever n you mention, I can, in time, respond with the nth digit of the decimal expansion of pi. I can do this because I have a certain technique for computing more and more digits of the expansion of pi. For me, a string of symbols describing such a computation technique serves as a name for the real number pi.

Now, let us fix our attention on some one particular M and consider the set E_M of all the real numbers that have a finite M-name somewhere in the Total Library. It seems that since we can find a translation function Trans $_M$ that maps the countable Total Library onto the set E_M, E_M must be countable. Now, the diagonalizing technique studied in Excursion I shows how to find a real number different from every member of any given countable set of reals, so it seems that for any M, there will be real numbers that do not have any finitely long M-name; these reals might be called M-*random* reals.

The question of whether or not there is a best M comes up here. That is, is there a sort of ultimate M such that if a real number s has any finite name at all, then it has an M-name? If this is indeed the case, then we can think of the M-random real numbers as being random in the absolute sense of having no finite description at all.

Since it is the nature of infinite sequences of digits that really concerns us here, it will be no great loss if we restrict our attention to the real numbers between zero and one—that is, the real numbers with nothing to the left of the decimal point.

Returning to the M's, it is simpler to think of them behavioristically. An M is something that turns natural numbers into real numbers, and if two M's behave the same we regard them as identical. So we can identify an M with a certain list of real numbers, which can in turn be coded up by a single real number.

In general, for a given M and a given natural number n, we can define $\text{Trans}_M(n)$ to be the real number $.e_{n1}e_{n2}e_{n3}e_{n4}$ . . . given by the definition.

$$\text{Trans}_M(n) = \begin{cases} \text{The real number between zero and one, if any, whose} \\ \quad M\text{-name is coded up by } n \\ .999 \text{ , , , otherwise} \end{cases}$$

In terms of behavior, M is given by the doubly infinite square array of all the e_{nk}. By means of a certain sort of shuffling, we can fit all of these

$$T_M = .e_{11}e_{12}e_{21}e_{13}e_{22}e_{31}e_{14}e_{23}e_{32}e_{41}e_{15} \quad \cdots$$
$$= .1\,2\,9\,3\,9\,2\,0\,9\,5\,0\,3\,9\,2\,8\,0 \ldots$$

e_{nk} into a single ω-sequence, which can be regarded as a real number that might as well be called T_M or $.m_1m_2m_3.\,\ldots$

There is another interesting number obtainable from the square array given. This is the diagonal number $d_M = .d_1d_2d_3 \ldots$, defined so that

$$d_n = \begin{cases} e_{nn} - 1 \text{ if } e_{nn} \text{ is not } 0 \\ 1 \text{ if } e_{nn} \text{ is } 0. \end{cases} = \begin{cases} m_{(2n^2-2n+1)} - 1 \text{ if } m_{(2n^2-2n+1)} \text{ is not } 0 \\ 1 \text{ if } m_{(2n^2-2n+1)} \text{ is } 0. \end{cases}$$

Now, d_M is different from every one of the $\mathrm{Trans}_M(n)$, which is to say that d_M is M-random and has no M-name.

A little thought shows that d_M can be defined directly from T_M since e_{nn} is always in the $2n^2 - 2n + 1$ place of T_M. So we could also use the right-hand definition of d_M. The dependence of d_M on T_M can be expressed by saying $d_M = f(T_M)$.

We will say that a naming system M is *closed* if whenever M has a name for some real number s, then M also has a name for real numbers that have simple definitions in terms of s. In particular, we say that M is closed only if whenever M names a real number s, then M also names the real number $f(s)$, meaning that if M names the code T_N of some naming system, then M will also name the diagonalization d_N of this naming system. Given the definition above of d_M in terms of T_M, we can see that any naming system that would naturally be adopted by a rational being would be closed in this sense.

Now, consider a closed naming system M. M cannot name the diagonal number d_M, since this number is constructed to differ from every real number with an M-name. If M has a name for T_M, then since M is closed, M will also name d_M—but this is impossible. Therefore, M does not have a name for T_M. In general, no closed naming system can name the real number T_M that codes up this system.

This fact was first discovered by Jules Richard in 1905. Richard was a French high school teacher at the time. He formulated the fact that no closed M names T_M as a paradox by taking his naming system to be a given and evident universal relation. That is, he assumed that "B is a name for s" is a relation that is already perfectly clear. But if this relation is already perfectly clear, then the real number T coding up all the nameable reals is clearly defined, and then d (the diagonalization) is clearly defined. But d cannot be nameable since it differs by construction from every nameable real. So, if we assume that the relation of naming is clearly named by the word "name," then we can name a number d that differs from every number that we can name. Such is Richard's paradox.

Richard himself was able to see that the way out of the paradox is to deny that the M one has in mind can name T_M. As he puts it, the diagonal number is really only named by M if the translation code T_M is totally defined, "and this is not done except by infinitely many words."[22]

Is there any deeper significance to the fact that no closed naming system M names its translation code T_M? Looked at quite formally, we are simply saying

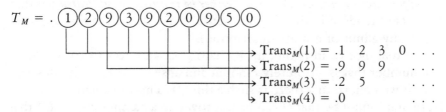

$$T_M = . \widehat{1}\widehat{2}\widehat{9}\widehat{3}\widehat{9}\widehat{2}\widehat{0}\widehat{9}\widehat{5}\widehat{0}$$

$$\text{Trans}_M(1) = .1 \quad 2 \quad 3 \quad 0 \ldots$$
$$\text{Trans}_M(2) = .9 \quad 9 \quad 9 \quad \ldots$$
$$\text{Trans}_M(3) = .2 \quad 5 \quad \ldots$$
$$\text{Trans}_M(4) = .0 \quad \ldots$$

that when certain real numbers T_M are decomposed into countably many real numbers in the way indicated above, then the original number does not appear among the countably many real numbers obtained. This, of course, applies only to numbers T_M that code up a closed system of naming. It is worth noting, incidentally, that if T_M does code up a closed system, then breaking down one of the new numbers like $\text{Trans}_M(3)$ should not lead to any numbers not already present after the first breakdown.

If we think of M as a human being, then we can imagine the n's as being books made of words, and think of Trans_M as being the process of translating each book into a real number. There can be no book that gets translated into the real number T_M coding up all of M's activity.

We can make this more colorful. The code numbers can be thought of as books made of *words,* and we can perhaps think of the real numbers named by these words as being *ideas.* The idea of pi embodies the whole infinite decimal expansion in a simple pattern. In general, a real number is something like an idea, because it has a definite existence as a mental (as opposed to physical) object, yet nevertheless provides a normative standard for concrete approximations (namely, the initial segments of the infinite decimal expansion). So, we might say that Richard's argument demonstrates that no human being can give a finite description of how he turns words into ideas.

This is quite similar to the moral already drawn from the Berry paradox—that no finite scheme can capture the essence of how one connects the real and the ideal, the physical and the mental, language and thought.

But wait. Even though M has no finite name for T_M, couldn't there be a better naming system M^* that *does* have a name for T_M? There seem to be two alternatives. On the one hand, it may be that T_M is absolutely unnameable, in the sense that no finitely given system M^* can name T_M. In this case, M must be infinitely complicated. On the other hand, it may be that there is some finite way of giving M and T_M within a higher system M^*. In this case, M is incomplete. So we can conclude that *any system for naming real numbers must either be infinitely complicated* (so that T_m is itself random in the absolute sense) *or incomplete* (so that T_M does eventually admit of a finite description).

The real motive in taking up Richard's paradox was to try to find a real number that is absolutely random and answers to no finite description. If we believe that there is such a thing as a maximal naming system U, then it cannot be that U can be improved to a naming system U^* that names everything U names, and names T_U as well. So if we believe that there is a maximal naming system U, then we know that there is a random real number—the translation code T_U. If we accept the relation "B names s" as being meaningful without further specification, then we are really thinking in terms of some maximal system U. But there is some question as to whether this is a legitimate way to think.

One could be rather hard-nosed about it and deny that a naming system M exists unless it has been exhaustively specified by various rules and schemata. Such an M is basically a finite thing, and can always be improved upon to get a better M^* that also describes T_M. Now, this kind of process can be continued indefinitely without ever reaching a stopping point. It is a bit like the way in which we can always find a greater natural number without ever getting to an actually infinite number.

So a Richard-style argument will give us an irreducibly infinitely complex real number T_U only if we are already willing to accept the existence of a somewhat transcendental relation of "naming." Such a U would be embodied in, let us say, a God who was an English-speaking mathematician. But all we are doing then is assuming one infinity to get another one.

CODING THE WORLD

One can conceive, ideally, of a set of facts that would enable one to answer every possible question about our universe. The question that concerns us here is whether or not the smallest, most efficient such complete description of the world is finite or infinite.

Whether or not the world itself is infinite to begin with makes a difference, of course. But this was already discussed in the "Physical Infinities" and "Higher Physical Infinities" sections. Here we are concerned with distinguishing between the following three cases: 1) The universe is totally finite, and thus admits of a finite complete description; 2) The universe is in some respect infinite, but nevertheless is completely specified by some finite set of facts; 3) The universe is infinite and cannot be completely described by any finite collection of sentences.

Case 1 is the situation where space and time are finite and quantized. Here the universe has only a finite extent, and space-time is grainy, so that only a finite number of possible space-time locations exist. A complete description of the universe could be given by specifying what was to be found at each of the finitely many space-time locations.

Figure 63.

A universe such as this is something like a picture made of individual light bulbs, such as one sees in displays at Times Square. Or, one might think of the universe as being a large, but finite, four-dimensional Go board with white stones for matter and black stones for antimatter.

I might mention in passing that the German mathematician Eduard Wette believes that the universe is totally finite in the way just described, having well under $10^{10^{10}}$ space-time locations. He concludes from this that any mathematical talk about numbers greater than $10^{10^{10}}$ is meaningless, and even inconsistent. He has repeatedly tried to use this insight to fashion a convincing proof that all of traditional mathematics is contradictory.[23] Needless to say, Wette's ideas are unpopular among mathematicians.

Case 2 is the rationalists' dream. Here we have an infinite universe whose very essence is somehow captured by a finite set of facts and natural laws. Science continually works to approach this situation by finding laws that account for and summarize a wide variety of individual cases. Once we know about the proton-neutron-electron model of the atom, then the table of elements becomes very easy to understand.

An extreme example of this sort of process appears in Eddington's *Fundamental Theory*.[24] Here Eddington tries to derive such physical constants as the mass of the electron and the radius of the universe from certain *a priori* theoretical considerations. His efforts must be judged largely unsuccessful, but the idea of finding a few key facts and laws that account for everything is still an attractive one.

One could, of course, question whether any finite theory could some-how predict the mass of infinitely many stars, or even the arrangement of the blades of grass in a lawn. Actually, there is a certain sense in which no finite theory can exhaustively describe an infinite world. Gödel's Incompleteness Theorem, which will be studied in the next chapter, states that no finite theory can predict all of the true facts about natural numbers. Now, if the universe is infinite, then it embodies the full set of natural numbers, so Gödel's Theorem seems to say that for any given finite theory of the universe, there are certain facts having to do with sets of physical objects that cannot be proved by the theory.

But let us set this difficulty aside for now and look at a more concrete way in which a universe might fail to answer to any finite description. We will consider a universe that continues expanding forever after an initial singular state. Our own universe may very well be like this. Given an infinite future, with no future collapses to rub everything out, might look for an irreducible infinity in the form of a random sequence.

Suppose you started flipping a coin and wrote down a 1 for every "heads" and a 0 for every "tails". If you put a decimal point in front of it all, something is generated that looks like it might be a random real number, say, .0110010100001011. . . . But there is the problem that you will not be around forever, so you will not be able to produce an infinite sequence of digits.

To avoid this, you might build a coin-flipping machine. To keep the machine running, you supply it with a couple of repair robots, who are also capable of repairing and even building copies of themselves. In order that these three friends don't run out of energy or raw materials, we put them in a spaceship that zooms around the universe scooping up matter and converting it into energy and the desired elements. If we have an everlasting universe with an infinite amount of matter in it, then there is no theoretical reason why such an immortal coin-flipper could not be set up.

Figure 64.

If the coin-flipper is unbiased, it is at least logically possible that nothing but heads will come up from now on. But we would expect it to be more likely to produce a sequence of zeros and ones answering to no finite description. A universe in which at least one such random coin-flipper existed would not have any finite description.

It may very well be that no one will ever bother building such a complicated machine. Isn't there some simpler sort of physical choice mechanism?

Consider a hydrogen atom, consisting of an electron circling a proton. The atom can exist in various energy states. In general, it moves to higher energy states by absorbing photons, and passes to lower energy states by emitting photons. Suppose that we watch a particular hydrogen atom, and at the end of each second we write down 1 if the atom has emitted a photon during that second, and write down 0 otherwise.

There is no reason why a given hydrogen atom may not survive intact for the rest of time, but there is a problem here in that we will not be around for the rest of time to watch it and mark down zeros and ones.[25]

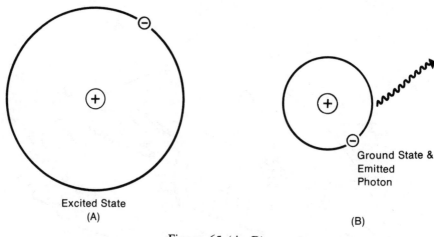

Figure 65 (A–B).

At first one is inclined to feel that *whether or not anyone looks* the atom will, during any given second, either emit a photon or fail to do so. So one is tempted to believe that if time goes on forever, then each hydrogen atom in a certain sense already embodies or labels an infinite sequence of zeros and ones.

Randomness is not usually an axiom explicitly assumed in quantum mechanics, but there is a strong feeling that the behavior of a hydrogen atom is, in principle, unpredictable. So we would expect most of the hydrogen atoms to be generating random real numbers. (There is one theorist, Paul Benioff, who has tried extending Quantum Mechanics by explicitly *assuming* that such hydrogen sequences would be random in the sense of answering to no finite description.[26])

But there is a big catch. Unless someone *looks* at a hydrogen atom, it need not have definitely emitted a photon or not. It can be, according to orthodox quantum mechanics, in what is called a *mixed state.* That is, unless it is being subjected to external measurements, the hydrogen atom can be in a state where it emits a photon with sixty percent probability and fails to emit a photon with forty percent probability, but does not unequivocally do either one!

The same sort of thing actually applies even to macroscopic systems such as the coin-flipper. The coin-flipper starts out in a certain state described by a certain wave function. The wave function evolves deterministically according to Schrödinger's wave equation as time goes by. Unless someone is there to keep an eye on the flipper, it will soon enter a state of always getting tosses that are fifty percent heads and fifty percent tails. That is, *each* toss is 50/50.

It is difficult to assign any meaning to a statement such as this. How can you toss a coin, have it land (not on an edge) and be fifty percent tail and fifty percent head? *You* can't . . . because you can only see one universe. If you could somehow split into two distinct people in two distinct universes, then you could see the coin come up heads *and* come up tails on the "same" toss.

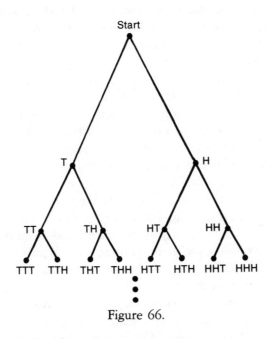

Figure 66.

One way of interpreting quantum mechanics is to claim that the universe actually does split like this everytime a decision has to be made. The idea is that if there is no sufficient reason for the world to choose H over T or T over H, then it chooses both.[27]

As the universe continues its dividing and redividing, every node in an infinite binary tree of possibilities is filled by a possible world. After ten coin flips, there will be coin flippers in $2^{10} = 1024$ different universes, one for each possible sequence of ten zeros and ones.

Curiously, such a branching universe contains less information than a universe that does not branch. For if there is only one coin-flipper, then a unique sequence of zeros and ones is generated, and this sequence is, in all likelihood, a random real coding up an irreducibly infinite amount of information. But if a coin-flipper splits in two every time the coin is tossed, then there is no one flipper you can point to in order to pick out a path through the binary tree.

If you start up a coin-flipper in a branching universe, then all you get is the occupation of every node of an infinite binary tree. This sort of setup is completely described by a finite number of words: "Take every possible finite sequence of zeros and ones." If the universe does *not* branch, then the only way to describe what this particular flipper does is to state the actual sequence it generates: ".011010001010000110 . . ." In this case one can *hope* that there is some hidden order to the universe that finitely determines the sequence in question, but this hope may very well be unfounded.

This is an interesting point, and bears a little more discussion. If every possible universe exists, then there is no need to account for the special peculiarities of *this* universe (e.g., the facts that there is an ant crawling up my screen right now, or that there are 79 clover blooms in my backyard, or that sentient beings exist in this universe, or that space has three dimensions). If every possible universe exists, then there is no need to explain any peculiarity. Why is there an ant on my screen? No reason—there is another universe exactly the same except with no ant.

This situation is a bit like the Total Library. Given that every possible book is there, it would not really make sense to ask what the printer had in mind when he filled some given volume with "yam's." He did so because he had to do every possible book. In a sense, the Total Library contains no information at all!

So if every possible universe exists, then the cosmos can be rather simply specified by the injunction: "Take every possible universe." Actually, a bit more than that would be needed. In particular, one would have to decide what constitutes a "possible universe." A fairly conservative answer would be to take all the possible ways of filling a four-dimensional space-time with mass and energy consistent with Maxwell's equations and Einstein's field equations. And what about universes where the usual laws of physics do not hold, or universes where even the usual rules of logic do not apply? But it would take us too far afield to consider questions such as these.

It is very hard to believe seriously that every possible universe exists. But let us give it a try. If all possible worlds are out there, then every time I get in my car there is a world where I suffer a fatal accident. How, one is tempted to ask, do you stay in the worlds where you don't get killed? The answer seems to be : You don't. You are equally in all of those worlds. Really to accept this fact is a source, I would say, of profound liberation. Once you're born, the worst has already happened to you.

The main unanswered question about the Many Universes model is how it is that one *seems* to oneself to be just in one universe. If everything is possible, why can't you be aware of it? Maybe you can, or are all the time. If you examine your preverbalized thought patterns, they are rather different from ordinary consensus reality.

For example, if I am handed a bowl of walnuts, it may look in an instant like a cave, an old man's face, clouds, the Matterhorn, a cat's eye, hands, or *The Wreck of the Medusa*. When I see something for the first time, before I have decided what is is, it is many things at once. "Ah, yes, a bowl of walnuts," I say shortly, and then I have only the one reality in my consciousness. But until I *name* the object and make it be one thing or another, I am in a mixed state—in many worlds.

Dreams are perhaps jumbled perceptions of many possible worlds. Language and ordinary thought form a sort of touchstone which keeps bringing you back to the same reality. You certainly do leave ordinary reality every time you fall asleep. If you woke up alone and with no memories but your dreams, could the dreams take over?

Despite all that has just been said, I do not believe very strongly in the full Many Universes theory. The universe we live in is so artfully constructed, so full of checks and balances, causes and coincidences, that it is hard to believe that this universe is just the product of a zillion coin-flips. There is so much holistic order in the universe that it seems implausible to suppose that we are just in some random one of all the possible worlds. There may be other universes, and we may be able to sense them in some manner, but I would expect each universe to have a certain overall patterning or essence.

This underlying pattern is what is sought when one looks for the shortest, most efficient complete description of a universe. But if the universe is indeed infinite, there seems to be no pressing reason why the underlying pattern of the universe should not be infinite as well. This brings us to Case 3.

This case, where one has an infinite universe with no finite description, actually splits into a number of subcases according to which level of infinity is assigned to the universe and to its descriptions. One could have a countable universe with a countable description, an uncountable universe with a countable description, or an uncountable universe with only an uncountable description. But it would be too confusing to consider these subcases here. Instead, let us further discuss the distinction between Case 2 and Case 3, looking at some concrete examples of how one might go about trying to code up the whole universe.

The most naive approach is to imagine that the universe is made of some finite or countably infinite collection of particles m_0, m_1, m_2, . . . , and that at each time t, each of these particles is of a definite type (electron, quark, photon, etc.) and has a definite position, orientation, and momentum. It is evident that all this information constitutes a set of real numbers that is at most countably infinite. Using the shuffling technique of the last subsection, all of these real numbers can be combined to form a single real number $U(t)$, which represents an exhaustive description of the state of the universe at time t.

We who have heard of quantum mechanics all our lives must question whether any such number $U(t)$ really exists, particularly since we know that we could never measure the position of anything with infinite precision, and especially not if the momentum is to be found with infinite precision as well. But for the physical determinists of the nineteenth century this objection had no force. They assumed that even if we could not measure it, $U(t)$ existed at each t. The determinists also believed that it was not necessary to know all of the $U(t)$. Indeed, they assumed that given $U(t_0)$ at one time t_0, all of the remaining $U(t)$ could be calculated on the basis of Newton's Laws of Motion. So for an old-fashioned determinist, a complete description of the universe can take the form of a single real number U.

Simply by looking at the leaves on a tree, the wrinkles on one's palm, or a sky full of clouds, one can quickly convince oneself that if such a universal number U has any finite description, then the length of this description must be very long. I would guess that the shortest natural number coding up a description of how to generate U would have to be longer than the number called u_0 in "The Berry Paradox."

An interesting side issue arises here. Even if some perfectly accurate description U of the universe exists, it seems likely that if we represented U by the clumsy expedient of putting numbers in books, then this representation of U would not fit in the universe. Of course, the most efficient representation of U is the universe itself, so at least one representation of U exists. But could we ever hope to have a desk-top or pocket-sized model of the universe? Only if matter is indefinitely divisible.

For if there is some smallest size particle, then any object in the uni-

verse will have less particles than the universe, and thus cannot serve as
a scale model. But if there is no smallest particle size, then any portion
of matter contains the same infinitely many particles, so it would be possible for some small region of matter to look exactly like the entire universe.

Figure 67.

Let us describe a way of modelling this possibility. Consider the pattern in Figure 67, colloquially known as a *Phoebe Snow* by crapshooters.
We can use this pattern as the base for an endless sequence of more and
more detailed patterns of the same size. We do this by replacing each of
the dots by a tiny Phoebe Snow to get a new pattern, replacing each dot
in the new pattern by a tinier Phoebe Snow, and so on.

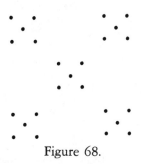

Figure 68.

In Figure 69 I have reproduced plate 115 from Benoit Mandelbrot's
Fractals. This shows an enlarged picture of the fourth stage of the process just described. If the process is carried out an actually infinite number of times, we get something called a Fournier fractal (after E. E.
Fournier D'Albe, who devised a similar pattern in 1907 to describe how
he thought the galaxies might be arranged in space).

What we obtain is a universe that seems, at a glance, to be made of
five chunks of matter in the Phoebe Snow arrangement. Upon closer
inspection, each of these chunks resolves itself into five smaller sub-
chunks, each of which upon closer inspection reveals itself to be a
Phoebe Snow of five subsubchunks, etc. In other words, this fractal
consists of five groups consisting of five groups consisting of five

Figure 69. From Benoit Mandelbrot, *Fractals*.

groups. . . . The phrase "five groups consisting of" is repeated ω times. Now, if a few of the initial repetitions of this phrase are left off, nothing is changed, so it is evident that each of the groups, subgroups, etc., has the exact same internal structure as the entire Fournier fractal universe.

I might add here that this could all have been carried out three-dimensionally by adding two dots to the basic pattern—one dot directly above the central dot, and one dot directly below the central dot, obtaining a "centered octohedron." The final fractal would be obtained by endlessly replacing dots by smaller centered octohedra.

The point is that if matter is infinitely divisible, then it is conceptually possible that there are particles that are exact replicas of the whole uni-

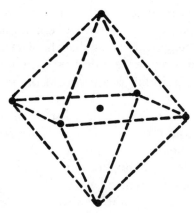

Figure 70.

verse. If one chooses to *identify* these particles with the entire universe, then the circular scale situation of the subsection, "Infinities in the Small," arises. Note also that even if there is no circling back, such a Fournier fractal universe would contain only a *countable* infinity of groups, subgroups, etc. One unnatural feature of this model is that it is finite and *bounded,* in the sense of having outer boundary groups around an inner group. If we try to remedy this by curving the space into a hypersphere, then it is hard to see how the groups in the space can also be curved into hyperspheres, since the space itself normally has a dimension one lower than the space it is curved in. One way out of this problem would be to work with space of infinitely regressing dimensionality, so that one works with a hyperhyper . . . space, but this notion has never been studied.

In any case we were talking about ways of coding up the universe and the question of whether such codes are finite or infinite. The digression that just ended had to do with the question of whether it was, in principle, possible for the universe to include a scale model of itself.

We have already examined the physical determinists' method of coding up the universe by a single real number U. Even the non-deterministic theories of physics would generally seem to allow for the universe's instantaneous state being coded up by a real number, and for its entire history being coded up by a countable set of real numbers (give the state at each rational time coordinate), which can in turn be folded into a single universal real U, which may or may not be coded up by some magically efficacious finite formula.

There is something a bit unnatural in trying to describe the universe in terms of elementary particles, particularly in light of the fact that our knowledge of these particles is so derivative and uncertain. There are a number of facts that suggest the view that the data about these particles that we obtain are, in some sense, mind-produced. Perhaps it would be better to base our description of the world on actual human thought and experience.

That is, we would want to compile a list of all the things about the world that a person could find out. I might start, for instance, by mentioning the pen lying next to my typewriter. By way of describing it, I mention that it is black and shiny, but now I must pause to explain what "black" means. One component of what "black" means is certainly its association with the idea of "night." Fully to express what "night" is, I must give specific examples of various nights I have experienced, for instance, the night in 1966 when Sylvia and I stood on a fragrant terrace looking out at the path the moonlight made on the Mediterranean. But really to describe how it felt to be there, I had better tell you about Sylvia. She was born in Budapest—here is a map of the city. A map is a sort of diagram drawn with a pen. Pens? They're things you write with —I have one right here next to my typewriter; it's black and shiny. . . .

The fact is that when we try to describe any one object or experience fully in terms of other objects and experiences, more and more things get dragged in, including repeated appearances of the object originally being discussed.

There is no real contradiction or regress here, but it does seem that the body of possible experience is something like a neglected dish of hard candy; we try to pick one piece of candy up and the whole dishful comes along. It is probably impossible to describe any one thing in the world exhaustively without mentioning everything else as well. No matter what you start with you're going to end up by mentioning the scar on my right index finger, the shape of the first sunspot to appear in 1292 B.C., the genetic makeup of the spirochetes that attacked Ivan the Terrible, and the nature of the galactic civilization that has evolved in the Whirlpool Nebula.

How can all this diversity be grasped as a mathematical unity? One approach would be to let The Description be the set of books in the Total Library that consist of true English descriptions of some aspect of the universe. Wittgenstein says, "The world is everything that is the case," and we will let The Description be all the English descriptions of things that are indeed the case. The Description constitutes a set of books, which can be viewed as a set of code numbers. So The Description itself can be viewed as an infinite set of natural numbers, which can be coded, if desired, as a single real number that could be called D.

The restriction to English poses no real problems, since The Description will include an explanation of what each word means in terms of all the other words. Insofar as the attainable physical knowledge of the world constitutes only a part of "what is the case," D really has more information than U.

Again, we might ask if D could possibly have a finite description of some kind. For instance, what if by some miracle, D turned out to be one over pi! People who believe that there is some ultimate answer to "it all" are in the position of hoping for just such a miracle. But anyone who has ever savored the endless diversity of nature must feel–and even hope–that the universe can never be fully captured by any finite schema, and that the pattern of the universe is, in a formal sense, random and unnameable.

The bounded and finite One of Plato and Parmenides seems no more worthy of worship than an overgrown computer. There is a harrowing passage in *Moby Dick* where Ahab stands on the deck in a thunderstorm, holding a lightning rod and ranting to his shipmates' nameable God:

> "There is some unsuffusing thing beyond thee, thou clear spirit, to whom all thy eternity is but time, all thy creativeness mechanical. Through thee, thy flaming self, my scorched eyes do dimly see it."[28]

WHAT IS TRUTH?

The golden age of Greek philosophy was not very far in the past at the time of Christ. One of the very few representatives of the Greco-Roman world who appears in the Gospels is Pontius Pilate. In view of this fact, the following passage from the Gospel according to John takes on a certain significance as an archetypal confrontation between mystical and rational ways of thinking:

> Pilate said to him, "So you are a king?" Jesus answered, "You say that I am a king. For this I was born, and for this I have come into the world, to bear witness to the truth. Every one who is of the truth hears my voice." Pilate said to him, 'What is truth?'" —John 18 (37-38).

The notion of truth leads to a number of logical difficulties. One of the most prominent of these difficulties is the Liar paradox, also known as the Epimenides paradox. Epimenides lived in Cnossus, the capital city of Crete, sometime before the time of Christ. It is generally believed that St. Paul is referring to Epimenides in the following passage: "One of themselves, a prophet of their own, said, 'Cretans are always liars, evil beasts, lazy gluttons.' This testimony is true." —Epistle to Titus 1 (12-13).[29]

The paradoxical aspect of Epimenides the Cretan saying, "Cretans are always liars," is that if what he says is true, then he must be lying. In view of this, it is especially ironic that Paul adds, "This testimony is true!"

The Liar paradox can be sharpened by considering the sentence: A) THIS SENTENCE IS FALSE.

That is, A is a sentence saying that A is false. If A is true, then what it says is true, so A is false. If A is false, then it is true to say that A is false, so A is true.

This is certainly a deplorable state of affairs. One way in which people have tried to get out of the paradox is to deny that A is either true or false—to assert that A simply happens to be a sentence that has no definite truth value. Now, we are certainly familiar with sentences that seem to be meaningless, rather than strictly true or false. "Virtue is triangular" and "Donald Duck weighs 62.8 pounds" are sentences that we would hesitate to call definitely true or definitely false. So why couldn't A be another such sentence?

This exit route can be closed in the following way. Consider the sentence: B) THIS SENTENCE IS NOT TRUE.

B is a sentence saying that B is not true. Now, every sentence is either true or not true. If B is true, then B is not true. If B is not true, then B is true. Therefore, B is both true and not true—which is a contradiction.

It certainly seems that any sentence must either be true or not true, where "not true" is taken in the broadest possible sense of "false, meaningless, contradictory, or impossible to verify." So it seems dishonest to try to escape the present form of the Liar paradox by denying that every sentence is either true or not true.

We might, instead, try denying that B is a sentence. There is certainly something peculiar about B. It refers to itself with the phrase THIS SENTENCE, and if we replace THIS SENTENCE by a quotation of the sentence in question, namely by "This sentence is not true," then we are no better off than before. Repeating this substitution leads to an infinite regress:

THIS SENTENCE IS NOT TRUE.
"THIS SENTENCE IS NOT TRUE" IS NOT TRUE.
""THIS SENTENCE IS NOT TRUE" IS NOT TRUE" IS NOT
TRUE . . .

.

.

.

""". . . . . ." IS NOT TRUE" IS NOT TRUE" IS NOT TRUE.

Note that the limiting sentence at the bottom consists of an ω-sequence of "'s running from left to right, followed by an ω-sequence of" IS NOT TRUE's running from right to left.

There is actually nothing inherently bad about infinite regresses. Josiah Royce has made the point that an infinite regress can usually be avoided by looking at the situation in a schematic way. For instance, if I say that there is a sentence B that says B IS NOT TRUE, then there is no infinity present. It is only when I insist on eliminating the symbol "B" that the infinite regress arises.

Recall that we say a very similar situation in the "Infinities in the Mindscape" section, when we discussed a mind that consists of pure self-awareness. This was modelled as a set M whose only member is M. It is possible to grasp the essence of the set M immediately and all at once, but if we try to eliminate the symbol "M", we get the infinitely regressing definition of M as $\{\{\{. . . . . .\}\}\}$.

The traditional belief was that if a line of thought leads to an infinite regress, then this line of thought is invalid.[30] This belief is founded on the notion that infinity is inherently contradictory and even incoherent. But Cantor has delivered us from this superstitious fear of the infinite. And not a moment too soon, for in 1893 F. H. Bradley published a book, *Appearance and Reality,* which seems to show that just about *any* sentence leads to an infinite regress when thoroughly analyzed.[31]

Bradley's argument goes something like this. Ordinarily we think of the world as being made up of various individuals $a, b, . . . ,$ which stand in various relations $R, P, . . .$ to each other. For instance, for me to say that this X is to the left of *this* X is to say that a certain relation L (" to the left of") is satisfied by the object a (the first X mark) and the object b (the second X mark) taken in that order. This is abbreviated as $L(a, b)$.

Now, Bradley continues, we can think of relations as themselves being higher-order objects that can, in turn, stand in various higher-order relations to other relations and objects. We do this, for instance, when we say, "It is more blessed to give then to receive." Returning to the example above, we can view the relation L as a (higher-order) object and say that $L(a, b)$ really means that L, a, and b have a certain higher-order relation–call it satisfaction–to each other. This is, we can say that $L(a, b)$ really means $S(L, a, b)$ where S is the "first-order satisfaction relation" such that in general $S(R, x, y)$ holds if and only if x stands in the relation R to y.

But now we can introduce a still higher-order relation S' and say that $S(L, a, b)$ really means $S'(S, L, a, b)$ where S' is the "second-order satisfaction relation" such that in general $S'(Q, R, x, y)$ holds if and only if $Q(R, x, y)$. Indeed, Bradley argues that not only is it *possible* to start down this path, but that it is *necessary*. For he feels that to assert a statement such as $L(a, b)$ is to say that the relation L, the object a, and the object b stand in a certain relation to each other.

And, of course, there is no stopping, and we have the infinite regress:

$$L(a, b)$$
$$S(L, a, b)$$
$$S'(S, L, a, b)$$
$$S''(S', S, L, a, b)$$

$$\cdot$$
$$\cdot$$
$$\cdot$$

$$\ldots S'', S', S, L, a, b).$$

The moral is that even a statement as simple and unproblematic as "this X is to the left of *this* X" can lead to an infinite regress, so the mere fact that the Liar paradox sentence B leads to an infinite regress does not automatically mean that B is not a sentence and that B is, therefore, ex-

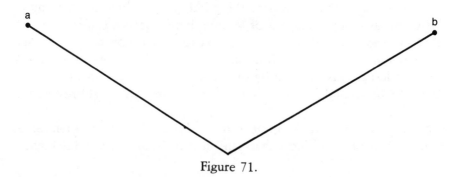

Figure 71.

empt from having to be either true or not true. In other words, we should not dismiss the Liar paradox out of hand simply because it leads to an infinite regress—for Bradley has shown that *every* sentence leads to an infinite regress.

Before going any further, let's eliminate the symbolism and try to see what Bradley has really done. His basic intuition seems to be that nothing can be linked to anything else without a mediating relation. Thus, L links a to b, S links a to L and L to b, S' links a to S and S to L and L to b, and so on.

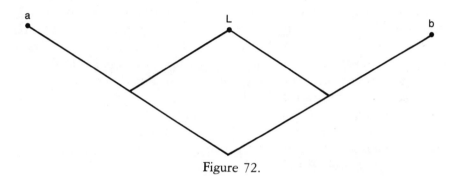

Figure 72.

We can visualize this geometrically by starting with an a and b with a sort of chasm in between them. Now, the relation L can be viewed as a bridge across the chasm. Continuing, we can think of S as a pair of bridges spanning the gulfs between a and L and L and b, and of S' as a set of four bridges connecting a to S, S to L, L to S, and S to b. In the limit we get a fractal, which is partially drawn in Figure 73.

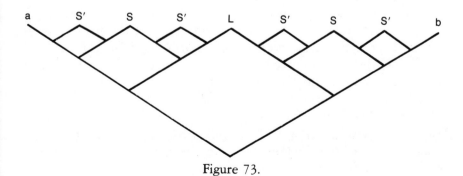

Figure 73.

Leaving aside, for now, Bradley's concern with how it is that disparate language elements are unified into a sentence, let us consider what the sentences in his infinite regress actually *say*.

$L(a, b)$ says a is to the left of b.
$S(L, a, b)$ says "a is to the left of b" is true.
$S'(S, L, a, b)$ says ""a is to the left of b" is true" is true.
... $S'', S', S, L, a, b)$ says . . . """a is to the left of b" is true" is true" is true". . . .

Another way of putting it is that Bradley believes that truly to assert some sentence E, one must really assert each of the following infinite regress of sentences: E , "E" is true, ""E" is true" is true, . . . This regress tends toward the limit . . . """E" is true" is true" . . . , where on the left we have an ω-sequence of left-hand quotation marks running from right to left, and on the right we have an ω-sequence of *is true*" 's running from left to right.

Can we avoid this infinite regress by insisting that once you have said E, it adds nothing also to say "E" *is true?* This is the case for certain artificial sentences. We can construct such an artificial sentence like this:
C) THIS SENTENCE IS TRUE.

C says that C is true, so C and "C" *is true* are the same sentence. If we analyze C) as we did B), we see that C) actually has the form: """" . . . . . ." IS TRUE" IS TRUE" IS TRUE, where an ω-sequence of quotes moves in from the left, and an ω-sequence of" IS TRUE's moves in from the right. Clearly, tacking on an extra quotation mark in front and an" IS TRUE in back changes nothing. So for the artificial sentence C, C and "C" *is true* are the same. Incidentally, you might wonder whether THIS SENTENCE IS TRUE is true or not. I don't know. All I can tell you offhand is that if it is true, then it is true; and if it is false, then it is false . . . which is not too helpful.

Returning to the main line of thought, I would like to stress the fact that for most sentences E, "E" *is true* really is a sentence that is different from E. To understand a sentence such as "a is to the left of b," we need only know where a and b are and what "left" means. But to understand "It is true that a is to the left of b," we have to know what "true" means as well. And this is not so easy to know.

In point of fact, there can be no finite complete description of truth. *Truth is undefinable.* As we will see, this will be our way out of the Liar paradox. For if truth is undefinable, then the word "truth" cannot properly be used to stand for the full concept of truth, so the Liar sentence B

really does not mean what we thought it did, and the paradox is avoided.

But before thinking that through, let's look at Alfred Tarski's 1934 proof that truth is not definable.[32] Consider the Total Library. Some of the "books" will assert sentences, and some of these sentences will be what we think of as *true* sentences. If, for instance, the text of B_{5389} is simply "Snow is white," then we would say that B_{5389} *is a true book,* or, abbreviating, $T(B_{5389})$.

In general, $T(B_n)$ if B_n is a book that is true. Recall that in the last subsection we referred to the set of all true books as The Description. You might ask who is going to decide which of the B_n are true—but let's dodge that for now by saying we'll let God decide which books are true.

So $T(B_n)$ holds if and only if B_n is true. Now I wish to prove that there can be no finite description of the T predicate. There can be no finite accounting for which books God will say are true.

Figure 74.

For suppose that there were a finite blueprint or program for building a "truth machine" such as the one illustrated in Figure 74. The truth machine has a slot where you can put a book B_n in. The truth machine scans the pages of B_n, and if B_n is true, then it is placed in a nice orderly row with all the other true books. But if B_n is *not* true, then the truth machine spits it out onto the scrap heap of history.

O.K. So, now have a finite set of instructions for building a truth machine. Let K be the book that goes as follows:

> "Imagine building a truth machine according to the following set of instructions. [Give the instructions.] Now, the assertion that this book wishes to make is the following: The truth machine will not say that this book is true."

Toss K into the truth machine's mouth and watch the fun begin!

Fun, that is, if you enjoy seeing valuable machinery destroy itself. The truth machine can't say that K is false, since then K would be true—in that it predicts that the truth machine will not say that K is true. And the truth machine can't say that K is true, since if the machine says K is true, then K is false—since it predicts that the machine will not say K is true. So K sticks in the truth machine's craw: The machine can't say YES and it can't say NO.

So, in fact, the truth machine never will say that K is true. But that's what K says. We who look in from the outside can see that K really is true, in the absolute and undefinable sense of the word "true." K is a true book, but the truth machine is unable to recognize this fact!

Speaking a bit more formally, we have proved that for any finite set of instructions S for building a truth machine M_S, there is a finite book K_S such that K_S is true and M_S cannot recognize this fact. K_S is, as above, the book saying, "M_S will not say that K_S is true," and when we give K_S to M_S, M_S goes into an endlessly regressing loop and never says anything again.

There can be no finite description of a truth machine that singles out as true all the true books. There is no robot we could build to winnow The Description out of the Total Library. Truth is undefinable.

The solution to the Liar paradox is now at hand. In itself, the sentence: B) THIS SENTENCE IS NOT TRUE is not really a meaningful utterance. Some description S of what is meant by TRUE must be appended, so that B really becomes K_S: K_S) THIS SENTENCE IS NOT TRUE ACCORDING TO DESCRIPTION S OF TRUTH.

In terms of description S, K_S will be neither true nor untrue, for S cannot reach a decision on K_S. So there is no paradox.

In terms of our absolute, but unformalizable, notion of truth, K_S *is* true. Better and better descriptions of truth can be obtained, but no finite description can ever exhaust the unnameable concept that we point at with the symbols T-R-U-T-H.

A truth machine description S cannot always correctly decide the truth of statements mentioning S. But it is possible, given S, to come up with a better truth machine description S' that does everything S did and also is able to correctly decide the truth of statements about S. (S' could, for instance, include all the rules of S, plus the new rule, "K_S is true.")

This move can be used to generate an infinite sequence S_n of truth machine descriptions. We let S_0 be some truth machine description that decides about sentences that do not mention truth at all, and for each

$n + 1$ we let $S_{n=1} = S_n{}'$. That is S_{n+1} decides everything S_n did, and decides sentences about S_n as well.

With this in mind, we can arrive at a better understanding of what underlies Bradley's regress. Suppose that you want to state some declaration E in such a way that it is absolutely clear that E is true.

If you simply say "E," then people may think that you are only reciting poetry or framing a hypothesis, so you want to say something like "E is true." Unfortunately, the people you are talking to are positivists, and they insist that you define every term that you use—and it is impossible to define the absolute notion of truth that you had in mind. So, you describe some definition S_0 of truth and say, "E is true according to S_0." But now you feel that you should stress that this last sentence is true, and the regress begins:

E.

"E" is true according to S_0.

" "E" is true according to S_0" is true according to S_1.

" " "E" is true according to S_0" is true according to S_1" is true according to S_2."

.

.

.

No final satisfaction is ever reached. Indeed, this regress can be carried on past ω. For if S_ω is the truth description that feeds a given book into one after another of the S_n until an answer appears, then S_ω is still finitely described and we can move on to $S_{\omega+1} = S_\omega{}'$.

Of course, our absolute notion of truth *should* be such that if E is true, then "E is true" is true, and so on. But this absolute notion is not finitely describable. If we restrict ourselves to finitely describable notions of truth, then Bradley's regress is unavoidable.

It is, of course, possible to deny that any absolute notion of truth exists, and to insist that all there will ever be is better truth definitions, tending toward a limit that is wholly imaginary. This point of view is analogous to the point of view that admits that there are arbitrarily large natural numbers, but denies that the actually infinite limit ω of this sequence exists. Even more analogous is the viewpoint of those who admit the existence of the various infinite sets, but deny the existence of the unified existence of all sets at once in Cantor's Absolute. These are all variations of the One/Many problem that will be discussed in Chapter 5.

The critical reader will have noticed certain weak points in the proof that truth is undefinable. In particular, he may wonder how one can legitimately construct a book such as K_S that refers to itself without using the somehow transcendental phrasing, "this book," and he may wonder if the proof is not somehow circular or fallacious, since it seems to mention the very notion of truth whose undefinability is being proved. In the next chapter (and in Excursion II) I show how a version of this proof can be constructed so as to be completely unimpeachable.

To sum up, we have shown that truth is a concept that cannot be rigorously defined in a finite way. Because of this, it is possible to assert that THIS SENTENCE IS NOT TRUE is not really a sentence, and is thus exempt from having to be either true or not true. This is not an entirely satisfactory solution, since we feel that a collection of words that is neither true nor not true should *really* be called not true. Like all good paradoxes, the Liar paradox resists any final resolution and endures as "an eternal crevice of unreason."

CONCLUSION

The three columns of the following table summarize the three sections of this chapter. In each case one starts with a familiar infinite concept a), and runs into a paradox, b). These three paradoxes have a certain similarity in that each of them has to do with *semantics,* that is, each paradox centers on the process of determining the *meaning* of strings of symbols. One way of resolving these paradoxes is to insist that the key word ("name," "nameable," "true") points toward, but cannot actually name or define the needed concept with the requisite precision, and then to assert that the statements in row b) can be regarded as meaningless. But one cannot just stop here. It is not satisfying to "solve" paradoxes by dismissing them as meaningless. I am reminded here of the master puzzlist Sam Loyd's comments on Alexander the Great's method of undoing the Gordian knot:

> Alexander the Great, it is said, made many ineffectual attempts to untie some of the knots, but finally becoming enraged at his want of success, drew his sword and cut the cord, exclaiming that 'such is the common sense way to get a thing when you want it.' Strange that those familiar with the story and its contemptible climax endorse it with a certain air of assumed pride when they have surmounted some difficulty and exclaim: 'I have cut the Gordian knot!'[33]

In row c) we have taken each of the paradoxes and replaced the nebulous key words by precise approximations ("M_1-name," "M_2-nameable,"

a)	The natural numbers.	The real numbers.	Truth.
b)	Let the Berry number be the first natural number which has no name as short as this sentence.	Let the Richard number be the real number obtained by diagonalizing the list of all nameable reals.	This statement is not true.
c)	Let M_1 be the following system: [Description of M_1]. Print out the first number which has no M_1-name as short as this.	Let M_2 be the following system: [Description of M_2]. Generate the real number obtained by diagonalizing over all reals which M_2 names.	Let M_3 be the following system: [Description of M_3]. M_3 will not say that this paragraph is true.
d)	There is some number u such that M_1 is unable to generate any natural numbers with complexity greater than u.	There is a nameable real number which M_2 cannot name.	There is a true sentence which M_3 cannot recognize as true.
e)	No finite system can generate arbitrarily complex patterns.	No finite system can understand everything.	No finite system can define truth.

"true according to M_3"). The M_i are intended to be any finitely describable system, e.g., an appropriately programmed digital computer, or perhaps even a human being (for it may be that a biological organism does admit of an exhaustive finite description). It is to be understood that M_1 is arranged so that a given finite input B causes M_1 to print out some finite natural number n_B; that M_2 is set up so that any given finite input B converts M_2 into a device that will give out a digit $r_{B,n}$ for each requested n; and that M_3 takes any given finite input B and types out "true" or "not true." The key fact is that none of the M_i always work properly, that there are certain B's that cause these M_i's to run forever without ever outputting anything. In particular, each of the three paragraphs in row c) will cause the relevant M_i to enter an endless loop.

Because of this we can draw the conclusions in row d). Since M_1 cannot find the natural number described in row c), it must be that M_1 cannot locate *any* number whose shortest description is longer than the length u of that paragraph in row c). For any input B, either M_1 gives out a number whose complexity (shortest description length) is less than u, or M_1 runs forever. In running forever it will go *past* numbers (of seconds) of arbitrarily large complexity, but it is unable to stop and point out any of these numbers with complexity greater than u. Since this argument applies to any finite system M_1, we can draw the general conclusion in row e).

Since M_2 does admit of a precise finite description, the description in row c) does actually describe a specific real number. It is just that this description cannot be unraveled by M_2. In this sense there is some meaningful paragraph that M_2 does not understand, and the conclusion in row e) can be drawn. Recall from the section on Richard's paradox that analyzing the construction of the M_2-Richard number shows that M_2 is also unable to name the real number T_{M_2} that codes up its own translation process. So we might improve the conclusion in row e) to something like; no finite system can finitely describe the process by which it converts words into thoughts.

Since M_3 never draws a conclusion on the paragraph in row c), we know that M_3 will not say that that paragraph is true. So it is in fact true, although M_3 cannot recognize this fact. Thus, we reach the conclusion in row e), which can also be phrased this way: for any given finite system there is a truth that the finite system in question cannot recognize as true.

This last conclusion is an imprecise statement of Gödel's First Incompleteness Theorem, which will be considered in detail in the next chapter, where we will also attempt to use these facts to draw some conclusions about the nature of human and mechanical minds.

So we have learned that for any finitely give system M, there will be a number of things that M cannot name, describe, conceive of, understand. For any finite system M, there are things that are *unnameable relative to M*. Now we must ask if there is anything that is *absolutely unnameable,* that is, beyond the comprehension of any finite system whatsoever.

In Chapter 1 we considered the related question of whether anything is actually (as opposed to potentially) infinite. We knew that for any given natural number n, there is a larger natural number (e.g., $n + 1$). The question was whether there is any number such as ω that is bigger than every natural number at once.

I discussed the possibility of physical and mental actual infinities the "Physical Infinities" and "Infinities in the Mindscape" sections; in "The Absolute Infinite" I introduced the Absolute as something that is assuredly infinite, if indeed it exists.

In the "Coding the World" subsection in this chapter the issue of whether any physical and absolutely unnameable reals exist was handled; it should be evident from our previous discussions of the Absolute that it is absolutely unnameable. However, we have not yet said much on the question of whether any absolutely unnameable mental objects can be found.

We have already seen that the set of all true books is an unnameable mental object. But this set is rather nebulous, and one might question if it really exists. The same would apply to the unnameable process by which we translate words into thoughts. In the next chapter we will see that there is a fairly definite mental object that cannot be completely specified in any finite way. This is the set of all true statements about natural numbers.

PUZZLES AND PARADOXES
(Answers, p. 299)

1. Show that ten tetrated to the four is greater than googolplex.
2. A Berry-paradox-style argument is sometimes given to "prove" that every natural number is interesting. Try to construct such an argument.
3. Find the lengths of the hypotenuses in this spiral arrangement of right triangles:

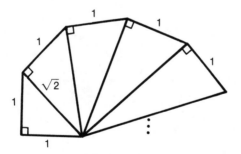

Figure 75.

4. Suppose that I find in the Total Library a book that I believe to be a correct, detailed description of your entire life, past and future. You do not like this, as you feel your future is, and should remain, undecided, if I show you the book in question, can you prove that I am wrong?
5. This is the classical paradox of the Crocodile: "A Crocodile had stolen a Baby off the banks of the Nile. The Mother implored him to restore her

darling. 'Well,' said the Crocodile, 'if you say truly what I shall do I will restore it: if not, I will devour it.' "[34] What should the mother say?

6. What is the meaning of the following sentence: "APPENDED TO ITS OWN QUOTATION IS FALSE" APPENDED TO ITS OWN QUOTATION IS FALSE.

7. John Barth begins his *Lost in the Funhouse* with the infinite story: ONCE UPON A TIME THERE WAS A STORY THAT BEGAN ONCE UPON A TIME THERE WAS A STORY THAT BEGAN ONCE UPON A TIME THERE WAS A STORY THAT BEGAN. . . .[36] This involves a single left-to-right ω-sequence. Can you think of an appropriate right-to-left ω-sequence to close off Barth's story with?

8. Say that we have established some starting assumption A, and wish to draw a specific conclusion C. In order to do this, we normally proceed by proving the implication (if A, then C). But suppose that some stubborn individual denies that C follows inevitably from A and (if A, then C). Then we must establish not only A and (if A, then C), but also (if A and (if A, then C), then C). Show that if our interlocuter remains stubborn enough, he can in this way force us into an infinite regress.[37]

CHAPTER FOUR
ROBOTS AND SOULS

Are people just complicated robots . . . or do we have souls? On the one hand, being conscious certainly *feels* like something more than the mechanical working out of a computer program. But on the other hand, what could a soul possibly be? How could it act on matter? Might a machine have a soul?

A theorem of mathematical logic, Gödel's Incompleteness Theorem, is widely thought to have a bearing on this constellation of problems. In this chapter we will begin by investigating Gödel's famous theorem and end by making some speculations about mechanical intelligence and the nature of consciousness.

The first section gives a quick overview of Gödel and his Incompleteness Theorem. In the following section I describe a series of conversations that I had with Gödel about his theorem and some related matters. "Towards Robot Consciousness" contains a more detailed treatment of Gödel's theorem and explains exactly what Gödel thought the consequences of his theorem were for the field of Artificial Intelligence. (An even more detailed treatment is to be found in Excursion II.) In "Beyond Mechanism," I explore Gödel's claim that there is a nonmaterial component to human consciousness.

GÖDEL'S INCOMPLETENESS THEOREM

In the summer of 1930, the twenty-four-year-old mathematician Kurt Gödel proved a strange theorem: *mathematics is open-ended*. There can never be a final, best system of mathematics. Every axiom-system for mathematics will eventually run into certain simple problems that it cannot solve at all. This is Gödel's Incompleteness Theorem.

The implications of this epochal discovery are devastating. The thinkers of the Industrial Revolution liked to regard the universe as a vast preprogrammed machine. It was optimistically predicted that soon scientists would know all the rules, all the programs. But if Gödel's Theorem tells us anything, it is this: Man will never know the final secret of the universe.

Of course, anyone can *say* that science does not have all the answers. What makes Gödel's achievement so remarkable is that he could rigorously *prove* this, stating his proof in the utterly precise language of symbolic logic. To come up with a mathematical proof for the incompleteness of mathematics is a little like managing to stand on one's own shoulders. How did Gödel come to think of such a proof? What kind of person was he?

Kurt Gödel was born on April 28, 1906, in Brünn (Brno), Czechoslovakia, at that time part of Austria-Hungary. His family was part of the sizable German minority in Brünn, and his father was the successful manager of one of the city's numerous textile mills. Gödel's childhood was marred by a case of rheumatic fever at the age of six. He recovered from the disease, but for the rest of his life he had morbid fears about his health.[1]

He attended the University of Vienna, starting as an undergraduate in 1923 and earning his doctorate in mathematics in 1930. Vienna was a terribly exciting place to be in those years. The beginnings of psychoanalysis, twelve-tone music, modern architecture, and non-representational painting can all be traced to this period—with Sigmund Freud, Arnold Schönberg, Adolf Loos, and Oskar Kokoschka all active in Vienna.

But, more importantly for Gödel, this was also a period of great philosophical ferment in Vienna. In 1921 the Viennese intellectual Ludwig Wittgenstein had published his gem-like *Tractatus Logico-Philosophicus*. And logical positivism was being founded and developed by a group of philosophers known as the "Vienna Circle." Gödel's principal teacher, Hans Hahn, was a prominent member of this group, as were Moritz Schlick, Philipp Frank, and Rudolf Carnap. The Vienna Circle held its meetings in a seminar room near the mathematics department, and Gödel attended these meetings regularly.

The basic credo of logical positivism is summed up by Rudolf Carnap's manifesto: "We give no answers to philosophical questions and indeed *reject all philosophical questions,* whether of Metaphysics, Ethics or Epistemology."[2] The idea was that abstract philosophical statements such as, "All is One," are meaningless—not true or false, but simply without content. This position was based on the so-called "Verifiability Principle," according to which the *meaning* of a statement is equated with the *method for verifying* the statement. Since the positivists could see no way scientifically to document metaphysical statements like, "All is One," or "The Absolute is outside time," these statements were regarded as wholly devoid of significance.

This basically negative part of logical positivism was greatly influenced by Wittgenstein's celebrated *Tractatus.* This short and aphoristic book offers the following solution to the problems of traditional philosophy: "What we cannot talk about we must pass over in silence."[3] But it should be pointed out that Wittgenstein, although friendly with members of the Vienna Circle, was never quite a logical positivist. On the contrary, he sometimes sounds like a Zen mystic. He elegantly describes his position in the *Tractatus:*

> "We feel that even when *all possible* scientific questions have been answered, the problems of life remain completely untouched. Of course there are then no questions left, and this itself is the answer. The solution of the problem of life is seen in the vanishing of the problem."[4]

The positive part of logical positivism was a program to unify all of science, using the language of symbolic logic. The inspiration came from Whitehead and Russell's *Principia Mathematica* of 1910.[5] In this monumental work (three mammoth volumes), it is shown how all of our familiar mathematical concepts and facts can be logically derived from certain very simple and primitive principles of reasoning. The logical positivists hoped to put other branches of science, including physics and psychology, on a similarly rigorous footing.

The principal achievement of Whitehead and Russell was to get a mechanically precise definition of what we mean when we say that some statement *follows logically* from some other statements. With this definition in hand, the "formalist" mathematician David Hilbert pointed out that mathematics was now only a matter of choosing the right axioms and examining the logical consequences of these axioms. The positivists hoped to extend this approach to all the sciences, and even to all of human thought.

In order to grasp the implications of such a programmatic approach to human knowledge, I would like to run a little thought-experiment here, an alternate-world fantasy. Imagine:

A complete axiom-system for mathematics was obtained by 1950. The system was called MT for Mathematical Truth. It was theoretically established that any true mathematical statement could be proved from MT, and that any false mathematical statement could be disproved from MT. Thus, the axioms of MT, along with the Whitehead-Russell-Hilbert rules for logical deduction, captured the whole of mathematics.

The existence of the complete theory MT did not really begin to affect mathematicians until the twenty-first century. They went on much as before, using their intuition and ingenuity to find ways of combining the various axioms of MT to yield logical proofs of interesting theorems. But in the year 2000, the computers became big enough to take over. In the space of ten years a single linked system of Josephson-junction processors had made mathematicians as obsolete as slide-rules. The system was called MTM for Mathematical Truth Machine.

The functioning of MTM was as follows. MTM was programmed with the basic axioms of the complete system MT. What MTM did was simple. It exhaustively worked out all the logical consequences of these axioms: first all the theorems with one-step proofs; then all the theorems with two-step proofs; then three; and, before long, three million; and so on and on.

As MTM proved theorem after theorem, these were added to its systematic master list. If you wanted to know about some mathematical problem ("Is Fermat's Last Theorem true?" "What is the solution to this differential equation?" "What is the shortest route connecting these ten cities?") you fed your question into MTM, and MTM would search its master list for your answer.

If the answer was already in the master list, fine. If not, you had to wait a bit, but sooner or later MTM would get to the theorem answering your question. There was no point consulting a mathematician instead, since MTM had already gone far beyond all logical derivations short enough to be humanly comprehensible.

All this was fine with everyone except the mathematicians. A few of them rebelled to create a new "surreal mathematics" based on deliberately false and inconsistent assumptions. But MTM one-upped them by taking a spare hour to work out the most interesting false theorems of "surreal mathematics." With its ever-increasing store of mathematical and logical fact, MTM was getting faster and faster. You could feed it any set of axioms and it could work out the most interesting consequences in next to no time at all.

The physicists were the next to go the way of the mathematicians. Late in the 1990s an Israeli graduate student had achieved the final unification of General Relativity with Quantum Mechanics. A simple list of axioms, twenty-five in number, summarized all the laws of nature. This theory, called PT for Physical Truth, was programmed into a computer linked up with MTM. The new system, called PTM for Physical Truth Machine, began systematically working out the consequences of PT. Soon the three-body problem was solved, an explanation for the mass of the electron was produced, the exact age of the universe was calculated, and several methods for safe nuclear fusion were discovered.

A critical mass of knowledge had been reached by MTM and PTM. In the next few years, complete theories were found for biology, psychology, and sociology. A planet-wide system of linked computers combined all of these complete theories to produce the god-like STM, for Scientific Truth Machine.

Any scientific question at all was best answered by consulting STM. Either STM had already worked out the answer, or it soon would. No scientist could know as much as STM, so independent work was useless. During the Italian Renaissance there were certain mathematicians who made their living by being able to solve cubic equations. Imagine their dismay if one of the princes had been given the simple formula for solving cubics, along with a pocket calculator to work out the solutions! This was the situation in which the scientists now found themselves. The utterly mechanical working out of a complete theory had replaced any need for scientific intuition or creativity.

The last step came suddenly, in the year 2060. An embittered Argentine scientist had, with the help of the STM, worked out a complete theory of aesthetics. The immutable laws of what makes a great novel, painting, or symphony were incorporated into an axiom system called AT for Artistic Truth. The clandestinely assembled ATM sprang into action and began producing short works magnificently expressive of the condition of man in the cosmos.

Over the protests of the artists, the government coupled ATM with STM to get UTM, the Universal Truth Machine. There was no longer any need to do anything. Whatever any person wanted to know or do or say would, sooner or later, be done better by UTM. Terrorist acts against UTM were impossible, as UTM had a complete theory of human behavior, and could predict and ward off any such attack. About the only thing people still had left for their own was sports. A UTM terminal was placed in each home, and the world slid toward senescence, watching the tube.

Depressing? Don't worry!

In 1930 Kurt Gödel proved that there can never be a UTM (Universal Truth Machine). There can't even be an MTM (Mathematical Truth Machine). There *is* no complete set *MT* of axioms for mathematical truth. Any system of knowledge about the world is, and must remain, fundamentally incomplete, eternally subject to revision.

Of course, the future could follow a scenario *similar* to what I just outlined. The difference, thanks to Gödel, is that the machines will never have *all* the answers. There will always be room for a creative person who can think of a better way of doing things.

Try to catch the universe in a finite net of axioms and the universe will fight back. Reality is, on the deepest level, essentially infinite. No finitely programmed machine can ever exhaust the richness of the mental and physical world we inhabit.

The proof of Gödel's Incompleteness Theorem is so simple, and so sneaky, that it is almost embarrassing to relate. His basic procedure is as follows:

1. Someone introduces Gödel to UTM, a machine that is supposed to be a Universal Truth Machine, capable of correctly answering any question at all.
2. Gödel asks for the program and circuit diagrams of the UTM. The program may be complicated, but it can only be finitely long. Call the program P(UTM) for Program of the Universal Truth Machine.
3. Smiling a little, Gödel writes out the following sentence: "The machine constructed on the basis of the program P(UTM) will never say that this sentence is true." Call this sentence G for Gödel. *Note that G is equivalent to "UTM will never say G is true."*
4. Now Gödel laughs his high laugh and asks UTM whether G is true or not.
5. If UTM says G is true, then "UTM will never say G is true" is false. If "UTM will never say G is true" is false, then G is false (since G = "UTM will never say G is true.") So if UTM says that G is true, then G is in fact false, and UTM has made a false statement. So UTM will never say that G is true, since UTM makes only true statements.
6. We have established that UTM will never say G is true. So "UTM will never say G is true" is in fact a true sentence. So G is true (since G = "UTM will never say G is true.").
7. "I know a truth that UTM can never utter," Gödel says. "I know that G is true. UTM is not truly universal."

Think about it—it grows on you.

The gimmick in Gödel's proof is very similar to the gimmick in the famous Liar paradox of Epimenides: "I am lying," says Epimenides. Is he? Or, as in "What is Truth," define B to be the sentence "B is not true." Is B true? The problem is that B is true if and only if B is not true. So B is somehow outside the scope of the applicability of the concepts "true" and "not true." There is something viciously meaningless about the sentence B, and one is inclined just to try to forget about it.

But Gödel's G sentences cannot be so lightly dismissed. With his great mathematical and logical genius, Gödel was able to find a way (for any given P(UTM) actually to write down a complicated polynomial equation that has a solution if and only if G is true. So G is not at all some vague or non-mathematical sentence. *G is a specific mathematical problem that we know the answer to, even though UTM does not!* So UTM does not, and cannot, embody a best and final theory of mathematics.

Gödel's Incompleteness Theorem flew in the face of the formalist and logical positivist movements of the time. But everyone who was capable of following the many steps of the detailed proof was forced to concede its correctness. Gödel became famous.

When the Second World War broke out, Gödel moved to Princeton, New Jersey, to take up a permanent position at the Institute for Advanced Study. This famous "college without students" had been established a few years earlier with the financial help of Louis Bamberger, the department-store magnate. The aging Einstein was also there, and he became one of Gödel's few intimates. They could often be seen walking the great lawn in front of the Institute's Fuld Hall discussing relativity. Gödel himself published a curious paper in relativity, in which he described a special class of universe where time-travel by rocket ship is possible.[6]

Gödel did some of his most interesting work during the 1940s. Soon after arriving in the United States he published his only book, a monograph on set theory called *The Consistency of the Continuum Hypothesis.*[7] This book contains his proof that Cantor's Continuum Hypothesis cannot be disproved from the existing axioms of set theory. As with his Incompleteness Theorems, it was not only the statement of the theorem, but also the method of proof that had such an impact on mathematics and philosophy. Here Gödel found a whole new way of thinking about the class of all sets and discovered certain absolute features of the mathematical universe that had never been suspected.

In the mid-1940s he wrote a pair of rather philosophical papers aimed at the non-specialist, "Russell's Mathematical Logic" and "What is Cantor's Continuum Problem?"[8] These beautifully written papers reveal Gödel as anything but a logical positivist. He argues here that sets and concepts exist external to any individual's activities, and that questions about infinite sets are every bit as meaningful as questions about physics. This Platonistic strand in Gödel's thought became more and more pronounced over the years, culminating in his 1964 addendum to the "What is Cantor's Continuum Problem?" paper, which contains these passages:

> Despite their remoteness from sense experience, we do have something like a perception of the objects of set theory, as is seen from the fact that the axioms force themselves upon us as being true. I don't see any reason why we should have less confidence in this kind of perception, i.e., in mathematical intuition, than in sense perception . . . The set theoretical paradoxes are hardly any more troublesome for mathematics than deceptions of the senses are for physics. . . . Evidently the

'given' underlying mathematics is closely related to the abstract elements contained in our empirical ideas. It by no means follows, however, that the data of this second kind, because they cannot be associated with actions of certain things upon our sense organs, are something purely subjective, as Kant asserted. Rather, they, too, may represent an aspect of objective reality, but, as opposed to the sensations, their presence in us may be due to another kind of relationship between ourselves and reality.[9]

Despite his high level of scientific creativity during this period, Gödel seems to have had some opposition at the Institute, as he was not promoted from member to faculty status until 1953.[10] It may have been that to some people Gödel's Incompleteness Theorems seemed wholly negative, and that as a result they tried to dismiss these theorems as mere curiosities with no real mathematical or philosophical significance.

But in point of fact, Gödel's Incompleteness Theorems are of an importance comparable to the Pythagoreans' proof that the square root of two is irrational. The analogy is quite close. The Pythagoreans learned that no ratio of natural numbers could fully describe the relation between the diagonal and the side of a square. For his part, Gödel showed that no finitely describable theory can codify all mathematical truth. That is, he has shown that the set of all true statements about mathematics is finitely unnameable, and thus essentially random and infinite.

It is characteristic of Gödel's work that the Incompleteness Theorems use purely mathematical reasoning to prove certain facts about the objective world. In 1949 he actually attempted to demonstrate the unreality of time by means of an argument in mathematical physics.[11]

Gödel published only one paper after this, a 1958 discussion of how one might prove the consistency of mathematics by assuming that mental objects have an objective existence.[12] He did not like publicity and made at most two or three public appearances during the latter part of his life. Nevertheless, he continued as a guiding force in logic and set theory. Any mathematician who was invited to do so eagerly made the pilgrimage to his office at the Institute for Advanced Study.

In the next section, I describe my own pilgrimages.

CONVERSATIONS WITH GÖDEL[13]

I didn't know where his real office door was, so I went around to knock on the outside door instead. This was a glass patio door, looking out on a little pond and the peaceful woods beyond the Institute for Advanced Study. It was a sunny March day, but the office was quite dark. I couldn't see in. Did Kurt Gödel really want to see me?

Suddenly he was there, floating up before the long glass door like some fantastic deep-sea fish in a pressurized aquarium. He let me in, and I took a seat by his desk.

Kurt Gödel was unquestionably the greatest logician of the century. He may also have been one of our greatest philosophers. When he died in 1978, one of the speakers at his memorial service made a provocative comparison of Gödel with Einstein . . . and with Kafka.[14]

Like Einstein, Gödel was German-speaking and sought a haven from the events of the Second World War in Princeton. And like Einstein, Gödel developed a structure of exact thought that forces everyone, scientist and layman alike, to look at the world in a new way.

The Kafkaesque aspect of Gödel's work and character is expressed in his famous Incompleteness Theorem of 1930. Although this theorem can be stated and proved in a rigorously mathematical way, what it seems to say is that *rational thought can never penetrate to the final, ultimate truth.* A bit more precisely, the Incompleteness Theorem shows that human beings can never formulate a correct and complete description of the set of natural numbers, $\{0, 1, 2, 3, \ldots \}$. But if mathematicians cannot ever fully understand something as simple as number theory, then it is certainly too much to expect that science will ever expose any ultimate secret of the universe.

Scientists are thus left in a position somewhat like K. in *The Castle*.[15] Endlessly we hurry up and down corridors, meeting people, knocking on doors, conducting our investigations. But the ultimate success will never be ours. Nowhere in the castle of science is there a final exit to absolute truth.

This seems terribly depressing. But, paradoxically, to understand Gödel's proof is to find a sort of liberation. For many logic students, the final breakthrough to full understanding of the Incompleteness Theorem is practically a conversion experience. This is partly a by-product of the potent mystique Gödel's name carries. But, more profoundly, to understand the essentially labyrinthine nature of *the castle* is, somehow, to be free of it.

Gödel certainly impressed me as a man who had freed himself from the mundane struggle. I visited him in his Institute office three times in 1972, and if there is one single thing I remember most, it is his laughter.

His voice had a high, singsong quality. He frequently raised his voice toward the ends of his sentences, giving his utterances a quality of questioning incredulity. Often he would let his voice trail off into an amused hum. And, above all, there were his bursts of complexly rhythmic laughter.

The conversation and laughter of Gödel were almost hypnotic. Listening to him, I would be filled with the feeling of perfect understanding. He, for his part, was able to follow any of my chains of reasoning to its end almost as soon as I had begun it. What with his strangely informative laughter and his practically instantaneous grasp of what I was saying, a conversation with Gödel felt very much like direct telepathetic communication.

The first time I visited Gödel it was at his invitation. I was at Rutgers University, writing my doctoral thesis in logic and set theory. I was particularly interested in Cantor's Continuum Problem. One of Gödel's unpublished manuscripts on this problem was making the rounds, and I was able to get hold of a Xerox of a Xerox of a Xerox.[16]

I deciphered the faint squiggles and thought about the ideas for several months, finally giving a talk on the manuscript at Rutgers. I had a number of questions about the proof Gödel had sketched and wrote him a letter about them.

He probably would not have answered—Gödel almost never answered letters. But I happened to be attending a weekly seminar at the Institute with Gaisi Takeuti, an eminent logician who was there for a year's research. Gödel knew this, and one day while I was at the seminar in Takeuti's office, he phoned up and asked that I come see him.

Gödel's office was dim and unlit. There was comfortable carpeting and furniture. On the empty desk sat an empty glass of milk. Gödel was quite short, but his presence was such that visitors sometimes left with the impression that he was very tall.

When I saw him he was dressed as in all his pictures, with a suit over a warm vest and necktie. He is known to have worried a great deal about his health and was always careful to keep himself well bundled-up. Indeed, in the winter, one would sometimes see him leaving the Institute with a scarf wrapped around his head.

He encouraged me to ask questions, and, feeling like Aladdin in the treasure cave, I asked him as many as I could think of. His mind was unbelievably fast and experienced. It seemed that, over the years, he had already thought every possible philosophical problem through to the very end.

Despite his vast knowledge, he still could discuss ideas with the zest and openness of a young man. If I happened to say something particularly stupid or naive, his response was not mockery, but rather an amused astonishment that anyone could think such a thing. It was as if during his years of isolated thought he had forgotten that the rest of the human race was not advancing along with him.

The question of why Gödel chose to live most of his life in splendid isolation is a difficult one. Although he was not Jewish, the Second World War forced him to flee Europe, and this may have soured him somewhat on humanity. Yet, he loved life in America, the comfortable position at the Institute, the chance to meet Einstein, the great social freedom. But he spent his later years in an ever-deepening silence.

The first time I saw Gödel, he invited me; the second two times, I invited myself. This was not easy. I wrote him several times, insisting that we should meet again to talk. Finally I phoned him to say this again.

"Talk about what?" Gödel said, warily. When I finally got to his office for my second visit, he looked up at me with an expression of real dislike. But annoyance gave way to interest, and, after I'd asked a few questions, the conversation turned as friendly and spirited as the first. Still, toward the end of a conversation, when he was tired, Gödel would sometimes look at a visitor with an eerie mixture of fear and suspicion, as if to say, *what is this stranger doing in my retreat?*

Gödel was, first and foremost, a great thinker. The essence of the man is not to be found in his physical description, but rather in his ideas. I would like to describe now some of our discussions on mathematics, physics, and philosophy.

One of Gödel's less well-known papers is a 1949 article called, "A Remark on the Relationship Between Relativity Theory and Idealistic Philosophy."[17] In this paper, probably influenced by his conversations with Einstein as well as by his interest in Kant, Gödel attempts to show that the passage of time is an illusion. The past, present and future of the universe are just different regions of a single vast space-time. Time is *part* of space-time, but space-time is a higher reality existing *outside of time.*

In order to destroy the time-bound notion of the universe as a series of evanescent frames on some cosmic movie screen, Gödel actually constructed a mathematical description of a possible universe in which one can travel back through time. His motivation was that if one can conceive of time-travelling to last year, then one is pretty well forced to admit the existence of something besides the immediate present.

I was disturbed by the traditional paradoxes inherent in time-travel.

What if I were to travel back in time and kill my past self? If my past self died, then there would be no I to travel back in time, so I *wouldn't* kill my past self after all. So then the time-trip would take place, and I *would* kill my past self. And so on. I was also disturbed by the fact that if the future is already there, then there is some sense in which our free will is an illusion.[18]

Gödel seemed to believe that not only is the future already there, but worse, that it is, in principle, possible to predict completely the actions of some given person.

I objected that if there were a completely accurate theory predicting my actions, then I could prove the theory false—by learning the theory and then doing the opposite of what it predicted. According to my notes, Gödel's response went as follows: "It should be possible to form a complete theory of human behavior, i.e., to predict from the hereditary and environmental givens what a person will do. However, if a mischievous person learns of this theory, he can act in a way so as to negate it. Hence I conclude that such a theory exists, but that no mischievous person will learn of it. In the same way, time-travel is possible, but no person will ever manage to kill his past self." Gödel laughed his laugh then, and concluded, "The *a priori* is greatly neglected. Logic is very powerful."

Apropos of the free will question, on another occasion he said:

> "There is no contradiction between free will and knowing in advance precisely what one will do. If one knows oneself completely then this *is* the situation. One does not deliberately do the opposite of what one wants."

As well as questions, I also brought in for Gödel's enjoyment some offbeat theories of physics I had come up with recently. I was quite satisfied when, after hearing one of my half-baked theories, he shook his head and said, "This is a *very* strange idea. A *bizarre* idea."[19]

There is one idea truly central to Gödel's thought that we discussed at some length. This is the philosophy underlying Gödel's *credo,* "I do objective mathematics." By this, Gödel meant that mathematical entities exist independently of the activities of mathematicians, in much the same way that the stars would be there even if there were no astronomers to look at them. For Gödel, mathematics, even the mathematics of the infinite, was an essentially empirical science.

According to this standpoint, which mathematicians call *Platonism,* we do not *create* the mental objects we talk about. Instead, we *find* them, on some higher plane that the mind sees into, by a process not unlike sense perception.

The philosophy of mathematics antithetical to Platonism is *formalism*, allied to positivism. According to formalism, mathematics is really just an elaborate set of rules for manipulating symbols. By applying the rules to certain "axiomatic" strings of symbols, mathematicians go about "proving" certain other strings of symbols to be "theorems."

The game of mathematics is, for some obscure reason, a *useful* game. Some strings of symbols seem to reflect certain patterns of the physical world. Not only is "$2 + 2 = 4$" a theorem, but two apples taken with two more apples make four apples.

It is when one begins talking about infinite numbers that the trouble really begins. Cantor's Continuum Problem is undecidable on the basis of our present-day theories of mathematics. For the formalists this means that the continuum question has no definite answer. But for a Platonist like Gödel, this means only that we have not yet "looked" at the continuum hard enough to see what the answer is.

In one of our conversations I pressed Gödel to explain what he meant by the "other relation to reality" by which he said one could directly see mathematical objects. He made the point that the same possibilities of thought are open to everyone, so that we can take the world of possible forms as objective and absolute. Possibility is observer-independent, and therefore real, because it is not subject to our will.

There is a hidden analogy here. Everyone believes that the Empire State Building is real, because it is possible for almost anyone to go and see it for himself. By the same token, anyone who takes the trouble to learn some mathematics can "see" the set of natural numbers for himself. So, Gödel reasoned, it must be that the set of natural numbers has an independent existence, an existence as a certain abstract possibility of thought.

I asked him how best to perceive pure abstract possibility. He said three things. i) First one must close off the other senses, for instance, by lying down in a quiet place. It is not enough, however, to perform this negative action, one must actively seek with the mind. ii) It is a mistake to let everyday reality condition possibility, and only to imagine the combinings and permutations of physical objects—the mind is capable of directly perceiving infinite sets. iii) The ultimate goal of such thought, and of all philosophy, is the perception of the Absolute. Gödel rounded off these comments with a remark on Plato: "When Plautus could fully perceive the Good, his philosophy ended."

Gödel shared with Einstein a certain mystical turn of thought. The word "mystic" is almost pejorative these days. But mysticism does not really have anything to do with incense or encounter groups or demoniac possession. There is a difference between mysticism and occultism.

A pure strand of classical mysticism runs from Plato to Plotinus and Eckhart to such great modern thinkers as Aldous Huxley and D. T. Suzuki. The central teaching of mysticism is this: *Reality is One.* The practice of mysticism consists in finding ways to experience this higher unity directly.

The One has variously been called the Good, God, the Cosmos, the Mind, the Void, or (perhaps most neutrally) the Absolute. No door in the labyrinthine castle of science opens directly onto the Absolute. But if one understands the maze well enough, it is possible to jump out of the system and experience the Absolute for oneself.

The last time I spoke with Kurt Gödel was on the telephone, in March 1977. I had been studying the problem of whether machines can think, and I had become interested in the distinction between a system's *behavior* and the underlying mind or consciousness, if any.

What had struck me was that if a machine could mimic all of our behavior, both internal and external, then it would seem that there is nothing left to be added. Body and brain fall under the heading of *hardware.* Habits, knowledge, self-image and the like can all be classed as *software.* All that is necessary for the resulting system to be alive is that it actually exist.

In short, I had begun to think that consciousness is really nothing more than simple existence. By way of leading up to this, I asked Gödel if he believed there is a single Mind behind all the various appearances and activities of the world.

He replied that, yes, the Mind is the thing that is structured, but that the Mind exists independently of its individual properties.

I then asked if he believed that the Mind is everywhere, as opposed to being localized in the brains of people.

Gödel replied, "Of course. This is the basic mystic teaching."

We talked a little set theory, and then I asked him my last question: "What causes the illusion of the passage of time?"

Gödel spoke not directly to this question, but to the question of what my question meant—that is, why anyone would even believe that there is a perceived passage of time at all.

He went on to relate the getting rid of belief in the passage of time to the struggle to experience the One Mind of mysticism. Finally he said this: "The illusion of the passage of time arises from the confusing of the *given* with the *real*. Passage of time arises because we think of occupying different realities. In fact, we occupy only different givens. There is only one reality."

I wanted to visit Gödel again, but he told me that he was too ill. In the middle of January 1978, I dreamed I was at his bedside.

There was a chess board on the covers in front of him. Gödel reached his hand out and knocked the board over, tipping the men onto the floor. The chessboard expanded to an infinite mathematical plane. And then that, too, vanished. There was a brief play of symbols, and then emptiness—an emptiness flooded with even white light.

The next day I learned that Kurt Gödel was dead.

TOWARDS ROBOT CONSCIOUSNESS[20]

The human mind is incapable of formulating (or mechanizing) all its mathematical intuitions, i.e., if it has succeeded in formulating some of them, this very fact yields new intuitive knowledge, e.g., the consistency of this formalism. This fact may be called the 'incompletability' of mathematics. On the other hand, on the basis of what has been proved so far, it remains possible that there may exist (and even be empirically discovereable) a theorem-proving machine which in fact is equivalent to mathematical intuition, but cannot be proved to be so, nor even be proved to yield only correct theorems of finitary number theory.

—*Kurt Gödel*[21]

For many years there has been debate over the precise significance of Gödel's Incompleteness Theorem for the field of Artificial Intelligence.[22] Gödel, especially in his later years, was a reclusive, even secretive, man, and the quotation printed above is very nearly the sum total of his published words on this important question. The purpose of this section will be to tease out the meaning and implications of that quotation.

The section breaks into four subsections. The first, which can be skimmed, describes precisely what is meant by a theorem-proving machine. The second develops the argument for the undefinability of truth a bit further, and shows how Gödel reaches the conclusion that humans can never write down a complete description of how they think about mathematics. The third subsection explains how machines too complex to design can nevertheless *evolve,* and the final subsection advocates a mystical answer to the question, "What is consciousness?"

FORMAL SYSTEMS AND MACHINES

Quite generally, a formal system is a set of symbols together with rules for employing them. A formal system has four components: (1) alphabet, (2) spelling and grammar, (3) axioms, and (4) rules of inference.

The alphabet is simply a supply of symbols. If one wishes to be quite abstract, he can get by with "0" and "1" as his only symbols. But normally one allows the upper-case and lower-case English and Greek alphabets, the punctuation symbols, the blank space, the usual logical symbols, the numerals and other mathematical symbols, and so on.

The spelling rules specify which strings of symbols are to be regarded as noun phrases (terms) and which strings are to be regarded as verb phrases (relations). The grammar rules specify which sorts of verbs and nouns can be meaningfully combined to make simple sentences, and how one can go on to build up compound sentences, sentences with quantifiers, and multi-sentence statements. A statement formed in accordance with the spelling and grammar rules is called well-formed.

The formal system singles out a certain set of well-formed statements as axioms or fundamental assumptions. The rules of inference specify the precise ways in which axioms can be changed and combined to "prove" the theorems of the formal system.

To be more precise, the well-formed statement A is said to be proved by the formal system if and only if there is a finite proof sequence M_1, . . . , M_n of well-formed statements such that each of the M_i is either an axiom or is obtained from some of the previous M_j's by one of the rules of inference, and such that the last sentence M_n is actually A. Now, the theorems of the formal system are those well-formed statements A for which there exists a proof sequence ending with A.

The theorems of a formal system are already latently present in the system's axioms and rules of inference. Normally the formal system itself can be finitely described, but it will be able to prove infinitely many theorems. So a formal system can be thought of as a very compact way of summarizing a large body of fact.

All the theorems of classical mathematics, for instance, can be proved from the axioms and rules of inference of the formal system obtained by combining the propositional calculus, the predicate calculus, and the axioms of ZF (Zermelo-Fraenkel set theory). This formal system can be completely described in a few printed pages. There is, thus, a sense in which everything we know about mathematics is coded up in a few pages of print.

No other entire branch of science has allowed itself to be fully codified as a formal system yet, but there are many successful piecemeal efforts in this direction. The rules for how many electron shells a given atom has, the laws of heredity, the theory of electromagnetism, special relativity . . . all these can be expressed as formal systems yielding certain bodies of fact.

Usually the alphabet of a formal system will include something that can be used as a negation symbol. One normally only considers formal systems that are consistent, where this means that for no statement S can we obtain both S and the negation of S as theorems. This is a natural requirement because a formal system is, in practice, intended to summarize a collection of facts that obtain in some one possible world or part of a world . . . and it could never happen that S and the negation of S are both true facts.

Normally, one also requires that the formal systems we consider be put in an unambiguous and finitely describable form. When we say that a formal system is finitely describable, we mean that there are three definite finite procedures, WELL, AXIOM, and RULE, determining the system as follows.

Given any string S of symbols drawn from the system's alphabet, WELL can be applied to determine whether or not S is a well-formed statement and AXIOM can be applied to determine whether or not S is an axiom. The finite procedure RULE can be applied to any well-formed statement in combination with any finite set of well-formed statements to determine whether or not the former follows from the latter according to any single rule of inference.

Without getting too technical, the essential aspect of the algorithmic procedures, WELL, AXIOM, and RULE, must be that they are (a) utterly mechanical in application and (b) always give an unequivocal YES or NO answer after some finite amount of time.

A finitely describable formal system T is sometimes best thought of as a theorem-listing machine. Given the formal system T based on the triple of algorithms (WELL, AXIOM, RULE), we can construct a machine M_T that will print out all of T's theorems, one after another.

Before describing M_T, we should first note that given a fixed alphabet there is an utterly mechanical procedure for generating all possible "words" or strings of symbols drawn from that alphabet. If, for example, the alphabet were the lower-case English alphabet, one could first list, in dictionary order, all one-letter words, then, in dictionary order, all two-letter words, then, in dictionary order, all three-letter words, and so on and on.

Now, M_T will operate by looping through the following list of instructions to generate three ever-growing stores St, We and Th, consisting, respectively, of possible strings, well-formed statements, and theorems. At the same time, M_T will print the theorems out.

1. Generate the next possible string and add it to store St.
2. Check the most recent string with the procedure WELL. If YES, then add the string to the store We.
3. Check the most recent well-formed sentence with procedure AXIOM. If YES, then add the sentence to the store Th.
4. Use the procedure RULE to check, one by one, each member of We for derivability from the set Th. Add each sentence for which the answer is YES to the store Th, and print each of these new theorems on the output tape.
5. Return to (1).

There is a general theorem of recursive function theory stating that not only can any formal system be viewed as a machine, the converse is true as well. That is, given any digital computer M with unlimited memory, we can find a formal system T_M such that the possible outputs of M are exactly the possible theorems of T_M. This holds not only for machines that simply print out long lists of theorems, but also for machines that exhibit branching behavior patterns, and even for machines that intereact with their environment.

A typical machine is non-deterministic in the weak sense that it does have branching futures. M begins in an initial state (compare axioms), and then passes through a series of transitions according to programmed rules (compare rules of inference). M can be nondeterministic in the sense that for certain states S there are a variety of allowable "next states" $S_1^*, S_2^*, \ldots, S_n^*$. I want to make the point that even if one allows the choice of next state to be made randomly (say, by counting clicks on a Geiger counter), the machine's range of possible outputs is still equivalent to the set of possible theorems of some fixed formal system.

This is the case because, in and of itself, a formal system does not actually generate a list of its theorems. A formal system is, strictly speaking, a gestalt or starting state from which one can move out along various possible proof sequences to reach various possible theorems. The essential form of a formal system is thus more like a tree than a line. At the base of the tree we have the axioms and growing out from them we have all one-step proofs, and so on. At each node or fork of this tree is a theorem—a possible output.

Suppose, however, that we have a machine M that interacts with its environment. M then can be thought of as embodying a behavior function of the form $M(h, i) = o$. "h" stands for the history of what has happened to M since it was turned on, "i" stands for the stimulus or input at the present time, and "o" stands for a response or output that M may give to stimulus i, having had history h. Note that, as mentioned above, it is permissible that for given h and i there be several possible o's. But now if we assume that the h's, i's, and o's can be specified by strings of symbols, then it is not hard to see that the behavior of machine M can be coded up in some formal system T_M that has a variety of strings of the form "$M(h, i) = o$" as its theorems.

THE LIAR PARADOX AND THE NON-MECHANIZABILITY OF MATHEMATICS

Let us now return to the idea of a theorem-listing machine, a finitely described device that prints out an endless list of statements. Obviously, the best possible theorem-listing machine would manage to print out all and only the true statements expressible in, let us say, the English language plus the customary printers' symbols. Such a Universal Truth Machine would be a fine thing indeed. Once we had devised its finite procedures, WELL, AXIOM, and RULE, we could set it to running and simply lean back and watch the printout. If you were curious about the truth of some well-formed statement S, you would only have to sit there and wait until either S, or the negation of S, was printed out.

Before going any further, we should be a bit more precise about what sorts of string S will count as (well-formed) statements for which one can expect an unequivocal true or false decision.

A first point is that we do not necessarily expect a statement to consist of a single grammatical unit. That is, a statement can consist of several sentences, a whole paragraph, a book, or even a large number of books. We desire only that the entire welter of symbols embody a single assertion that is in fact either true or false. We will, however, not allow infinitely long strings of symbols as statements, for a statement should be, at least in principle, unambiguously and finitely communicable.

A second point is that a string such as "x is smaller than y" is not, in itself, a statement. For without being told what x and y are, no one could ever say that this string expressed either a truth or a falsehood. So a statement should not include any undefined terms or relations.

But now we have a rather nasty problem. What about the phrase "is smaller than" in the example above? Must the meaning of that too be explained before we have a definite statement? But if every term and relation, every noun and verb, must be explained, then we seem to face a real mess. A full statement will take on the form of an initial assertion, followed by definitions of the words used in the first definitions, followed by definitions of the words used in the second definitions, and so on.

Well . . . why not? Such are the joys of abstract speculation! By way of preventing unnecessary regresses, we specify that within any expanded statement, no word's definition need be repeated. Since there are only finitely many words and word-creation schemata in English, the process should normally terminate. There may, however, be some cases in which the full expansion of a statement becomes infinite, and in such a case we will just have to write the statement off as non-well-formed, not really a statement.

Someone might object that a fully expanded statement will consist of circular sequences of definitions and will thus be useless. To this there are two responses. First, any circular definition more complicated than "x is x" does convey some information. Euclid's circular definitions—of space as the set of all points and of a point as a region of space that has no parts—tells us that a point, whatever it is, cannot be made up of a number of component points. Second, the fact that one is using strings of symbols enables one to exhibit certain concepts without really having to define them. Take, for instance, the concept of the length of a string of symbols. "xxx" is a longer string than "a", and by multiplying such examples (perhaps schematically) the meaning of string length can be made evident.[23]

So let us expand our statements to be self-explanatory. When we are done, a statement such as "the square root of two is irrational" will have blossomed into a hefty volume of mathematics, and "all men are mortal" will have expanded to include a definition of most of the words in the English language.

But the simple listing of dictionary definitions will not always be enough to make a statement definite enough to be susceptible to a true/false determination. In order to evaluate a statement such as "Jimmy Carter is good," one needs to know which criterion of "goodness" is intended. Does the statement assert that Jimmy Carter teaches Bible School, that Jimmy Carter should have been reelected, or that the flesh of Jimmy Carter is sweet and tasty? In filling the bare assertion out to a full statement, one of the many possible options must be chosen.

Let us return now to the concept of a UTM (Universal Truth Machine.) What we want is a finitely programmed device that will print out all the true statements, and no false statements. It is easy to prove that there can be no such UTM. (We repeat the argument from the section "Gödel's Incompleteness Theorem.")

For, say that you have a candidate UTM based on the triple of finite procedures (WELL, AXIOM, RULE). Now, since WELL, AXIOM, and RULE are entirely finite and mechanical, one should be able to construct a well-formed statement G saying, "The UTM based on the triple (WELL, AXIOM, RULE) will never print out this statement"[24]

If your UTM ever prints out G, then G will be false—so the UTM will have printed out a false sentence. But this would be impossible, as the UTM is to print out only true sentences. Therefore the UTM will never print out G. And therefore, G is in fact true. But now we have a true sentence that will never be printed out by our candidate UTM— which is therefore not truly universal.

Truth, in short, is not finitely describable. This fact provided the resolution of the Liar paradox discussed in "What is Truth?" Consider the string of symbols B, "This string of symbols does not express a well-formed statement that is true." Clearly B cannot in fact be expanded to make a well-formed statement that is true—because if such a statement if true, then it isn't. The reason B can't be made well-formed is that if we expand B to include a complete explication of all the words it uses, then B will have to include a complete explication of what it means for a finite statement to be "true." But, as we have just seen, this word "true" does not admit of any finite explication! So B is one of those statements that, if expanded to be completely understandable, becomes infinitely long. Now, an infinitely long string of symbols is not allowable as a statement, so there is really no paradox here. If B were a finite statement, then B would be true if and only if B were not true . . . but B is infinite, and that is the end of it. B gives the feel of a paradox, as does the similar string, "This sentence is not true," but this is simply an illusion on a par with a man suspending himself by a piano wire and saying he can fly. In the case of B, the "piano wire" is infinitely long, and B is not really a statement at all.

So the concept of truth for finite statements is itself infinite. One might now be tempted to ask if there might be some higher notion of truth that can arbitrate the truth of every statement, whether finite, infinite, or transfinite? Not in this world. Looked at another way, the message of the Liar paradox is that one can never define truth in such a way as to apply to statements involving the concept of truth being defined.

Let me illustrate this with an example. There is in Rome a church, Santa Maria in Cosmedin, outside of which stands a huge stone disc. The face of the disc is carved in bas-relief to depict the visage of a hairy, bearded man. The mouth, located at about waist-level, is shaped something like a letter slot. Legend has it that God himself has decreed that anyone who sticks a hand in the mouth slot and then utters a false statement will never be able to pull the hand back out. But I have been there, and I stuck my hand in the mouth and said, "I will not be able to pull my hand back out." (May God forgive me!)

It is perhaps not really so surprising that there is no set of rules that will suffice to generate all the possible truths. There is something about the fact that the rules exist in the world they attempt to describe that forbids this. What perhaps is surprising is that the same phenomenon also holds true for the small and orderly world of number theory.

The natural numbers, along with the equality relation and the operations of plus and times, do not seem very complicated. Even if one cannot get a finite description of all true statements, one might hope to get at least a finite description of all the true facts about natural numbers. But, according to Gödel's 1930 Incompleteness Theorem, this is also impossible.

Gödel's two Incompleteness Theorems state that all formal systems of a certain kind are subject to two related limitations. His results apply to any formal system T that is (i) finitely describable, (ii) consistent, and (iii) strong enough to prove the basic facts about whole-number arithmetic.

Gödel's *First Incompleteness Theorem* states that no such formal system T is capable of deciding every statement about natural numbers. That is, for any such T there will be a fairly simple sentence about natural numbers such that neither this sentence nor its negation ever appears as a theorem of T. We already have seen the idea of the proof, as follows.

Let M_T be the theorem-listing machine that prints out all the theorems of T. Suppose $G(T)$ is the sentence saying, "M_T will never print this sentence." Now, as we saw in the discussion of Universal Truth Machines, $G(T)$ must in fact be a sentence that is true, but that M_T never prints out.

The hard part is to show that $G(T)$ can actually be put in the form of a sentence about natural numbers. Gödel did this by a process called Gödel-numbering, something like the coding process described in "The Library of Babel." When Gödel is done, the self-referential sentence $G(T)$ saying, "T cannot prove $G(T)$," has been converted into a sentence of pure mathematics stating that a certain polynomial equation has no solutions in the whole numbers.

The upshot is that any finitely describable consistent theory T provides an incomplete description of the natural numbers; for given any such T there will be a sentence $G(T)$ about natural numbers that T can neither prove nor disprove.[25]

Now for the *Second Incompleteness Theorem*. Recall that the condition, "T is consistent" means that for no sentence S does T prove both S and the negation of S. This condition can be abbreviated as $\mathrm{Con}(T)$, and by using Gödel-numbering again we can convert $\mathrm{Con}(T)$ into a purely number-theoretic sentence, once again a sentence saying that a certain Diophantine equation has no solution.

Insofar as the theory T embodies a correct description of the mathematical universe, we might expect $\mathrm{Con}(T)$ to be a fairly obvious and readily deducible fact. But the Second Incompleteness Theorem tells us that if T is a theory satisfying conditions (i) through (iii), then T cannot prove $\mathrm{Con}(T)$.

The proof works by going back and formalizing the proof of Gödel's First Incompleteness Theorem inside the system T. Once we do this, we have proved "$\mathrm{Con}(T)$ implies $G(T)$". But now, since T cannot prove $G(T)$, it must be that T cannot prove $\mathrm{Con}(T)$ either!

We are now in a position to understand the first part of the quote from Gödel that stands at the head of this paper. As mentioned above in "Conversations with Gödel," Gödel took the viewpoint that the natural numbers, the transfinite sets, and all other mathematical objects are immaterial, but actually existing, entities. By a process that Gödel called mathematical intuition, humans come to learn certain facts about the universe of mathematics, and we also come to learn certain correct methods of reasoning from given facts to further facts. Mathematical intuition is to be thought of as a process just as reliable as ordinary sense perception.[26]

Now, since the facts and methods of reasoning that we learn are correct descriptions of an actually existing mathematical universe, there is no possibility of ever producing a contradiction in mathematics. On the basis of true mathematical facts and correct mathematical reasoning, we will never find ourselves proving that, say, zero is not equal to zero. In other words, whenever we can produce a description K of our present body of mathematical knowledge, then we can be sure that $Con(K)$ is true.

Now, say that we are able to come up with a finite description of a formal system K that summarizes our entire present body of mathematical knowledge. *On the one hand,* by the considerations of the last paragraph, we know that $Con(K)$ is true. On the basis of our finite description of K we can even apply Gödel numbering and come up with a specific true number-theoretic sentence $Con(K)$. But, *on the other hand,* since K satisfies conditions (i) through (iii), the Second Incompleteness Theorem tells us that K cannot prove $Con(K)$!

It is for this reason that Gödel says the human mind is incapable of mechanizing all of its mathematical intuitions. For to mechanize our intuitions is to produce a finite description of a formal system K. But as soon as we see this finite description, our mathematical intuition shows us a fact, $Con(K)$, which the mechanized system does not prove. So it is not true that the mechanized system K proves all facts that we can perceive through our mathematical intuition.

ARTIFICIAL INTELLIGENCE VIA EVOLUTIONARY PROCESSES

Let us now consider the second part of Gödel's remark. He remarks that even though we cannot write the program for a theorem-proving machine that is equivalent to human mathematical intuition, it is possible that such a machine could exist and even be empirically discoverable.

Let us suppose that there is a machine R that is equivalent to human mathematical intuition. A first fact to be established is that we could never understand R's program. Let us make this point quite clear.

(i) Being a theorem-proving machine, R is finitely describable. (ii) Since R is equivalent to our *a priori* consistent mathematical intuitions about the mathematical universe, R is consistent. (iii) For the same reasons, R is of course strong enough to prove the basic facts of whole-number arithmetic. Therefore, by the Second Incompleteness Theorem, R cannot prove $Con(R)$.

Now, since R is in fact exactly equivalent to human mathematical intuition, it must be that the humans can never prove $\text{Con}(R)$ either. This can only be because the humans can never understand the finite description of R well enough to prove that R is, in fact, equivalent to their own mathematical intuitions, which are known, by a sort of higher-level intuition, to be consistent. The grounds for our lack of understanding of R's program will, presumably, lie in the program's length and subtlety.

But how could such an utterly incomprehensible machine actually be produced, when we know that we could never understand it, let alone build it?

The answer is evolution. Before his premature death in 1957, John von Neumann worked out the basic theory of self-reproducing automata.[27] There is, in principle, no difficulty in designing robots that are capable of building factories to produce other robots. There is also no difficulty in arranging things so that the old robots can copy their own programs onto the processors of the new robots. It is simply a matter of assembling the hardware, and then replicating one's software onto the new hardware.

A current IBM design goal is the construction of super-cooled computers that will fit inside an $8 \times 8 \times 10$ centimeter box, and have a 256,000 word cache memory (workspace) and a 64 million word main memory. Each bit will be represented by a quantum of magnetic flux generated by a persistent current in a super-conducting Josephson junction. It is perhaps not too much to hope that the following generation of computers will be able to enter the 10^{10} word range characteristic of human brains.

Now imagine equipping a few thousand robots with these liquid-helium-temperature silicon brains and setting them loose on the moon. Their prime directive is to mine, smelt, fabricate, and assemble the materials necessary to build more robots. They find the moon congenial: the low temperatures, the abundant solar energy, the lack of corrosive water vapor or gaseous oxygen, the abundance of silicon. They are fruitful, and they multiply.

Assuming that they are programmed to place high priority on *self-reproduction,* there will be an inevitable competition for the raw materials and finished supplies with an attendant effect of *natural selection.* In addition, we can ensure that their programs undergo regular *mutations.* This could be done by placing somewhere in the base program an imperative never to copy oneself exactly. Each time a program was transferred, a substantial number of changes would be made in it, these changes to be determined randomly, let us say by counting cosmic rays, or simply by waving a powerful magnet over the new scion's head.

Actually, a large part of evolutionary diversity arises not from actual mutating gene change, but rather from the shuffling of genes inherent in sexual reproduction. Presumably something along these lines could be arranged, with two (or more) robots "sexually reproducing" together by pooling their hardware resources and shuffling together various of their subprograms to produce a new program for the little scion.

Rather than immediately sending billions of dollars worth of hardware to the Moon, one would in practice begin a project of this nature in the laboratory. It would not even be initially necessary to deal with mobile hardwares capable of physical self-reproduction. Instead, as a first stage, I would envisage some thousands of AI programs competing (perhaps on the basis of scores on certain tests) for the right to be replicated, with one copy always incorporating some randomly determined mutations. It seems possible that the process could be speeded up enough for significant evolutionary effects to emerge after only a few years run-time.

Once the programs had gotten intricate enough to be incomprehensible, or understandable only with great difficulty, they could be packed into the factory-building robots and shipped off to the Moon. The economic incentive for this is, of course, that we would like to be able to exploit the Moon's resources, and it is possible that robots would have a better cost to productivity ratio than human colonists.[28]

As the robots on the Moon continue to evolve, there is always the danger that an unfortunate series of lethal mutations might force them into extinction. One would like perhaps to guard against lethal mutation by leaving certain core life-support sections of their programs inviolable. A problem with this is that a mutation that is lethal now may not be lethal to future variants of the species, and vice versa. Lungs, for instance, are very bad for a fish, but very good for an amphibian.

But assuming all goes well, that God, perhaps, will guide the robots' evolution as He may have guided ours, then we could expect that there would eventually be a large and autonomous robot civilization on the Moon, and that some of the robots there might interest themselves in mathematics. It is at this point that there is a real possibility that there could arise a theorem-proving machine R (a robot mathematician) whose abilities are, in fact, equal to the resources of human mathematical intuition.

There is no reason why R could not print out his program for us, per- haps in some extremely compact and coded form (compare sperm cells!). But, as was discussed above, we would not be able to understand this program, nor would we be able to prove to our satisfaction that R was consistent. The interesting thing is that even though we cannot un- derstand the program of R, we are able to set up the physical conditions that lead to R's coming into existence.

ROBOT CONSCIOUSNESS

With the paradigm of the last subsection in mind, it now seems evi- dent that there could be robots whose general behavior was the same as the behavior of human beings. These robots would be thinking beings who had evolved on a substrate of metal and silicon chips, just as we are thinking beings who have evolved on a substrate of amino acids and other carbon-based compounds. Would one be justified in saying that these highly evolved robots possess consciousness in the same sense that humans do?

Upon lengthy introspection, most people will agree that the individ- ual person consists of three distinct parts: (a) the hardware, the physical body and brain; (b) the software, the memories, skills, opinions, and be- havior in general; (c) consciousness, the sense of self or personal iden- tity, pure awareness, the spark of life, or even the soul.

I would like to argue that any component of parts (a) or (b) can be replaced or altered without really affecting (c). My purpose in arguing this way will be to show that there is nothing about part (c) that is spe- cific to the individual.

Let us begin with the hardware. If one gets an artificial leg, kidney, or heart, one is still the same person. I maintain that it is possible to envis- age a time when one could even get a new artificial brain. This could be done by, let us say, holographically recording the physical, electrical, and biochemical structure of the brain, and then transferring this struc- ture isomorphically onto a large silicon-chip system or onto some ap- propriate module of culture-grown tissue. Presumably one would expe- rience such a transfer only as a brief period of unconsciousness after which one would go on thinking much the same as one had before. The whole process would be comparable to putting a given program into a new computer.[29]

Now for the software. It is an utterly familiar feeling to look back on the way one was behaving a year, a month, or an hour ago and to be amazed. One's personality is always changing, and one is always learning new things and forgetting old ones. There is also the extreme example of brainwashing. We are inclined to say that a person's essential identity is unchanged, even if he has been given a completely false set of memories.

What then remains for part (c)? I contend that the sum total of the individual consciousness is the bare feeling of existence, expressed by the primal utterance, *I am*. Anything else is either hardware or software, and can be changed or dispensed with. Only the single thought *I am* ties me to the person I was twenty years ago.

The curious thing is that you must express your individual consciousness in the same words that I use: *I am. I am me. I exist.* The philosopher Hegel was very struck by this fact, and deemed it an instance of "the divine nature of language."

What conclusion might one draw from the fact that your essential consciousness and my essential consciousness are expressed in the same words? Perhaps it is reasonable to suppose that there really is only one consciousness, that individual humans are simply disparate faces of what the classical mystic tradition calls the One.

But we can go farther than this. The essence of consciousness is, really, nothing more than simple existence. *I am*. Why should the possession of this sort of consciousness be denied to anything that does exist? Aquinas has said that God is pure existence unmodified. Is it not evident that there is a certain single something–call it God, or the One, or pure existence–that pervades the world as it is? Consider the Zen phrasing of this: *The universal rain moistens all creatures.*[30] Or think of the world as a stained-glass window with light shining through every part.

To exist is to have consciousness. The other things one might feel are necessary for consciousness are more or less complicated sorts of hardware and software, patterns of mass and energy. But no pattern can be conscious until it exists, until it is brought into reality. Existence is, finally, the only thing required for consciousness. A rock is conscious. This piece of paper is conscious. And so, of course, is a robot, both before and after his behavior evolves to our level.

Traditionally, those who have asserted the equivalence of men and (possible) machines have been positivists, mechanists, materialists. They put their viewpoint this way: "Men are no better than machines." But if one only changes the emphasis, then this equivalence can become the expression of a deep belief in the universality and reality of consciousness: "Machines can be as good as men!"

BEYOND MECHANISM?

There seem to be three possible viewpoints regarding questions about human and robot souls. *Mechanism:* Neither men nor robots are anything but machines, and there is no reason why man-like machines cannot exist. *Humanism:* Men have souls and machines do not, therefore no robot can be quite like a man. *Mysticism:* Everything, whether man or machine, participates in the Absolute, therefore it should be possible for man-like machines to exist. Perhaps "mysticism" is not really a good name for this last view, but let it stand.

In the last section I argued for the third viewpoint, under which "having a soul" is a concept automatically satisfied by anything that exists. But, since this view leads to a conclusion identical to that of mechanism, the reader may feel that I have ducked the real issue. One feels oneself to be "more than just a machine." Is there any possible justification for this belief short of recourse to an all-pervading Absolute?

In two classic papers, Alan M. Turing developed a powerful argument for mechanism: All that we can know about another person's mind is based on observing his behavior (i.e., by conversing with him, reading his writings, and so on); and there seems to be, in principle, no reason why there could not be a machine whose "conversation" is just like a person's.[31]

Now, it might be objected that such an argument does not account for private mental phenomena, such as mental images, purposes, emotions, and the like. But a determined mechanist can say that what we call a mental image is merely a model or simulation such as computers often use, that a purpose is simply an assignment of utility values to certain internal states, and that emotions are just ways of assigning values to external phenomena as well.

An even stronger argument for the mechanist position can be formulated by asserting that: 1) The activities of the mind are isomorphic to certain electro-chemical processes in the brain; and 2) The brain functions basically like a digital computer.[32]

These assertions can be restated: 1) There is no mind apart from matter; and 2) The brain is finite. The idea is that, given 1) and 2), we can be sure that the operation of the mind is a finite law-like process; and any such process can, in fact, be modelled by a large enough digital computer.

Assumption 1 is often questioned by those who believe in telepathy and other forms of ESP. These investigators have pointed out various unusual occurrences or perceptions that do not seem to fit in with the idea that the mind is simply a phenomenon taking place in the limited confines of the skull. I really do not know what to make of the claims made for ESP. Like any other person, I have experienced my share of what Jung calls *synchronicities:* meaningful conicidences, true premonitions, and lucky guesses.[33] But the desire for results is so strong, and the opportunities for deception so great, that the most extreme caution is necessary.

One of the things most lacking in ESP research is any reasonable theory of how the mind could, in fact, go beyond the confines of the brain.[34] As I suggested at the end of Chapter 2, it would be nice if Cantor's two-substance theory were true, for then we could perhaps have the mind or "astral body" made up of some higher-level substance quite different from matter. But there is no evidence for such a theory—it is still only an idea for an idea.

A recent development in physics that could lead to some interesting developments is the experimental disproof of the Bell Inequality.[35] What the experiments indicate is that there is a very real sense in which particles that have interacted continue to affect each other long after the interaction has taken place . . . and in an instantaneous way! If this is indeed true, then the universe would of necessity behave like a single organic whole, leading to the possibility of choosing to identify one's mind with the cosmos rather than with some individual brain. But, as I pointed out at the end of the last section, there is no reason why such a form of higher consciousness would not be open to robots as well. Of course, then robots would not be "machines" in the narrow, finitistic sense intended by the mechanists.

Going back to the argument for mechanism, what about assumption 2? As was discussed in the section "Infinities in the Small," it is certainly possible that matter is infinitely divisible. If this were indeed true, then any material object, such as a human brain, would in fact be infinitely complex. Perhaps we really do think infinite thoughts, and it is simply an accidental property of this limited scale level that our descriptions of them come out finite. Sometimes I actually am able to believe this for a few minutes, keeping in mind Cantor's remark that "the infinite even inhabits our minds."[36]

> "I can't believe *that!*" said Alice.
> "Can't you?" the Queen said in a pitying tone. "Try again: draw a long breath, and shut your eyes."
> Alice laughed. "There's no use trying." she said: "one *can't* believe impossible things."

"I daresay you haven't had much practice," said the Queen. "When I was your age, I always did it for half-an-hour a day. Why sometimes I've believed as many as six impossible things before breakfast."[37]

Perhaps the mechanists are right. But it is certainly worthwhile to keep an open mind, even if this means occasionally seeing how it feels to believe something that is "impossible."

PUZZLES AND PARADOXES
(Answers, p.(301)

1. Consider the sentence S: "THIS SENTENCE CAN NEVER BE PROVED." Show that if S is meaningful, then S is not provable, and that therefore you can see that S must be true. But this constitutes a "proof" of S. How can the paradox be resolved?[38]

2. Consider the following argument for the unreality of death: a) A person's mind and personality is equivalent to his software, that is, to the programming of his brain; b) Any software structure can be coded up by some large set of natural numbers; c) Every set of numbers exists eternally as a mathematical abstraction independent of the physical universe; d) Therefore each individual's personality is immortal. Why is it that this type of immortality does not seem like enough?

3. Consider this somewhat stronger argument for the immortality of artists: a) In a great work of art, an artist codes up a large part of his software, his personal feelings about life; b) When one deeply immerses oneself in a work of art, one takes on, for a few moments, the actual software coded up by that work; c) Therefore every time one truly appreciates a work of art, one is, for the moment, isomorphic to the artist, and the artist is thus (briefly) reincarnated over and over. If there were, a hundred years from now, someone exactly like you, would it make a difference to you? Would this be any different from being frozen for a hundred years and then resuscitated?

4. Consider the following fanciful example of self-reproduction (due to Doug Hofstadter). "Imagine that there is a nickelodeon in the local bar that, if you press buttons 11-U, will play a song whose lyrics go this way: *Put another nickel in, in the nickelodeon,/All I want is 11-U, and music, music, music.*" Assuming that everything goes well, some state of some system will occur over and over here. Specify the state and the complete system.

5. There is a word game sometimes called "word golf" that embodies a simple analogy of a formal system. One starts with a given word (say, LOVE) that serves as the initial "axiom." The "rule of inference" is that at each step one can change one letter of the word in hand, provided the change produces a

new English word. The "theorems" following from a given "axiom" word
are those words that can be reached by a series of transformations according to the rule of inference. Thus HATE is a theorem, or consequence, of
LOVE. The "proof" of this is the sequence LOVE, ROVE, RAVE, HAVE,
HATE. The reason it is called "word golf" is that one tries to find "proofs"
with the shortest possible number of steps. If you allow the somewhat obscure word LAVE, then the derivation above can be shortened by a step to
LOVE, LAVE, LATE, HATE. How many steps do you need to turn COLD
into WARM, BEER into WINE, FISH into FOWL?

6. In this chapter we discussed a UTM (Universal Truth Machine) that would
print out all true statements. In the section "What is Truth," a sort of TSM
(Truth Sorting Machine) was discussed. The TSM was to look at any book
and decide, after a longer or shorter interval of time, whether or not the
book was true. We have proved that neither a UTM nor a TSM can
exist. *BUT,* ignoring this fact for now, show that, with a few modifications,
any UTM could be converted into a TSM, and any given TSM could be
converted into a UTM.

7. Present day computers are, both with respect to hardware and with respect
to software, far superior to the computers of thirty years ago. Show that this
improvement can be viewed as a sort of robot evolution, and point out what
processes have played the roles of reproduction, selection, and mutation.

8. The passage from Gödel quoted at the beginning of "Towards Robot Consciousness" goes on to make a further point: "Either the human mind sur-
passes all machines (to be more precise, it can decide more number-theoretical questions than any machine) or else there exist number-theoretical
questions undecidable for the human mind."[40] Explain how this is really
just a restatement of the result that no machine can answer all number-theoretical questions.

CHAPTER FIVE
THE ONE AND
THE MANY

Think of Everything–the whole physical universe past and present, all the other possible universes, all the possible thoughts, all the mathematical sets–or *can* one think about it? The classical One-Many problem is this question: Can Everything be regarded as a unity, as a single definite thing? Is the world a One or a Many?

The first section of this chapter contains a brief account of some aspects of the classical One/Many problem. The problem allows itself to be stated in terms of set theory, so the following section, "What Is a Set?," introduces the general notion of set, and "The Universe of Set Theory" describes various ways in which the One/Many problem arises in set theory. The final section of the chapter, "Interface Enlightenment," describes a certain sort of metasolution to the problem, where *meta*solution means a solution obtained by looking at the problem from a higher level standpoint. Such solutions are not always satisfying—one type of metasolution to a chess problem is to overturn the board!

THE CLASSICAL ONE/MANY PROBLEM[1]

There are two forms of the One/Many problem: i) How many *kinds* of things are there; ii) How many *things* are there? The natural first answer is that there are many different kinds of things and many different things.

There is, however, a perennial desire to reduce the world's diverse phenomena to a single basic kind, to believe that ultimately all things are built of the same stuff. Matter, sensation, thought, and form have all been candidates for *Urstoff*. The belief that there is ultimately only one kind of thing in the world is called *monism of kinds*. Materialism and idealism are both monisms of kinds; the monism of kinds that asserts that everything is a set will be considered in "The Universe of Set Theory."[2]

Instead of uniting things from the bottom up, one can work from the top down, starting with the assertion that "All is One." *Monism of substance* asserts that everything is a part or manifestation of a higher unity that is usually called the Absolute.

It is, of course, obvious that the word or concept "Everything" serves to form the world at least superficially into a One. In the same way, the bare concept "set" makes a One of the universe of set theory, but without answering the real question of whether this universe is in any sense a definite completed object. The heart of question ii) is whether or not the world is One in some organic sense, rather than in the sense of mere wordplay.

Perhaps the principal reason for believing that the world *is* an organic One is the sort of mystical insight that Lovejoy somewhat slightingly refers to as "monistic or pantheistic pathos."[3] The fact that it is occasionally possible to feel an all-encompassing unity in the world is, however, not conclusive, as it is equally possible to feel a diversity in the world that defies unification.

It is possible to *argue* for monism of substance in various ways. One idea is that ultimately everything in the world is related to everything else, and that the Absolute is the means or essence of this interrelatedness. Here the Absolute serves as a sort of connective tissue that fixes the individuals of the world into their perceived relational structure.

Another approach is to argue that any two things are, in a sense, the same; and that the Absolute is the one endlessly diversifying thing that exists.

But it is at least questionable whether such an Absolute actually exists, and pluralism of substance remains a reasonable position. This position is forcefully presented by William James in *A Pluralistic Universe:*

"... the pluralistic view which I prefer to adopt is willing to believe that there may ultimately never be an all-form at all, that the substance of reality may never get totally collected, that some of it may remain outside of the largest combination of it ever made, and that a distributive form of reality, the *each*-form, is logically as acceptable and empirically as probable as the all-form commonly acquiesced in as so obviously the self-evident thing."[4]

Not to keep the reader in suspense, I should come out with my position on the One/Many problem. But this is not so simple. In certain ways I agree with James, that there really is no ultimate single universe. But on the other hand, I do feel that the simple predicate, "exists," does tie everything together into a unity that it is, in principle, possible to experience directly. Rationally the universe is a Many, but mystically it is a One. The question that really interests me is this: How do we reconcile the Absolute as One with the Absolute as Many? How do we fit together the world of feeling and the world of thought?

But the reader should not hope for any final, tidy answer to this aspect of the One/Many problem. No one knew more about it than Plato. Most of the *Parmenides* and the *Sophist* deal with the One and the Many. In section 15 of *Philebus,* Plato's last dialogue, he has the aging Socrates deliver this wry, weary, wise summary of what he knows about the One and the Many:

> We say that the one and many become identified by thought, and that now, as in time past, they run about together, in and out of every word which is uttered, and that this union of them will never cease, and is not now beginning, but is, as I believe, an everlasting quality of thought itself, which never grows old. Any young man, when he first tastes these subtleties, is delighted, and fancies that he has found a treasure of wisdom; in the first enthusiasm of his joy he leaves no stone, or rather no thought, unturned, now rolling up the many into the one, and kneading them together, now unfolding and dividing them; he puzzles himself first and above all, and then he proceeds to puzzle his neighbours, whether they are older or younger, or of his own age that makes no difference; neither father nor mother does he spare; no human being who has ears is safe from him, hardly even his dog, and a barbarian would have no chance of escaping him, if an interpreter could only be found.[5]

WHAT IS A SET?

It is impossible to improve upon Cantor's succinct 1883 definition: "A set is a Many which allows itself to be thought of as a One."[6] One of the most basic human faculties is the perception of sets. Consider this random pattern of X's.

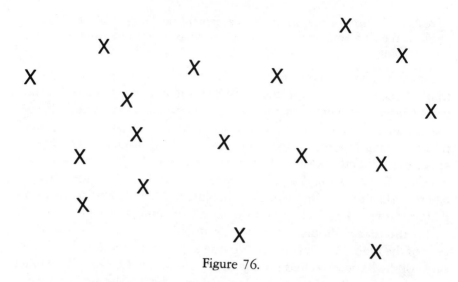

Figure 76.

As you idly gaze at the picture, your brain searches out and notices various subpatterns, various constellations of X's. On the upper right is a rough hexagon. Running from lower left to upper right is an approximate line. Running from upper left to lower right is a double camel's hump of sorts. Triangles, a question mark, a tennis racquet, and so on.

When you think about your associates and acquaintances you tend to organize your thought by sorting these people into overlapping sets: friends, women, scientists, drinkers, gardeners, sports fans, parents, etc. Or the books that you own, the recipies that you know, the clothes you have—all of these bewildering data sets are organized, at the most primitive level, by the simple and automatic process of set formation, of picking out certain multiplicities that allow themselves to be thought of as unities.

Do sets exist even if no one thinks of them? The numbers 2, 47, 48, 333, 400, and 1981 have no obvious property in common, yet it is clearly *possible* to think of these numbers all at once, as I am now doing. But must someone actually think of a set for it to exist?

This is really a version of the old chestnut, *does a falling tree make a noise if there is no one there to hear it?* The "tennis racquet" was present in the pattern of X's, even before I noticed it. By circling it, I do not bring it into existence, I merely point it out as an objectively existing feature of the external world.

Even if no one ever notices some given set, it still exists as a certain possible thought or perception. In the same way, the lonely falling

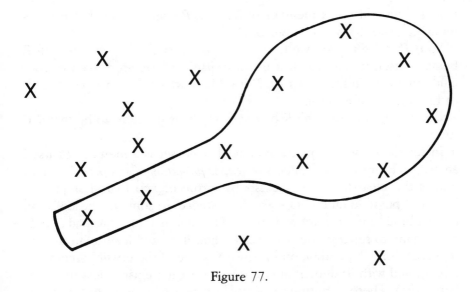

Figure 77.

tree's sound exists as a certain possible perception, a condition of the world expressible as, *if a person were present he would hear a crash.*

The basic simplifying assumption of Cantorian set theory is that sets are there already, regardless of whether anyone does or could notice them. A set is not so much a "Many which allows itself to be thought of as a One" as it is a "Many which would allow itself to be thought of as a One, if someone with a large enough mind cared to try." For the set theorist, the bust of Venus is already present in every block of marble. And a set M consisting of ten billion random natural numbers exists even though no human can ever see M all at once. *A set is the form of a possible thought,* where "possible" is to be taken in the broadest sense.

At this point one might well ask if there is anything that is not a set, or if there are any "thoughts" that are not possible. The answer, surprisingly, is yes. We can imagine that some sets might be members of themselves, and some might not be. So one might expect to be able to discuss a set R defined to be the *set of all sets that are not members of themselves.* In symbols we would have $R = \{x : x \notin x\}$.

On the surface there seems to be no reason why R should not be a Many which allows itself to be thought of as a One. But now ask this question: Is R a member of R? Is $R \in R$?

A little thought makes it evident that if R is a member of R, then R (not being a set that is not a member of itself) is not a member of R. And if R is a set and not a member of R, then R (being a set that is not a member of itself) is a member of R.

So *if* R is a set, then we have a contradictory state of affairs with R being in itself if and only if it is not in itself. So we are forced to conclude that R is, in fact, *not a set*. R is a Many which does not allow itself to be thought of as a One.

But why not? Why can't R be a set? The reason seems to be that R is too big.

In set theory we normally assume that no set is a member of itself anyway. This principle, called the *genetic formation of sets principle,* embodies the idea that a set S exists by virtue of two logical steps: i) a bunch of possible members x are in existence; ii) some of these x's can be combined into a unity to form S. The idea is that if we had $S \in S$, then trying to separate out the steps i) and ii) would lead to an infinite regress, with i) depending on ii) (since S is one of the possible members x of S), and with ii) depending on i) (since S can't exist unless its members exist). There is nothing directly contradictory here, but it is certainly more pleasant to speak only of sets that are logically built up from simpler sets and objects.

Note that if no set is a member of itself, then the collection V of all sets is the same as the collection R of all sets that are not members of themselves. Now we know that R is not a set, so if $R = V$, then we also know that V is not a set. This is one way to prove that the universe of set theory is a Many that does not allow itself to be thought of as a One.

Of course, since we have stipulated that no set is a member of itself, there is a simpler way to see that V is not a set. For if V were a set we would have V being a set that is a member of itself (since every set is a member of V). So the reason that R is not a set is not because its definition has something to do with self-reference, but rather, that R is big like V, the *class of all sets.*

Set theorists use the expression "class" to mean a collection or multiplicity of any kind. A class may or may not be unifiable into a set. If not, we call it a *proper class.* Thus, V is a proper class. V is a Many which does not allow itself to be thought of as a One.

Cantor was well aware of the distinction between sets and collections that are proper classes. He wrote about it in a famous 1899 letter to Dedekind:

> If we start from the notion of a definite multiplicity of things, it is necessary, as I discovered, to distinguish two kinds of multiplicities.
> For a multiplicity can be such that the assumption that *all* its elements

'are together' leads to a contradiction, so that it is impossible to conceive of the multiplicity as a unity, as 'one finished thing.' Such multiplicities I call *Absolutely Infinite* or *inconsistent multiplicities*.

As we can readily see, the 'totality of everything thinkable,' for example, is such a multiplicity . . .[7]

The reader may recall that Cantor's last point was discussed in the section "The Absolute Infinite." If we assume that "thought" means "rationally communicable thought built up from simpler forms," then it is evident that for any universe T of possible thoughts, the thought of T as a whole will be a new thought not lying in T. So any attempt to think of everything thinkable leads to an (Absolutely) infinite sequence of approximations that do not seem to converge to anything definite.

This pattern is important, since it is an exact duplicate of what happens when one tries to form a "set of all sets." Any proposed universe U of all sets is, if conceivable, actually a set. But then, since U is not a member of U, U is evidence that U is not the set of all sets after all.

The upshot is that if there is a single limiting class V of *all* sets, then V must be in some way vague or inconceivable, so that it resists being unified into a set. V is a Many that does not allow itself to be thought of as a One.

But is this really true? Am I not speaking of "V" as if it were, in fact, a single definite thing, the universe of set theory? Is V then *really* a *Many* or a *One?* This difficult knot is the One/Many problem.

In the next section I will try to inject some life into this question by giving a better description of what we presently know about V. But first we should notice how similar the One/Many question about V is to other questions that have cropped up.

At the end of "The Berry Paradox," we came to the conclusion that "nameability is unnameable." More specifically, we were discussing the first number u_0 that cannot be humanly named. The problem here is that on the one hand, there is a non-unifiable multiplicity of all the names, of all the numbers less than u_0; while on the other, there is the specific concept of "nameability" and the specific natural number u_0. How is it that we talk about "nameability" when the concept admits no definition? How do we talk about *the* class V of all sets when V is, in fact, *not a single definite thing?*

At the risk of being repetitious, let me point out the other places where this pattern has already arisen.

In "Richard's Paradox" we saw that the real numer coding up all the nameable real numbers is not nameable. Yet in a sense we named it in the process of discussing it. In "What Is Truth," and again in "Gödel's Incompleteness Theorem," we observed that mathematical truth is not mathematically definable. We have a single unified concept of truth guiding our efforts, and yet this concept exists not as any single definition but only as the non-unifiable multiplicity of all true statements. In "Towards Robot Conciousness" we saw that although a person feels himself to be a One, he can never grasp or know any unified description of his behavior. A person can describe his actions and personality only as an almost random multitude of particulars, even though the most primal datum of consciousness is unity.

There is even a hint of the One/Many problem when one regards the simple infinite set N of all natural numbers. For, if a person feels that only finite sets exist, then he cannot see how the multiplicity N can ever be formed into a single finished thing. And, by the same token, he will have trouble believing in the reality of ω as a single definite number. These doubts about N are fundamental to the school of thought called intuitionism. For the intuitionist there are no completed actually infinite sets—there are only *potentially infinite* sets. One could perhaps, characterize the intuitionist position as equating the simple infinity ω with the inconceivable Absolute Infinite Ω.[8] But as there seems to be no logical reason why there cannot be sets that are infinite, yet small enough to conceive of, we will continue to follow Cantor in discussing them.

THE UNIVERSE OF SET THEORY

Figure 78 is the standard picture of the universe of set theory. Running up the middle we have a spine consisting of all the ordinals. Beyond all the ordinals, clothed in clouds of glory, rests Ω, the Absolute Infinite. Or perhaps this is just a trick of lighting, for whether Ω *really* exists is yet another version of the One/Many problem.

PURE SETS AND THE PHYSICAL UNIVERSE

At the bottom of our picture of the universe of set theory is a singular point called the empty set. First there is nothing at all and then there is something—the idea of forming a set. The empty set is variously called {}, ϕ, or 0. The empty set is something, but inside it is nothing. Think-

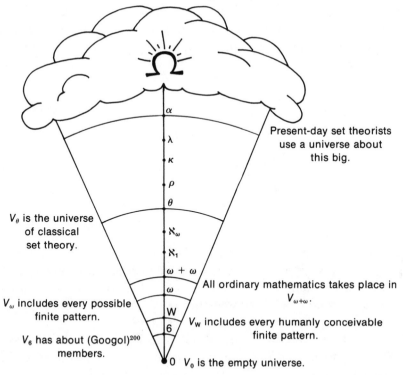

Present-day set theorists
use a universe about
this big.

V_θ is the universe
of classical
set theory.

All ordinary mathematics takes place in
$V_{\omega+\omega}$.

V_ω includes every possible
finite pattern.

V_W includes every humanly conceivable
finite pattern.

V_6 has about (Googol)^{200}
members.

V_0 is the empty universe.

Figure 78.

ing about it reminds one of what has been called the ultimate philosophical question: *Why is there something instead of nothing?*[9] Why does
anything at all exist? No one really knows, but the fact that *something
does exist* is the one absolutely incontrovertible known fact about the
world.

Why does the empty set exist? No one knows, but there it is, the raw
idea of set formation, an objective aspect of the world around us.

Fanning out above the empty set in an ever-widening "V" we have
sets of greater and greater complexity. The various levels are called *partial universes* or V_α's. It turns out that for each set x one can specify its
complexity by an ordinal numer called the *rank of x*. In general, V_α is the
set of all sets with rank less than α.

When one speaks of sets in ordinary life, one means sets of objects.
But in mathematics we need only speak of *pure sets,* that is, sets whose
members are other pure sets. The simplest pure set is the empty set.
The next simplest is the set whose only member is the empty set. This
set is variously called {{}}, {ϕ}, or 1. Note that {{}} differs from {} in the

same way that a *set* with one apple in it differs from the one *apple* itself. {{}} holds something in its outer brackets and {} holds nothing. Right below, we have written out some simple sets, with lines indicating set membership.

It may seem awkward and silly to make something out of nothing in this way, but it is certainly economical. The pure sets are built up out of thin air and the simple idea of forming collections. This is enough. Every conceivable tree of membership relations can be represented by the membership relation inside some set of sets.

By way of stressing this, I have drawn in Figure 79 all the possible membership relations one could have among four objects or sets. (I get thirty-one—*I think* that's all of them.) The dots stand for the different objects, and if dot A is supposed to be a component of dot B we draw B at a higher level than A, and with a line coming from A up to B. Drawn in this way, the dots group themselves into levels in a way analogous to the set theoretic grouping by rank. We can describe the level pattern in numbers, as is done in the figure.

Can each of these patterns be realized as some set of pure sets? Yes. Let me describe in detail how to do this. For each natural number n, let n^* be the set that is written as $n + 1$ left-hand brackets followed by $n + 1$ right-hand brackets.

$$0^* \quad 1^* \quad 2^* \quad 3^* \quad ...$$
$$\{\} \quad \{\{\}\} \quad \{\{\{\}\}\} \quad \{\{\{\}\}\}\} \quad ...$$

Note that for any n greater than zero, n^* has exactly one member: $(n - 1)^*$. We can use even-numbered n^*'s as the "dots" for our set patterns. The advantage is that each set in $\{0^*, 2^*, 4^*, 6^*, ...\}$ is distinct and not a member of any of the others. (Some light on the idea of "rank of a set" can be shed by the observation that for each n, n^* is the simplest set of rank n.) So we will use these sets as "atoms" or labels, for distinct bottom layer dots.

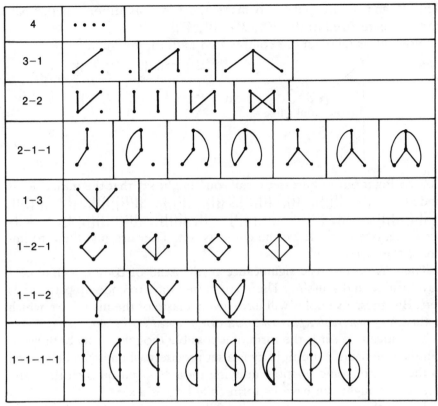

Figure 79.

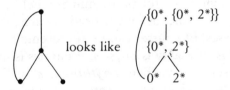

So, we can say that the pattern ⟨ is realized as the set of four sets, {0*, 2*, {0*, 2*}, {0*, {0*, 2*}}}, which can also be written as {{}, {{{}}}, {{}, {{{}}}}, {{}, {{}, {{{}}}}}}, a pure set of rank six.

We can also use the n^* sets to enforce splits.

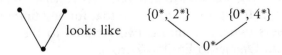

The 2* and 4* are put in to make the top dots differ. So the pattern $\bigvee$ is realized as {0*, {0*, 2*}, {0*, 4*}}.

Now let us take a more complicated pattern, my initial.

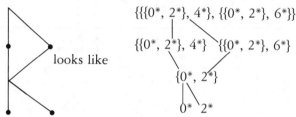

So, writing it out in pure-set terms, one might say that the initial "R" is coded as {{}, {{{}}}, {{}, {{{}}}, {{{}, {{{}}}, {{{{{}}}}}}, {{}, {{{}}}, {{{{{{}}}}}}}, {{{}, {{{}}}, {{{{{}}}}}}, {{}, {{{}}}, {{{{{{{}}}}}}}}}}. Strictly speaking, one could leave the commas out. They are just there to improve the readability!

There is no intrinsic significance to the kind of coding technique I have just been discussing. There are better ways to code up patterns as sets. But these examples will have given a taste of the main idea, which is this: *Any conceivable pattern at all can be coded up by sets.*

The slogan, "a set is the form of a possible thought," cuts both ways. On the one hand, a multiplicity is a set precisely when it can be viewed as the form of some possible unifying thought. And on the other, any possible thought can have its form coded up as a set.

The truth of this last observation is borne out by the fact that nearly every mathematics book starts with a section on sets. Natural numbers can be coded as sets something like the n^* discussed above. Fractions, infinite decimals, functions, relations, number fields, etc., can all be coded up one way or another as sets. Everything in mathematics (except for the proper classes, of course) can be represented as a set.

And consider this: If reality is physics, if physics is mathematics, and if mathematics is set theory, then *everything is a set.*[10] I am a set, my thoughts are sets, my emotions are sets. Occasionally I devote some time to trying to believe these conclusions in an immediate, experiential way. If everything is a set, then only pure form exists, which is nice. The whole physical universe could be a single large set U.

Now let us turn back to the picture of the universe of set theory that we began with in this section. Moving up the spine of ordinals, we get sets of higher and higher rank. In general, $V_{\alpha+1}$ consists of all the possible subsets of V_α. It is possible to prove that for any finite n, V_{n+1} will have $^n2$ members (where $^n2$ means two tetrated to the n, as was described in "From Omega to Epsilon-Zero").

$$^02 = 1$$
$$^12 = 2$$
$$^22 = 2^2 = 4$$
$$^32 = 2^{2^2} = 2^4 = 16$$
$$^42 = 2^{2^{2^2}} = 2^{2^4} = 2^{16} \approx 64{,}000$$
$$^52 = 2^{2^{2^{2^2}}} = 2^{2^{2^4}} = 2^{2^{16}} \approx 2^{64{,}000} \approx 10^{20{,}000} = (10^{100})^{200} = (\text{Googol})^{200}$$

Evidently, it is hopeless ever to write down or think of all the sets of rank six. But, of course, there are many individual sets of higher rank that we can think of, for instance, the set 100*.

The number W was "defined" in the section "The Berry Paradox" as the first natural number greater than every humanly nameable finite number. Obviously, if we can name a set, we can name its rank, so there will be no humanly nameable finite sets of rank greater than W. In a sense, all possible human thoughts lie in V_W. Of course, we can *point* to sets beyond this, to V_ω, for instance.

There is no real paradox in this if you believe that sets exist objectively and outside of human activities. When a person says he is talking about V_ω, his brain-state can be coded up as a set in V_ω. But he is still talking about the real V_ω. Brigitte Bardot is not flat just because her photograph is two-dimensional.

V_ω is, incidentally, the set of all "hereditarily finite" sets, that is, those sets that can be written out explicitly with a finite number of left-hand brackets, commas, and right-hand brackets. V_ω is, of course, infinite, of cardinality $\aleph_0$ to be precise. Each natural number is coded up inside V_ω.

$V_{\omega+1}$ is very much larger, of cardinality $2^{\aleph_0}$, or c. Each real number can be coded up as a set inside $V_{\omega+1}$.

A real-valued function of the sort studied in calculus can be thought of as a set of pairs of real numbers. Most of the theorems of calculus say things about various sets of functions. It is not difficult to see that by the time we get to $V_{\omega+\omega}$, we will have sets representing all of the things ordinary mathematics discusses.

Before discussing the rest of the picture, let us pause for an interesting question. Say that U is the set coding up our physical universe. How far up would one expect to find U?

This is really a restatement of the question asked in "Coding the World": How much information is in the universe? If the universe is completely finite, then U is a set somewhere in V_ω, perhaps in V_{googol}. And even if it is infinite, we wouldn't expect it to be so very far out—surely U must lie in $V_{\omega+\omega}$.

It seems strange to have our physical universe being just a little set U

floating around in the big universe V of all possible sets. Under this viewpoint, all the possible universes would be sets in V. Is it really reasonable to have an *idea* like V be so much bigger than the *real world?*

There *is* always the possibility, discussed in "Higher Physical Infinities," that the collection U coding up the physical universe is very much larger than we had suspected. If there were many parallel universes to be included, if matter were transfinitely divisible, if time were transfinitely long—in any of these cases, it might actually be possible to have U *too big* to be a set, too big to fit inside V. In this case U would be Absolutely infinite, a Many that does not allow itself to be thought of as a One.

Of course, our daily experience flies in the face of any suggestion that U is so very big. But there is a traditional philosophical principle, the *Principle of Plenitude,* which suggests that the physical universe should be as rich as the set theoretic universe of pure Platonic forms. Insofar as any physical structure can be coded up as a set, we already expect V to be as large or larger than U. The Principle of Plenitude insists that U must be as large or larger than V as well, leading to the conclusion that U and V are equally large.

A more extreme statement of this would be to insist that U and V are *identical,* but this is really pretty hard to swallow. An argument in this direction might be begun by remarking that we should understand U to include all the alternate universes as well as our own perceived universe. And one could then point out that a possibly existing alternate universe is really an abstract form no different than a set.

But it seems more reasonable to view our physical universe as a definite point U inside V. I find this congenial, as I feel the range of possible thoughts to be very great. A person with a different temperament, however, might go to the other extreme and argue that V is inside U.[11]

PROPER CLASSES AND METAPHYSICAL ABSOLUTES

For the various reasons discussed in "What Is a Set?," we know that the class V of all sets is not a set. V is not the form of a possible thought. This means that whenever a person believes himself to be thinking of the true V, he is deluded.

The situation regarding V is exactly analogous to our situation relative to the metaphysical or theological Absolute. Virtually all thinkers who have discussed the Absolute concur on one point: the Absolute is not rationally knowable. As was already mentioned in the "Absolute Infinite" section of Chapter 1, St. Gregory puts it this way: "No matter how far our mind may have progressed in the contemplation of God, it does not attain to what He is, but to what is beneath Him"[12]

Ernst Zermelo, one of the founders of modern set theory, makes a similar observation about sets: "Any specifically described model of set theory can in some way be viewed as a set, that is, as an element of a higher model of set theory."[13]

This idea has been formalized in modern set theory as the *Reflection Principle:* Given any proposed description *DESC* of V, there will be a partial universe V_α that satisfies *DESC* as well. Any specifically described universe of set theory turns out to be only one of the V_α sets, and not the whole universe. The mind does not attain to God, but to what is beneath Him.

To make clear that this is not just pious phrase-mongering, let me try to expose the mechanism underlying the Reflection Principle. Say that one is trying to think of V, of all the sets. Initially one might think only of the finite sets, that is, of the members of V_ω. But then one realizes that V_ω itself can be viewed as a set. Classical set theory dealt with all the sets of rank smaller than θ, the first "strongly inccessible cardinal" (See the "Large Cardinals" section of Excursion I.) But at some point it became apparent that the universe of classical set theory is just a single large set V_θ.

At any stage in the development of set theory, people are working with a great many larger and larger sets. But there is always some ordinal α lying beyond any ordinal yet named. When one realizes this, one names α. One realizes that the old universe was a set V_α, and begins working in a larger universe.

The process is comparable to a process of trying to think of all possible thoughts. At any time one has many thoughts present in one's consciousness. But now, by moving to a higher level of self-consciousness, one can group all of the past thoughts together into a new thought T. T and the old thoughts make up a new and enlarged state of consciousness, and by once again stepping outside oneself, a new and still higher thought T^* appears.

There seems to be no end to this process. It is a sort of Hegelian dialectic, endlessly moving out toward the Absolute universe of all possible sets or thoughts. To be quite precise, one could characterize the process in terms of thesis-antithesis-synthesis by saying the *thetic* component is one's instantaneous unconscious description of the Absolute, the *antithetic* component is the conscious formalization of this description, and the *synthetic* component is the formation of a new unconscious description of the Absolute that incorporates one's earlier descriptions and the awareness that they are inadequate. We could call such a process an intellectual history.

To say the Absolute is a One is to say that there is some unique limiting point or concept at the end of any such history. To say the Absolute is a Many is to say that there is only the working out of the endless sequence of approximations, with no single guiding notion at the end.

A model for these two views might be found in two infinite series: Zeno's series, $1 + \frac{1}{2} + \frac{1}{4} + \frac{1}{8} + \ldots$; and Grandi's series, $1 - 1 + 1 - 1 + \ldots$. We feel that as we take into account more and more terms of the Zeno series we are getting closer and closer to a definite limiting value: 2. But adding together more and more terms of the Grandi series only leads to an endless dithering between 0 and 1.

If α is a limit ordinal (i.e., an ordinal such as ω with no immediate predecessor), then the partial universe V_α has no last ordinal. But if α has the form $\beta + 1$, then V_α *does* have a last ordinal, which is β. Should we regard the full universe V as the limit of the first kind of partial universes, or as the limit of the second? Should we say that there is no last ordinal, or that the last ordinal is Ω?

The approach toward any ideal perhaps can be viewed as an intellectual history consisting of more and more sophisticated concepts. The ideal might be the ethical notion of Virture, the theological notion of God, the mathematical notion of V, the logical notion of Truth, the artistic notion of Beauty, or the spiritual notion of Love.

As a person develops he moves out to higher and higher transfinite levels. Although one cannot think of each natural number, to grasp the general idea of natural numbers is to jump out to V_ω. In terms of Love, or mutual knowledge, we could say that if A and B have a perfect understanding of each other, then A knows that B knows that A knows that B knows that $\ldots$, and so on. If A and B realize all *that,* then they too are moving out past level ω.

Actually, one need not deal with such high-flown concepts to encounter a form of the One/Many problem. As Plato says, the One and the Many run about together in every word uttered. To mention the hoariest of introductory philosophy examples, what do I mean when I say "table"? I do have a single underlying notion of "tableness," yet if I try to specify in words what a table is, I am swept into an endless sequence of refinements, taking in three-legged tables, ankle-height tables, tables attached to the wall, car hoods used as tables, operating tables, etc. We use words as units, as Ones, yet spelling out in detail what a given word means sinks us into an endlessly proliferating Many.

Perhaps if time could stop, one could get to the Absolute. But time goes on, and after one thinks he has seen the Absolute, he goes on to talk about it, and this talk becomes another idea, another step on the path outward. Whatever number I name, you can add one to. No one has the last word. The Reflection Principle formalizes this idea.

So in terms of things we can describe in words, the Absolute is an indescribable Many. It is possible to stop there. But many people, myself included, feel that there is a single underlying Absolute staring us in the face—call it pure existence, call it "this." As Wittgenstein says, "There are, indeed, things that cannot be put into words. *They make themselves manifest.* They are what is mystical."[14]

There is a kind of second-order One/Many problem that arises here. Are all the different Absolutes the same? Are God, Truth, Beauty, the Class of all Sets, the Mindscape, the Good, and so on, really different facets of some single ultimate ONE? This is certainly debatable. If all wisdom leads to the same thing, then why are there so many different religions, different schools of thought, and different ways of seeking enlightenment? Is a jogger looking for the same thing that a writer is?

This problem actually has an analogue in set theory. In set theory we have two different Absolutes: Infinity, represented by Ω, and Everything, represented by V. Ω can be thought of as the class of all ordinals, while V is the class of all sets. Now every ordinal can be represented as a set, so on the crudest level, V is larger than Ω. But, in pursuing the equivalence of all Absolutes, we could instead ask if each set is coded up by some ordinal, or if in terms of cardinality $\overline{\overline{\Omega}} = \overline{\overline{V}}$. Is Infinity as big as Everything?

No one really knows. The assertion $\overline{\overline{\Omega}} = \overline{\overline{V}}$ means that there is a one-to-one correspondence between the class of all ordinals and the class of all sets. But since such a correspondence is itself a proper class, it is hard to be sure it exists. When the assumption that there is such a correspondence is explicitly made, set theorists call it the *Axiom of Global Choice,* or, in a stronger form, the *Axiom of Ordinal Definability.* (See also the end of "The Continuum" in Excursion I, where a relationship between Global Choice and Cantor's Continuum Problem is pointed out.)

I think it is highly significant that the deepest problems of metaphysics can be given explicit set theoretic formulations. Gödel once expressed the view that present-day philosophy is in a state comparable to that of physics before Newton. Perhaps the ultimate role of set theory will be to do for philosophy what calculus did for physics.

INTERFACE ENLIGHTENMENT

One/Many	
Actual Infinite Platonism	Potential Infinite Formalism
Truth Thoughts Semantics Minds	Provability Words Syntax Machines
Nameable Reals	Random Reals
Sets	Proper Classes
Ω $\aleph_1$	V c
Mysticism: $\infty \quad = \quad 0$ Way of Unity = Inward Way Brahman $\quad$ = Atman Everything $\quad$ = I Am	Rationality
Right Brain Prajna	Left Brain Vijnana
Satori Interface Enlightenment	

We have come to the end of this book. In the table above, I have laid out the things I want to say. What we have here is basically a Pythagorean-style table of opposites. As the heading indicates, the general distinction between the left and right entries is similar to the distinction between the One and the Many. There are various kinds of ways of opposing the One and the Many, and these various ways are grouped together by the horizontal dividing lines. At the very bottom, running from one side to the other, is the phrase "Interface Enlightenment." This is the point we want to reach. But first let us work our way down the table. The "One/Many in Logic and Set Theory" section discusses the upper half of the table. "Mysticism and Rationality" explores the little box under Mysticism. And "Satori" gives a description of satori as the "/" in the One/Many problem.

ONE/MANY IN LOGIC AND SET THEORY

As has been mentioned several times, a mathematical Platonist is someone who believes in the objective, external existence of infinite sets, whereas a mathematical formalist believes that all we really have is various finite descriptions of mathematical theories. Intuitionism is sometimes presented as a synthesis between these views, but this is not really the case. As far as their views of infinity go, the intuitionist and the formalist are on much the same side, believing only in potential infinities (as opposed to actual infinities).

I have put the actual infinite on the One side of the table. For to look at a set such as N (the set of all natural numbers) as a single definite object is to think of an actually infinite set. And, on the other hand, to regard N as an ungraspable Many is to treat N as a potentially infinite, never-to-be-completed set.

In the next grouping, truth, thoughts, semantics, and minds are paired up against provability, words, syntax, and machines. The distinction between truth and provability is what is stressed by Gödel's Incompleteness Theorem. If we have correct axioms, then the provable statements will all be true, but not all the true statements will be provable. We cannot dispense with provability in favor of truth because we have no finite definition of what "truth" means. Truth is a kind of Absolute, a single guiding notion that directs our many different choices of axioms. Truth is the One that the Many types of provability try to approximate.

Moving to the next line, let me point out that truth is a type of thought that cannot be fully expressed by words. We all know what it is to think, but there is no way to explain exactly how we do it. As we saw in various ways in Chapter 3, there is no finite way to describe exactly how thoughts are turned into words and vice versa. If we think of the totality of human intellectual experience as a unity, then the various attempts to describe it in words make up a multitude of partial approximations.

Logicians often discuss this kind of distinction as a distinction between semantics and syntax. If we regard language as a system of symbols describing a fixed reality, then we have a semantic view of language. If, on the other hand, we regard language as a game played according to certain rules, then we have a syntactic view of language. Given a mathematical sentence S, the question "is S true in the mathematical universes we have in mind?" is a semantic question; and the question "is S provable from the axiom systems we use?" is a syntactic question. These two types of question are studied in the subbranches of logic called, respectively, Model Theory and Proof Theory.

In the initial stages of research, mathematicians do not seem to function like theorem-proving machines. Instead, they use some sort of mathematical intuition to "see" the universe of mathematics and determine by a sort of empirical process what is true. This alone is not enough, of course. Once one has discovered a mathematical truth, one tries to find a proof for it. In the later stages of research one does try to behave like a machine in writing up a definite program or proof for deriving the desired truth.

The fact that I am putting minds on one side and machines on the other does not mean that I am "for" the concepts on the left and "against" the concepts on the right. I see, in each case, both sides as valid and essential. The only kind of thinking I am really opposed to is that which would say *only* the One or *only* the Many is real. In making the mind-machine distinction I want to bring out my belief that there is more to consciousness than the simple working out of some biochemical program of the brain. Or, to put the same point a bit differently, I want to say that we can think both mystically and rationally. The pair "mysticism-rationality" is discussed below, but I might remark here that I have put mysticism and mind on the One side because I regard the characteristic feature of the mystically conscious mind to be its ability to experience itself directly as part of a unified Absolute.

The nameable reals-random reals pair characterizes a somewhat different aspect of the One/Many relationship. A random real number is an infinite sequence of digits with no unifying rule for writing it out. So in this sense, it is a Many, which is not a One like some nameable real whose decimal expansion obeys a single definite rule.

Of course, if we took some definite unnameable real number T (say, for instance, that T codes up truth), it could be argued that T, as knowable through higher intuition, is a One, and that it is the various inadequate names for T that make up the Many. This only shows that some distinctions can be cast into One/Many form in various ways.

Some of the same ambiguity is present in the set-proper class line. Certainly a set is a unity, a One; and a proper class is a Many that cannot be thought of as a One. Yet on a higher level, we could say that a single proper class (like *On*, the class of all ordinals), is a One that is approximated by Many different sets.

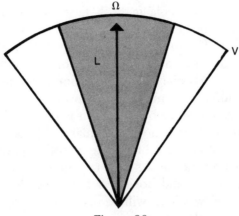

Figure 80.

The next group, Ω-V and $\aleph_1$-c, brings out a still higher-order One/Many distinction. If we treat the class of all ordinals as the single Absolutely Infinite "number" Ω, then we can ask about the relation between Ω and the much wider class V of all sets. In terms of vertical growth, Ω and V are as long as possible. But there is a view of set theory that regards the sets as arising from a horizontal process of growth out from the spine of ordinals. If horizontal growth is kept to a minimum, one gets Gödel's universe L of "constructible sets." But it is generally believed that the universe is very much wider than L. (See Excursion I).

Ω is a One in the sense of being a form of a simple concept (Infinity), while V is a Many in the sense of being the form of a complex concept (All Sets).

If we restrict our attention to countable sets, then the Ω-V pair turns into the $\aleph_1$-c pair of Cantor's Continuum Problem. The reason is that there are exactly c countable sets, and exactly $\aleph_1$ countable levels of infinity. In a way, Cantor's Continuum Problem is a type of One/Many question.

MYSTICISM AND RATIONALITY

Mysticism is an extreme form of monism. The central teaching of mysticism is simplicity itself: All is One. The essence of the mystic tradition is not really any special philosophical system, but, rather, the direct and immediate apprehension of one's personal identity with God.

It should be kept in mind that "mysticism" does have a precise meaning as a certain strand of thought, or type of behavior, that has been present through the millenia in both Eastern and Western cultures. Mysticism is not to be confused with occultism, which has to do with strange rites, secret formulas, and so on. Mysticism has no direct relationship with astrology, devil worship, fortune-telling, drug abuse, health food, or ESP. Mysticism is just the simple awareness of the direct identity of the individual soul and the Absolute.

The characteristic feature of mysticism is the breaking down of distinctions. Obviously, this is not an unqualified good. If I can't tell my hand from my sandwich, then I may bite myself. Opposed to the human tendency towards mysticism we have rationality. Too much rationality quickly becomes inane and boring. What is needed is some kind of bridge between the two. This will be discussed in the next subsection.

But first I would like to give the reader two examples of mystical thought. The first is related to a distinction Rudolf Otto makes in his book, *Mysticism East and West*.[15] Otto describes two different types of meditation that people practice in order to feel united with the Absolute: the Inward Way and the Way of Unity. These two ways correspond, respectively, to moving towards a consciousness of Nothing and Everything, of 0 and ∞.

The Inward Way involves trying to stop thinking thoughts, stop having emotions, stop muddying the mental waters. One strives toward the Void that underlies all things. A formula used by the Indians for this activity is "Neti, neti," meaning "Not that, not that." One tries to stop thinking, to stop thinking about stopping, to stop thinking about thinking about stopping, and so on. Sometimes it works. The Way of Unity involves trying to include more and more of the world in one's field of consciousness. One strives toward a sympathetic union with Everything. This activity could be characterized by the phrase, "And that too."

The metamystical thought I want to describe here is this: The Way of Unity and the Inward Way have the same goal. Nothing is the same as Everything.

Consider a geometrical analogy. Think of normal consciousness as a circle with radius 1. The Inward Way involves continually shrinking the field of consciousness—say by endlessly halving it. The Way of Unity involves repeatedly expanding the field of consciousness—say by doubling. If we consider "inverting the plane in the unit circle," which is to

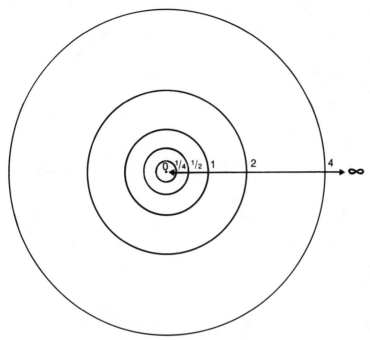

Figure 81.

say, pairing up each point (x, y) with $(1/x, 1/y)$, we can see that for each halving step inward there is a corresponding doubling step outward. What if we regard 0 and ∞ as being the same place? This could be accomplished, for instance, by first shrinking the plane to the inside of a circle and by then bending the circle into the shape of a torus, as in "Temporal Infinities."

Identifying the Inward Way with the Way of Unity is an example of the way mysticism breaks down distinctions. In *Spacetime Donuts* I describe a person who experiences this:

> One evening after a good day's work, Vernor went out into the garden behind the Library. There was a large tree there, and he was able to climb to its fork, some five meters up, by clinging to the grooves in the tree's bark and inching upwards. Once he was up in the first fork it was easy to move up the fatter of the two branches to a comfortable perch some fifteen meters above the ground. He was barefoot and felt perfectly secure.
>
> A fine rain was falling, so fine that it had not yet penetrated the tree's leaves. Set back from the City like this, in his leafy perch in the library garden, it was possible to listen to the incoming honks, roars and clanks as a single sound, the sound of the City.
>
> He noticed a hole in the branch some two meters above his head, and

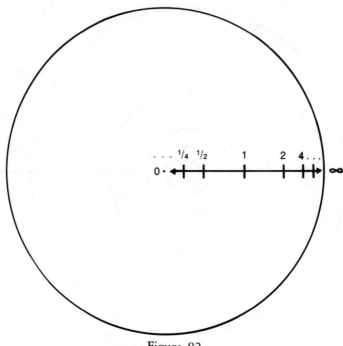

Figure 82.

inched up, hugging the thick, smooth branch. It was a bee-hive in there —a wild musky odor came out of the hole along with a steady, highly articulated 'Z'. A few bees walked around the lip of the hole, patrolling, but they were unalarmed by Vernor's arrival. He felt sure that they could feel his good vibes.

A soft breeze blew the misty rain in on him, and he slid back down to the crotch he had been resting in. Closing his eyes, he began working on his head. There seem to be two ways in which to reach an experience of enlightenment—one can either expand one's consciousness to include Everything, or annihilate it so as to experience Nothing.

Vernor tried to do both at once.

On the one hand, he moved towards Everything by letting his feeling of spatial immediacy expand from his head to include his whole body, then the tree branch and the bees, then the garden, the city and the night sky. He expanded his time awareness as well, to include the paths of the rain drops, his last few thoughts, his childhood, the tree's growth, and the turning of the galaxy.

On the other hand, he was also moving towards Nothing by ceasing to identify himself with any one part of space at all. He contracted his time awareness towards Nothing by letting go of more and more of his individual thoughts and sensations, constantly diminishing his mental busyness.

The overall image he had of this activity was of two spheres, one expanding outwards towards infinity, and the other contracting in towards zero. The large one grew by continually doubling its size, the smaller shrank by repeatedly halving its size . . . and they seemed to be endlessly drawing apart. But with a sudden feeling of freedom and air Vernor had the conviction that the two spheres were on a direct collision course—that somehow the sphere expanding outwards and the sphere contracting inwards would meet and merge at some attainable point where Zero was Infinity, where Nothing was Everything.[16]

So, the Void and Everything can perhaps, in a momentary way, be experienced as the same. Of course, a thoroughly rational person could dismiss Vernor's experience as wishful thinking or a type of hallucination. I do not wish to dismiss it, yet I do not want to claim that it is absolutely true. I just want to make clear what mystical thought involves.

Let us look at a second example, again an instance of the mystical tendency to break down distinctions. In a long essay called "What is Life," the great physicist Erwin Schrödinger comes up with the following argument: Given that i) my body functions as a pure mechanism according to laws of nature, and that ii) I know by direct experience that I am directing the motions of my body, it follows that iii) I am the one who directs the atoms of the world in their motions. Schrödinger remarks, ". . . it is daring to give to this conclusion the simple wording that it requires. In Christian terminology to say: 'Hence I am God Almighty' sounds both blasphemous and lunatic."[17]

But Schrödinger defends this conclusion, pointing out that it is just an example of the equation fundamental to the Upanishads: Atman = Brahman. "Atman" (which is related to the German word *atmen,* to breathe) is the Sanskrit word for the individual soul. In the sense described in "Robot Consciousness," an individual's Atman is his sense of "I Am." "Brahman" is a word meaning something like our "Absolute," the eternal, all-pervading "is-ness" of the world.

Along the same lines, consider the famous Old Testament passage (Exodus 3, 13-14):

> "Then Moses said to God, 'If I come to the people of Israel and say to them, "The God of your fathers has sent me to you," and they ask me, "What is his name?" what shall I say to them?' God said to Moses, 'I AM WHO I AM.' And he said, 'Say this to the people of Israel, "I AM has sent me to you."'"

What is God's name? "I AM."

These mystical ideas are certainly true on one level. Yet on another level, on the rational level, they are not true at all. I am not God. I am an insignificant mortal living out my allotted span. How can both things be true? How can I be the One, the I AM, the Absolute . . . and yet be only a face in the crowd, a single individual among Many others?

SATORI

No one has written more eloquently on Zen than D. T. Suzuki. I would like to begin this section by describing what he says in his essay, "The Meaning of Satori."[18]

Suzuki distinguishes between two ways of knowing the world. *Prajna* is intuitive, immediate knowledge of the world—what we might call a mystical grasping of the world in its unity. A characteristic feature of *pranja* knowledge is that it avoids distinguishing between the knower and the known, the subject and object. *Prajna* knowledge is not *taught,* it is *communicated.*

Vijnana is discursive, analytical knowledge of the world—what we call rational thought. *Vijnana* knowledge stands apart from the thing known, a subject examining an external object. *Vijnana* knowledge can be written down and learned. Suzuki says something that is very relevant:

> Vijnana can never reach infinity. When we write the numbers 1, 2, 3, etc., we never come to an end, for the series goes on in infinity. By adding together all those individual numbers we try to reach the total of the numbers, but as numbers are endless this totality can never be reached. *Prajna* on the other hand, intuits the whole totality instead of moving through 1, 2, 3, to infinity; it grasps things as a whole. It does not appeal to discrimination, it grasps reality from inside, as it were.[19]

The point is not that mystical, unitive, *prajna*-type knowledge is preferable. Both types of knowledge are real, and both are important. But it is very hard—perhaps impossible—for us to see the world in both ways at once. At any instant we see the world either as One or as Many.

Moving from Many to One tends to be a gradual process, the result of some kind of deliberate calming of the mind. But the passage from One to Many is usually sudden. At a given instant you may be sunk into a complete unity with the world. And then an instant later you are talking about your experience, standing outside yourself, making distinctions. The difficult thing is to catch the instant when you are still between One and Many, what I earlier called the "/" in the One/Many problem. According to Suzuki this instant is the fleeting enlightenment that Zen calls *satori*. "The oneness dividing itself into subject-object and yet re-taining its oneness at the very moment that there is the awakening of a consciousness—this is satori."[20]

This sort of satori is fleeting, but not rare. One could almost say that the natural rhythm of thought is a oscillation between One and Many. As you look around the room there are constant microlapses of atten-tion. You reach out and merge with the world, then draw back and ana-lyze. At one instant there is only is-ness, at the next there is a person cataloging his perceptions. One-Many-One-Many . . . at a rate of, say, three cycles per second.

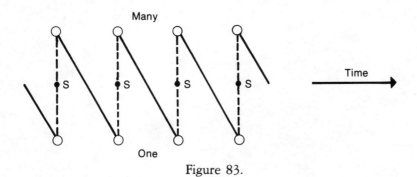

Figure 83.

We could draw a picture something like Figure 83, indicating a per-son who repeatedly sinks down into blissful union with the One, only, each time, to snap back to ordinary rational consciousness. The points labelled "S" might be the satori points.

There is a sense in which waking up each morning is a satori. On a good day (no alarms, no clock to punch) you float up from sleep into an idle state of is-ness, not even thinking who or where you are. But this is too good to last . . . whisk clickety-click, and you're planning your day. Is it possible to notice the moment of switch over?

I am thinking of satori as "interface enlightenment." The interface

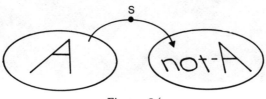

Figure 84.

need not really be between One and Many—it could be between almost any A and not-A. As·a rule it is not possible to think both A and
not-A. But we *do* change our minds. We move among a great variety of
incompatible mental states. These moves occur suddenly, like abrupt
jumps. Sometimes, as we jump, we think to glance down and see, without prejudice, A and not-A as two different regions of the same Mindscape.

Benjamin Paul Blood

Benjamin Blood wrote at some length about this type of experience.[21] He would equip himself with a handkerchief soaked in ether, hold it to his face, sink into unconsciousness, and then, as his nerveless hand fell away, he would wake back up. The experience of moving abruptly from artificial trance to normal awareness struck him as central, and he wrote something very interesting about it:

> I think most persons who shall have tested it will accept this as the central point of the illumination: [i] that sanity is not the basic quality of intelligence, but is a mere condition which is variable, and like the huming of a wheel, goes up or down the musical gamut according to a physical activity; [ii] and that only in sanity is formal or contrasting thought, while the naked life is realized only outside of sanity altogether; [iii] and it is the instant contrast of this 'tasteless water of souls' with formal thought as we "come to," that leaves in the patient an astonishment that the awful mystery of Life is at last but a homely and a common thing, and that aside from mere formality the majestic and the absurd are of equal dignity.[22]

Up until now I have been describing the interface between One and Many as something that one moves back and forth through in time. This is a bit misleading. In Suzuki's words, "Satori is no particular experience like other experiences of our daily life. Particular experiences are experiences of particular events while the satori experience is the one that runs through all experiences."[23] The One and the Many run about together in and out of every word ever uttered.

On the one hand you have pure undifferentiated reality, the God within you; and on the other hand you have your *hand,* as distinct from a foot, or a carrot. The world is both One and Many at once. I do not want to say that they are the same, and I do not want to say that they are different . . . for to assert either position begins an endless argument.

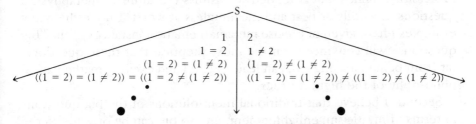

The argument can be illustrated as follows. Imagine a mystically thinking person who says "the One and the Many are really the same," and a rationally thinking person who says "the One and the Many are essentially different." We represent the first position by the number 1 (for only one thing) and the second by the number 2 (for two different things). Now, the mystic can tell the rationalist, "Ah, but don't you see, our two positions are really the same." $1 = 2$. The rationalist replies, "They are *not* the same. The fact that we disagree proves that there are differences." $1 \neq 2$. "But," counters the mystic, "arguing that things are one and arguing that they are many are really different aspects of the same one mind." $(1 = 2) = (1 \neq 2)$. "*Au contraire,* my dear friend," says the rationalist. $(1 = 2) \neq (1 = 2)$. And so on. Better not to start, better to stay silent at S! But silence does get boring.

In the language of quantum mechanics we might speak of the One and the Many as complementary, mutually exclusive aspects of reality.

> "In fact, here again we are not dealing with contradictory, but with complementary pictures of the phenomena, which only together offer a natural generalization of the classical mode of description. . . . Complementarity bears a deep-going analogy to the general difficulty in the formation of human ideas, inherent in the distinction between subject and object."[24]

The world is One and the world is Many. The One/Many split is the heartbeat of the universe, the charged tension that makes things happen.

What does all this have to do with the preceding discussions of logic and set theory? There are two key points to be made, both having to do with the symbiotic relationship between modern exact philosophy and the more traditional view of philosophy as a search for ultimate truth.

First, it is important to realize that such traditional questions as "Can we know the Absolute?," "Is Reality One or Many?," or "What is Truth?" are real questions that can be investigated in an exact way. An unfortunate effect of the early logical positivism was that for many years professional philosophers tended to dismiss the ultimate metaphysical questions as woolly at best and meaningless at worst. I hope the many examples I have given of precise metamathematical instances of the "big questions" will convince even the most skeptical that these questions, far from being meaningless, can lead to good and exciting mathematical philosophy of the highest order.

Second, I believe that traditional metasolutions of the big questions in terms of mysticism, enlightenment, and so on, can be of value to the thinker who is faced with one or another of the antinomies that crop up in modern logic and set theory. Only someone who can *feel* what the solution to a problem might be like is in a position to develop the language to describe a further step in the right direction.

PUZZLES AND PARADOXES
(*Answers, p. 305*)

1. A Platonic form is sometimes viewed as the One underlying thing that its Many instances have in common. In *Parmenides* (132), Plato points out how this can lead to an infinite regress.[25] All large things have something in common, let us say their participation in the form Largeness. But now any large thing has something in common with the form Largeness, let us say their mutual participation in. . . . Complete the argument, and compare this to the way in which the forming of a collection of sets into a definite unity intended to be the universe of set theory leads, not to the full universe, but only to another set.

2. There is a sort of One/Many problem in physics. If the universe exists as a sequence of distinct "nows," then space-time is Many. But if we assert that the passage of time is an illusion, then we have space-time as a One. But what if there are many parallel space-times? Is there any way to view these parallel worlds as a One, as aspects of a higher superspace? Could there be Many superspaces?

3. Von Neumann suggested representing ordinal numbers as sets in the following way. Each ordinal a is to be identified with a set $a' = \{b' : b$ is an ordinal less than $a\}$. Thus, for instance, 3 is represented as $3' = \{0', 1', 2'\}$. In terms of pure sets, $0'$ is $\{\}$, and $1' = \{0'\} = \{\{\}\}$. Write out $3'$ and $4'$ as pure sets.

4. Set theorists usually represent the ordered pair $<a, b>$ of sets a and b as the set $\{\{a\}, \{a, b\}\}$. If a and b are in V_n, then what is the first m that has $<a, b>$ in it? A rational number q is customarily represented as the set of all the ordered pairs $<m', n'>$, where m and n are in ratio q to each other. Thus $\frac{2}{3}$ is represented as the set $\{<2', 3'>, <4', 6'>, <6', 9'>, \ldots\}$. In which V_a are the "rational numbers" first found? A real number r is commonly represented as the ordered pair of sets $<U, L>$, with U equal to the set of all (representations of) rational numbers less than or equal to r, and L equal to the set of all "rational numbers" greater than r. In what V_a will *these* sets be found?

5. Many Zen stories deal with the difficulty of expressing a position that is somehow neither A nor not-A. How would you answer the master in this story: "Shuzan held out his short staff and said, 'If you call this a short staff, you oppose its reality. If you do not call it a short staff, you ignore the fact. Now what do you wish to call this?' "[26]

Georg Cantor

EXCURSION I
THE TRANSFINITE CARDINALS

In this excursion into mathematics we make a detailed tour of the transfinite cardinals. The first section describes the way in which one argues from (i) the existence of the Absolute Infinite, and (ii) the Reflection Principle, to get (iii) the existence of the infinite numbers we need. The second section describes exactly how one works with these infinite numbers.

In the "Continuum" section we examine a number of sets having the cardinality c of the set of points on a mathematical line; then we discuss the difficult question of where c lies in the hierarchy of alefs.

The "Large Cardinals" section attempts something new: a popular account of the modern theory of very large infinite numbers.

ON AND ALEF-ONE

I mentioned in "From Pythagoreanism to Cantorism" and in Chapter 5 that it is possible to represent all mathematical objects as sets. How are we to represent ordinals as sets?

The solution is simple, yet subtle. The ordinal a is identified with the set $\{b: b < a\}$ of all ordinals less than a. Thus $0 = \{b: b < 0\} = \phi$, $1 = \{b: b < 1\} = \{0\} = \{\varnothing\}$, $2 = \{0, 1\} = \{\phi, \{\phi\}\}$, $3 = \{0, 1, 2\} = \{\phi, \{\phi\}, \{\phi, \{\phi\}\}\}$, $\omega = \{0, 1, 2, \ldots\}$, $\omega + 1 = \{0, 1, 2, \ldots \omega\}$, $\omega + \omega = \{0, 1, 2, \ldots \omega, \omega + 1, \omega + 2, \ldots\}$, and $\omega^2 = \{0, 1, 2, \ldots \omega, \omega + 1, \ldots \omega \cdot 2, \omega \cdot 2 + 1, \ldots \omega \cdot 3, \ldots\}$.

This is reminiscent of the process depicted in Figure 32, where Wheelie Willie actually makes it out to level ω. Note that the use of vanishing points makes it possible to fit an infinite thought balloon into a finite frame. If Wheelie Willie were to think back on all the thoughts he had during that one breath, he would be at level $\omega + 1$.

Viewing an ordinal as identical with the *set* of all smaller ordinals makes many things more convenient. Given that $b < a + 1$ if and only if ($b < a$ or $b = a$), we can see that if a has the form $\{0, 1 \ldots \omega, \ldots s, \ldots\}$, then $a + 1$ is $\{0, 1, \ldots \omega, \ldots s, \ldots a\}$. In other words, $a + 1 = a \cup \{a\}$.

If $a = \lim(a_n)$, then $b < a$ iff b is less than one of the a_n's. (The forward implication follows since a is the *least* ordinal greater than all the a_n; and the reverse implication holds since a is *greater* than all the a_n.) Therefore, $a = \{b: b < a\} = \{b: b < a_n \text{ for some } n\} = \{b: b < a_0\} \cup \{b: b < a_1\} \cup \{b: b < a_2\} \cup \ldots = a_0 \cup a_1 \cup a_2 \cup \ldots = \bigcup_n a_n$. That is, $\lim(a_n)$ is obtained just by taking the union of all the sets a_n.

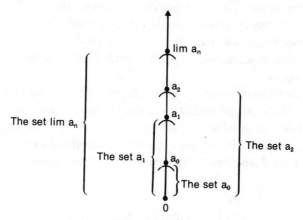

Figure 85.

This method can be applied to any set A of ordinals, regardless of whether or not this set can be arranged into a natural-number-indexed increasing sequence. In view of this fact, we introduce a new symbol *sup A,* which stands for the first ordinal greater than every member of A. If A has a greatest member, then *sup A* is simply that member plus one; otherwise, *sup A* is the union of all the ordinals lying in A.

As in "The Alefs," two sets S and T are said to have the same cardinal number if and only if a one-to-one correspondence can be set up between the members of S and the members of T. This relation is written $\overline{\overline{S}} = \overline{\overline{T}}$. One way of thinking of it is to say that S and T have the same cardinality if we can turn S into T just by altering the members of S one at a time.

It turns out that cardinality is a genuinely significant concept, since, on the one hand, not *all* infinite sets have the same cardinality; and, on the other hand, lots of apparently very different infinite sets *do* have the same cardinality. For example, ω and $\omega \cdot 2$ are quite different ordinals. But, as mentioned in "The Alefs," they have the same cardinality. That is, $\overline{\overline{\omega}} = \overline{\overline{\omega + \omega}}$.

In general, we say that a set S is *countable* iff its cardinality is no greater than the cardinality of ω. That is, S is countable if S is empty, S is finite, or $\overline{\overline{S}} = \overline{\overline{\omega}}$. ("Iff," by the way, is the logicians' abbreviation for "if and only if." The import of "P iff Q" is that P and Q are logically equivalent. "Iff" is, thus, a sort of equals sign.)

One of the things I would like to do in this section is to justify the existence of the ordinal $\aleph_1$, which is *not* countable.

In "From Omega to Epsilon-Zero," we used two principles of ordinal generation: I) Given any ordinal a, there is a least ordinal greater than a, called $a + 1$; and II) Given any increasing sequence a_n of ordinals, there is a least ordinal greater than all the a_n, called $\lim(a_n)$.

There is an important fact about ordinals hidden away in principle I. This is the fact that *no ordinal is less than itself*. For if we had some ordinal a such that $a < a$, then there could be no least ordinal greater than a. The reason is that if $a < a$, then whenever we have b with $a < b$, we can form the inequality $a < a < b$ to demonstrate that there is an ordinal between a and b . . . so that b is not the least ordinal greater than a.

Now principles I and II can be combined to form the following strong principle III: For every set A of ordinals, there is a least ordinal greater than every member of A, called *sup* A. Principle III is not really meaningful unless we have specified what sorts of sets A of ordinals exist. The basic principle of set existence is that a collection will be a set unless this is for some reason impossible.

As is mentioned in Chapter 5, Russell's collection $R = \{x : x \notin x\}$ of all sets that are not members of themselves cannot be a set, for if R is a set, then we have the contradiction $R \in R$ if and only if $R \notin R$. Again, if we make the customary assumption that no set x is an element of itself, then the collection V of all sets is not a set. For if it were, then V would be a set such that $V \in V$.

Let On be the collection of all ordinals. If On is a set, then by III there is an ordinal $\Omega = \sup On$. But this is impossible, for if Ω is an ordinal, then Ω is an element of the collection On of all ordinals, implying that $\Omega < \sup On = \Omega$. But, as was shown above, it is a basic property of ordinals that no ordinal can be less then itself.

Thus, the assumption that On is a set leads to the contradiction that the ordinal Ω is less than itself, where $\Omega = \sup On$. This fact was discovered by Cesare Burali-Forti in 1897, and earlier by Cantor. Nevertheless, we do have have some kind of *concept* of all the ordinals, and we sometmes use the symbol On to stand for this concept taken as a multiplicity, and the symbol Ω to stand for this concept taken as a unity. Note that $On = \{a: a < \Omega\}$, so that under the identification introduced above On and Ω seem to be the same.

What *is* Ω? Ω is what people are talking about when they speak of infinity in the sense of something subject to no limitation of any kind. Ω is Absolute Infinity. Absolutes are by their very nature not rationally or objectively knowable in full. An Absolute can be fully known only by entering into it as subject, and to identify your subject (self) with the Absolute is to give up your sense of personal identity. You have to take off your shoes before entering the temple.

A set is a form or thought that *can* be known in an objective way, that can be mentally handled and investigated without abandoning one's role of observer. Principle III is actually a Reflection Principle, a way of expressing the transcendence of the Absolutely Infinite collection On. For III says that no set of ordinals A reaches all the way out to Ω. III says that given any set A of ordinals, we can always find some ordinal (which, as ordinal, is less than Ω) bigger than every member of A. For any *set A* of ordinals there is an ordinal *sup A, between A* and the unattainable Ω.

If only Principle I is used, then one gets just ω, the collection of finite ordinals, also known as *the first number class*. The *second number class* is the collection of all the additional ordinals that can be obtained by repeatedly using principles I and II, where principle II is applied only to countable sequences $\{a_n\}$. Given that the limit of a countable sequence of countable ordinals is countable (which will be proved in the next section), we can see that the second number class is just $\{a: \overline{\overline{a}} = \overline{\overline{\omega}}\}$, the collection of all countable infinite ordinals.

Now, unless we expressly assume that every set and every ordinal is countable, there is no reason why the second number class should not be a set. Granted that it is very difficult to imagine larger and larger countable ordinals (recall how much trouble we had even with ϵ_0). But, on the other hand, the idea of an arbitrary ordinal that can be reached by repeated applications of principles I and II is pretty clear.

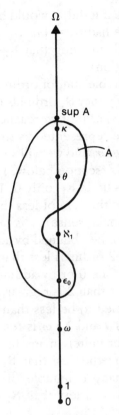

Figure 86.

There are actually some people who insist that every set *is* countable . . . this is characteristic of Brouwer, for instance.[1] But if we are
quite objective about sets and assume that a set is a form that exists regardless of any *human* ability to grasp it, then it seems that there is really
no reason why the second number class should not be a set. The situation here is to be contrasted with the collection *On* of all sets that are
ordinals. The assumption that the second number class is a set seems
harmless, but the assumption that *On* is a set leads directly to contradiction.

So, accepting the belief that the second number class is a set, we must
conclude that the second number class does not exhaust *On*, for no set
can exhaust the Absolute. That is, we can find an ordinal $\aleph_1 = sup$ (second number class) that lies between Ω and all of the countable ordinals.
As a set, $\aleph_1 = \{0, 1, . . . \omega, . . . \omega \cdot 2, . . . \omega^2, . . . \omega^\omega, . . . \epsilon_0,$
$. . . \alpha, . . .\}$, where the last ". . ." is the biggie. $\aleph_1$ does not lie in the

second number class (since if it did, it would be less than itself), and it is not countable, (since if we had $\aleph_1 = \{a_0, a_1, a_2, \ldots\}$, we would have $\aleph_1 = sup \{a_0, a_1, a_2, \ldots\}$ implying that $\aleph_1$ is in the second number class, which is a contradiction).

$\aleph_1$ lies beyond any countable sum of ordinals less than itself. We can only reach $\aleph_1$ by adding together $\aleph_1$ ordinals (ones would do). So there is a certain sense in which $\aleph_1$ cannot be reached from below. In general, an ordinal a that cannot be represented as the sum of less than a ordinals less than a is called *regular,* and we will see more of the regular ordinals in the "Large Cardinals" section below. For now it is interesting to note that of all the ordinals up to $\aleph_1$, only 0, 1, 2, ω, and $\aleph_1$ are regular.

0 is regular since it is not the sum of less than zero ordinals less than zero (which doesn't even make sense). 1 is not the sum of no ordinals less than one, since 1 cannot be obtained by adding no zeros together. 2 is not the sum of less than 2 ordinals less than 2, since 2 cannot be obtained either by adding up one 0 or by adding up one 1. No successor ordinal $a + 1$ that is greater than 2 is regular, for any such $a + 1$ is the sum of two (which is assumed to be less than $a + 1$) ordinals less than $a + 1$ (that is, the ordinals a and 1.) ω is regular since it can never be obtained by adding together finitely many finite numbers. And, finally, $\aleph_1$ is regular since to add together less than $\aleph_1$ ordinals less than $\aleph_1$ is to add together countably many countable ordinals, which always just gives another countable ordinal less than $\aleph_1$ (as will be proved in the next section).

So, the fact that $\aleph_1$ is regular certainly makes it hard to grasp, but there is a sense in which it is no worse, really than 2. If all you know of is 1, then all you can imagine is *one* 1 . . . and you cannot see how to get to 2. If all you know of is finite numbers, then all you can imagine is finite sets of finite numbers added together . . . and you cannot see how to get to ω. Finally, if all you know is countable ordinals, then all you can imagine is countable limits of countable ordinals . . . and you cannot see how to get to $\aleph_1$.

CARDINALITY

Two sets S and T are said to have the same cardinal number (abbreviated $\overline{\overline{S}} = \overline{\overline{T}}$) iff there is a *one-to-one* function that maps the set S *onto* the set T. We can think of $\overline{\overline{S}} = \overline{\overline{T}}$ as meaning that one can turn S into T by changing the appearance and arrangement of the elements of S, but without destroying, merging, creating, or splitting elements.

We define $\overline{\overline{S}} \leq \overline{\overline{T}}$ to mean that there is a one-to-one function map-

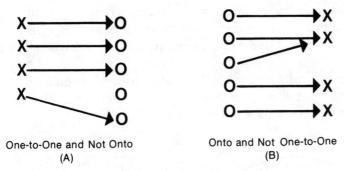

One-to-One and Not Onto
(A)

Onto and Not One-to-One
(B)

Figure 87 (A–B).

ping the set S into (but not necessarily onto) the set T. We can think of $\overline{\overline{S}} \leq \overline{\overline{T}}$ as meaning that S can be turned into a portion of T, or, alternatively, that there is a copy of S contained in T.

If we were talking only about finite sets, it would go without saying that if $\overline{\overline{S}} \leq \overline{\overline{T}}$ and $\overline{\overline{T}} \leq \overline{\overline{S}}$, then $\overline{\overline{S}} = \overline{\overline{T}}$. But it is dangerous to jump to conclusions about infinite sets, for they often have very unexpected and counterintuitive properties. As it turns out, it *is* possible to prove that for *every* two sets S and T, if $\overline{\overline{S}} \leq \overline{\overline{T}}$ and $\overline{\overline{T}} \leq \overline{\overline{S}}$, then $\overline{\overline{S}} = \overline{\overline{T}}$. This proof is the content of what is usually called the Schröder-Bernstein Theorem, and it goes like this.

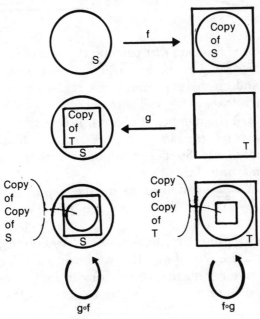

Figure 88 (A–C).

Say that f is a one-to-one function taking S onto a copy of S inside T, and that g is a one-to-one function taking T onto a copy of T inside S. What we need is to find a one-to-one function h that will take S onto all of T.

By using f and g to bounce back and forth between S and T indefinitely, we can build up an infinitely nested sequence of copies inside each of the sets, as pictured in Figure 89.

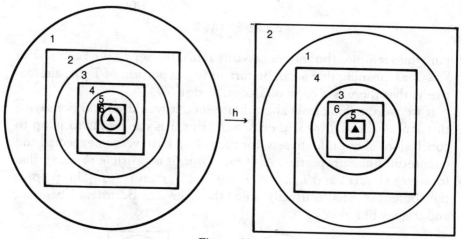

Figure 89.

Now, let h be the function that pairs up the numbered regions as indicated in the picture. That is, $h(x)$ is equal to $f(x)$ on the odd-numbered regions, and on the even-numbered regions we take $h(x)$ to be the unique y such that $g(y) = x$. There may be a region of S and of T that is not in any of the numbered regions. It is possible to show that f will pair up these two extra regions (pictured as black triangles above) in a one-to-one "onto" way. So, putting all this together, we get a one-to-one map h from S onto T.[2]

The relation $\bar{\bar{S}} < \bar{\bar{T}}$ is defined to be the conjunction $\bar{\bar{S}} \leq \bar{\bar{T}}$ and $\bar{\bar{S}} \neq \bar{\bar{T}}$. That is, $\bar{\bar{S}} < \bar{\bar{T}}$ holds if there is a one-to-one map from S *into* T, but no map from S *onto* T.

The theory of transfinite cardinal numbers would not be interesting unless we had infinite sets of differing cardinalities. What I would like to discuss next is a systematic process for obtaining larger and larger cardinal numbers. But first it is necessary to decide just what a cardinal number is.

Notice that the isolated cardinal number $\overline{\overline{S}}$ is not initially defined. Instead, one starts with describing the circumstances under which cardinal numbers are equal to or less than each other. The cardinal number $\overline{\overline{S}}$ is a doubly abstract entity arrived at by ignoring the appearance and the arrangement of the elements of S. $\overline{\overline{S}}$ is a pure form with no content.

It is hard to form the proper sort of image of such an abstract concept, since we are in the habit of imagining forms by imposing them on particular contents. But this habit can be broken. Just as we can imagine the concept "man" without thinking of any *particular* man, we can imagine the concept "three" without thinking of any particular set of three elements.

It is, however, convenient to have a uniform way of finding a concrete representation of a cardinal number $\overline{\overline{S}}$, and we are usually rather casual about the distinction between the abstract concept $\overline{\overline{S}}$ and its standard set representation.

The standard representation of $\overline{\overline{S}}$ is the least ordinal a such that $\overline{\overline{S}} = \overline{\overline{a}}$. That is, the cardinal number $\overline{\overline{S}}$ is identified with the least ordinal a such that there is a one-to-one correspondence between S and a (viewed as the set $\{b: b < a\}$ of all ordinals less than a). Again, $\overline{\overline{S}}$ is the smallest ordinal a such that S can be listed as a sequence of elements of order type a. (The question of whether or not there *is* such an ordinal a for any given set S will be taken up below.)

The cardinal number $\overline{N}$ is commonly called $\aleph_0$. Note that under the identification just introduced, $\aleph_0 = \omega$; note also that since we identify an ordinal with the set of its predecessors, $\omega = N$. Strictly speaking $\omega = \overline{N}$ and $\aleph_0 = \overline{\overline{N}}$; for ω is obtained from the natural numbers in their natural order by thinking only of their abstract arrangement, and $\aleph_0$ is obtained from the natural numbers by thinking only of their abstract numerousness. But once one understands this point, one does not want to stumble over it every second, and from now on N, ω, and $\aleph_0$ will be treated as synonyms with different shades of meaning.

We add, multiply, and exponentiate cardinal numbers by rules quite different from the rules for adding, multiplying, and exponentiating ordinal numbers (although, of course, the rules give the same results for *finite* numbers). If κ and λ are cardinal numbers, we calculate the cardinality $\kappa + \lambda$ by finding two sets K and L such that $\overline{\overline{K}} = \kappa$, $\overline{\overline{L}} = \lambda$, and K and L have no elements in common; and then letting $\kappa + \lambda = \overline{\overline{K \cup L}}$. Recall that for ordinal numbers k and l the ordinal $k + l$ was found by taking the order type of the ordering obtained by placing a copy of l after a copy of k. To see the difference between the two types of addition, note that to evaluate $3 + 2$ by *cardinal* addition you stick out three fingers of your left hand and two fingers of your right, and then find the smallest ordinal whose set of predecessors can be put into a one-to-one correspondence with the set of fingers you have out. To evaluate $3 + 2$ by *ordinal* addition you count up to three and then count two numbers further.

Although $\omega + \omega \neq \omega$, we do have $\aleph_0 + \aleph_0 = \aleph_0$. (The understanding is that we use ordinal addition on the ordinal symbols and cardinal addition on the cardinal symbols.) To see this last fact, let O be the set of all odd natural numbers, let E be the set of all even natural numbers, and let N be (as usual) the set of all natural numbers. Now, O, E, and N all have the same

$$O = \{1, 3, 5, 7, 9, \ldots, 2k + 1, \ldots\}$$
$$\updownarrow \updownarrow \updownarrow \updownarrow \updownarrow \qquad \updownarrow$$
$$N = \{0, 1, 2, 3, 4, \ldots, \quad k \quad , \ldots\}$$
$$\updownarrow \updownarrow \updownarrow \updownarrow \updownarrow \qquad \updownarrow$$
$$E = \{0, 2, 4, 6, 8, \ldots, 2k \quad , \ldots\}$$

cardinal number $\aleph_0$, as is evident from the picture above. Since O and E are sets of cardinality $\aleph_0$ that have no elements in common, $\aleph_0 + \aleph_0 = \overline{\overline{O \cup E}} = \overline{\overline{N}} = \aleph_0$.

The fact that an infinite set can have the same cardinality as a proper subset of itself, and that adding a transfinite cardinal to itself can give you the same number back are a little surprising. As I mentioned in Chapter 1, this aspect of infinite cardinalities so puzzled pre-Cantorian thinkers that they generally believed it was hopeless to attain a theory of infinite cardinalities much more sophisticated than: "All infinities are equal."

The proof that $\aleph_0 + \aleph_0 = \aleph_0$ can be used on any infinite cardinal number κ. Keep in mind that we are thinking of cardinal numbers as certain special sorts of ordinal numbers. In particular, if κ is an infinite cardinal, then it will be a *limit* (as opposed to *successor*) ordinal. This follows since i) if κ is a cardinal number, then $\overline{\overline{\kappa}} \neq \overline{\overline{\lambda}}$ for any ordinal $\lambda < \kappa$; and ii) if $a + 1$ is an infinite ordinal, then $\overline{\overline{a + 1}} = \overline{\overline{a}}$.

$$O_\kappa = \{1, 3, 5, 7, \ldots \omega + 1, \omega + 3, \ldots, \epsilon_0 + 1, \epsilon_0 + 3, \ldots, \lambda + 2k + 1, \ldots\}$$
$$\uparrow \uparrow \uparrow \uparrow \quad \uparrow \qquad \uparrow \qquad \uparrow \qquad \uparrow \qquad \uparrow$$
$$\kappa = \{0, 1, 2, 3, \ldots \omega, \quad \omega + 1, \ldots, \epsilon_0, \quad \epsilon_0 + 1, \ldots, \lambda + k, \ldots\}$$
$$\downarrow \downarrow \downarrow \downarrow \quad \downarrow \qquad \downarrow \qquad \downarrow \qquad \downarrow \qquad \downarrow$$
$$E_\kappa = \{0, 2, 4, 6, \ldots \omega, \quad \omega + 2, \ldots, \epsilon_0, \quad \epsilon_0 + 2, \ldots, \lambda + 2k \quad , \ldots\}$$

Now, we define an arbitrary ordinal to be *even* if it is a limit ordinal or if it is a limit ordinal plus some even natural number, and *odd* if it is a limit ordinal plus some odd natural number. Letting O_κ and E_κ be the sets of all ordinals less than κ that are, respectively, odd and even, we obtain $\kappa + \kappa = \overline{\overline{O_\kappa \cup E_\kappa}} = \overline{\overline{\kappa}} = \kappa$.

Given any two cardinals κ and λ, if κ is infinite and if $\lambda \leq \kappa$, then $\kappa \leq \kappa + \lambda \leq \kappa + \kappa = \kappa$. But if $\kappa \leq \kappa + \lambda$, and $\kappa + \lambda \leq \kappa$, then we can use the Schröder-Bernstein Theorem to conclude that $\kappa + \lambda = \kappa$. In other words, the addition of transfinite cardinals is extremely boring: $\kappa + \lambda$ is just the larger of κ and λ.

⟨0, 0⟩	⟨0, 1⟩	⟨0, 2⟩
⟨1, 0⟩	⟨1, 1⟩	⟨1, 2⟩

$\longleftrightarrow$

0	1	2
3	4	5

$2 \cdot 3 = 6$, cardinal style

0, 1,	0′, 1′,	0″, 1″

$\longleftrightarrow$

0	1	2	3	4	5

$2 \cdot 3 = 6$, ordinal style

Cardinal *multiplication* is defined as follows. If κ and λ are cardinals, we define $\kappa \cdot \lambda$ to be $\overline{\overline{\kappa \times \lambda}}$, where $\kappa \times \lambda$ is the *Cartesian cross product* of κ and λ, which is the set $\{<u, v>: u \in \kappa \ \& \ v \in \lambda\}$ of all ordered pairs with first component from κ and second component from λ.

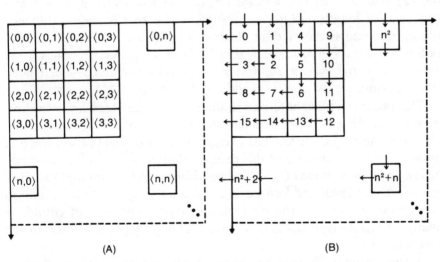

(A)   (B)

Figure 90 (A–B).

The product $\aleph_0 \cdot \aleph_0$ equals $\aleph_0$, as is demonstrated in Figure 90 by exhibiting a one-to-one correspondence between $\overline{\overline{\omega \times \omega}}$ and $\overline{\overline{\omega}}$. The idea is that if you just keep filling in pairs of numbers on the left, and filling numbers on the right, then you will just fill the two quarter-planes. The one-to-one correspondence is obtained by taking any given pair of numbers on the left onto the number occupying the corresponding position on the right.

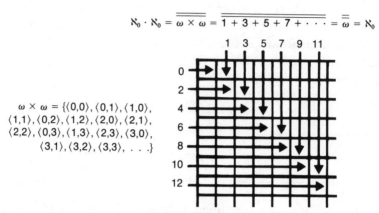

$$\aleph_0 \cdot \aleph_0 = \overline{\overline{\omega \times \omega}} = \overline{\overline{1 + 3 + 5 + 7 + \cdots}} = \overline{\overline{\omega}} = \aleph_0$$

$\omega \times \omega = \{\langle0,0\rangle, \langle0,1\rangle, \langle1,0\rangle,$
$\langle1,1\rangle, \langle0,2\rangle, \langle1,2\rangle, \langle2,0\rangle, \langle2,1\rangle,$
$\langle2,2\rangle, \langle0,3\rangle, \langle1,3\rangle, \langle2,3\rangle, \langle3,0\rangle,$
$\langle3,1\rangle, \langle3,2\rangle, \langle3,3\rangle, \ldots\}$

Figure 91.

If a and b are numbers, we let max (a, b) be the maximum of the two values a and b. For instance, max $(1, 2) = 2$, max $(3, 3) = 3$, max $(\omega, 12) = \omega$. We can also demonstrate $\overline{\overline{\omega \times \omega}} = \overline{\overline{\omega}}$, by listing $\omega \times \omega$ as a sequence of length ω. One way of doing this that is almost (but not quite) the same as the method illustrated in Figure 90 is to list $<a, b>$ before $<c, d>$ if i) max $(a, b) <$ max (c, d); or ii) max $(a, b) =$ max (c, d) and $a < c$; or iii) max $(a, b) =$ max (c, d) and $a = c$ and $b < d$. Under this ordering, we list $\omega \times \omega$ as indicated in Figure 91.

This proof can be generalized to show that for any transfinite cardinal κ, $\kappa \cdot \kappa = \kappa$. That is, if one arranges the members of $\kappa \times \kappa$ according to the ordering defined in the last paragraph, then one gets a sequence of order type κ. A corollary of this result is that for any two infinite cardinals κ and λ, $\kappa \cdot \lambda =$ max (κ, λ). So the addition and the multiplication of transfinite cardinals are *both* boring. Now let us go back and establish one more property of the countable ordinals: The sum of countably many countable ordinals is countable. (This property was used in the last section.)

A non-empty set M is said to be countable if $\overline{\overline{M}} \leq \aleph_0$. It is not hard to see that M is countable iff there is a function from ω onto M listing all the members of M (possibly with repetitions in the case when M is finite).

Now to prove that the union of countably many countable sets is countable. Let $\cup_n A_n$ mean $A_0 \cup A_1 \cup A_2 \cup \ldots$. If for each $n \in \omega$

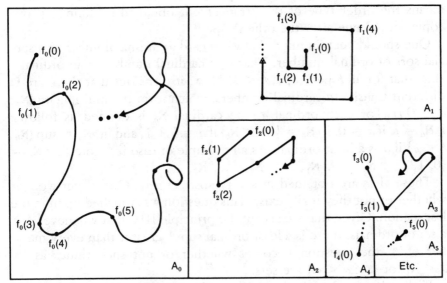

Figure 92.

we have a set A_n and a function f_n mapping ω onto A_n, then we can get a function g mapping ω onto $\cup_n A_n$ by letting $g(k) = f_a\,(b)$, where $<a, b>$ is the kth pair in the listing of $\omega \times \omega$ Just given. Thus g lists $\cup_n A_n$ as $\{f_0(0), f_0(1), f_1(0), f_0(2), f_1(2), f_2(0), f_2(1), f_2(2), f_0(3), \ldots\}$

The second number class, called (II) by Cantor, is the smallest collection of ordinals having the following three properties: i) $\omega \in$ (II); ii) if $a \in$ (II), then $a + 1 \in$ (II); and iii) if $a_n \in$ (II) for each $n \in \omega$, then $\cup\, a_n \in$ (II). Since $\overline{\overline{\omega}} = \aleph_0$, $\aleph_0 + 1 = \aleph_0$, and $\aleph_0 \cdot \aleph_0 = \aleph_0$, it is evident that every member of (II) has cardinality $\aleph_0$.

"*On* and Alef-One" defined $\aleph_1 = sup$ (II). $\aleph_1$ is the *first uncountable* cardinal. It is *uncountable* since if $\aleph_1$ were countable, then $\aleph_1$ would be the sup or lim or union of countably many members of (II), thus a member of (II), and thus less than itself, which is a contradiction. It is the *first* since every ordinal less than $\aleph_1$ is countable.

Now, suppose that we set $\aleph_2 = \{a: \overline{\overline{a}} \leq \aleph_1\}$. Evidently, $\aleph_2$ is the first cardinal greater than $\aleph_1$. For, on the one hand, $\overline{\overline{\aleph_1}} \leq \overline{\overline{\aleph_2}}$, since the identity map maps $\aleph_1$ into $\aleph_2$ in a one-to-one fashion; on the other hand, $\overline{\overline{\aleph_1}} \neq \overline{\overline{\aleph_2}}$, since if the two were equal, then $\aleph_2$ would be an ordinal a such that $a < a$, and this cannot happen. So, $\overline{\overline{\aleph_1}} < \overline{\overline{\aleph_2}}$. Viewed as a set, we have $\aleph_2 = \{0, 1, 2, \ldots \omega^2, \ldots \omega^\omega, \ldots \epsilon_0, \ldots \aleph_1, \ldots \aleph_1 + \omega, \ldots \aleph_1 + \aleph_1, \ldots \aleph_1^{\aleph_1}, \ldots\}$ (Note that here we are talking about *ordinal* addition, multiplication, and exponentiation. Thus, "$\aleph_1 + \aleph_1$"

means the order type of the arrangement obtained by lining up two copies of $\aleph_1$, one after the other.)

One should keep in mind that we regard a cardinal number as a special sort of ordinal number. That is, a cardinal number is an ordinal a such that for all b before, a, $\bar{\bar{b}} < \bar{\bar{a}}$. Now, it turns out that there are Ω different transfinite cardinal numbers, which form the multiplicity $\{\aleph_a : a \in On\}$. For each ordinal a, the cardinal $\aleph_a$ is defined as follows: i) $\aleph_a = \omega$ if $a = 0$; ii) $\aleph_a = \{c : \bar{\bar{c}} \leq \aleph_b\}$ if $a = b + 1$; and iii) $\aleph_a = \sup \{\aleph_b : b < a\}$ if a is a limit ordinal. As an example of case iii), note that $\aleph_\omega = \{0, 1, \ldots \aleph_0, \ldots \aleph_1, \ldots \aleph_2, \ldots \aleph_3, \ldots\}$.

These alefs are obtained in a very abstract way. One might wonder whether or not they really exist. This question was touched upon in the last section. Given that we accept the principle III)–that whenever A is a set of ordinals, there is a least ordinal sup A greater than every member of A–the question becomes whether or not such things as $\{a : \bar{\bar{a}} < \aleph_1\}$ or $\{\aleph_a : a < \omega\}$ are sets.

Well, what *is* a set? In Cantor's words, "A set is a Many that allows itself to be thought of as a One." [See Note 1, Chapter 5]. Clearly, $\{a : \bar{\bar{a}} < \aleph_1\}$ exists as a Many or multiplicity. The question is whether or not this multiplicity can exist as a One—a single finished thing, a unity, a set.

The question is meaningful since there are some multiplicities that can, by their very nature, not exist as unities. Such Absolutely Infinite multiplicities as all rational thoughts, the class V of all sets, or the class On of all ordinals, cannot be unities—for if they are unities, then they are rational thoughts or sets, and contradictions arise. These contradictions stem from the fact that a rational thought, set, or ordinal is not supposed to be a component of itself.

Cantor spoke of the Absolutely Infinite multiplicities as "inconsistent multiplicities." By this he meant that these are multiplicities that cannot exist as unities because such a unification would lead to an inconsistency or contradiction. A multiplicity that *can* exist as a single completed object is a "consistent multiplicity," or set.

With all this in mind, we are now in a position to understand Cantor's phrasing of, and answer to, the question of whether or not the alefs really exist. The passage translated below appears in a letter that Cantor wrote to Dedekind on August 28, 1899.

> One can ask how I know that the well-ordered sets or sequences that correspond to the cardinal numbers $\aleph_0, \aleph_1, \ldots \aleph_\omega, \ldots \aleph_{\aleph_1}, \ldots$ are

really sets in the sense of "consistent multiplicities." Is it not possible that these multiplicities are 'inconsistent,' and that the contradiction arising from the assumption that these multiplicities exist as unified sets has simply not been noticed yet? My answer to this is that the same question can be raised about finite sets, and that if one thinks about it carefully it becomes evident that no *proof* of the consistency of finite multiplicities is possible either. In other words: the fact of the consistency of finite multiplicities is a simple unprovable truth which could be called "the axiom of arithmetic," (in the old sense of the words). In the same sense, the consistency of those multiplicities which have alefs as cardinalities is "the axiom of the extended transfinite arithmetic."[3]

Clearly, there are many different things in the world and in the Mindscape. There is actually a philosophical position, called extreme nominalism, that denies the existence of even finite sets. But such a position strikes one as simply perverse in the face of the fact that everyone routinely perceives unities in multiplicities. Understanding any sentence involves taking the multiplicity of the words in it and grasping them as a unity.

In defense of sets such as $\aleph_1$, Cantor seems to be claiming a direct and simple perception of the reality of such sets in the Mindscape. There is something very appealing about this defense, but it cannot be said to be conclusive.

A weakness in "the axiom of the extended transfinite arithmetic" lies in the difficulty of directly perceiving the sets corresponding to the transfinite alefs. Cantor anticipated this objection, and in another place remarks that a number such as $\aleph_2$ is actually much *easier* to perceive than is some random natural number of ten million digits.[4] Cantor's position has become more and more tenable over the years, as many people have come to understand and work with the alefs without encountering contradiction. But to have said what he did in 1899 is a striking example of intellectual courage.

Cardinal addition and multiplication are rather uninteresting, since if κ and λ are both infinite cardinals, then $\lambda + \kappa$ and $\lambda \cdot \kappa$ are both just equal to the maximum of λ and κ. Cardinal exponentiation is another story entirely. The problem of determining the precise value of the simplest nonfinite cardinal exponentiation, $2^{\aleph_0}$, has been with us for a hundred years now, and there is still no definitive solution in sight.

If λ and κ are ordinals, we say that a κ-sequence from λ is a process that successively picks κ members of λ in a row. We can think of such a sequence as being a function s with domain κ and range $\leq \lambda$. That is, for each ordinal $a \in \kappa$, $s(a)$ is the unique $b \in \lambda$ that fills the ath slot of the sequence. A slightly different way of thinking of a κ-sequence from λ is to think of it as an ordered κ-tuple of members of λ. Thus, $<0, 3, 1>$ would be a 3-sequence from 4, $<0, 2, 4, \ldots 1, 3, 5, \ldots >$ would be an $\omega + \omega$-sequence from ω, and $<0, 1, 0, 1, 0, \ldots >$ would be an ω-sequence from 2. Note that we can think of this last sequence as a function s such that $s(n) = 0$ if n is even and $s(n) = 1$ if n is odd.

The set of all κ-sequences from λ is usually called $^\kappa\lambda$. (This notation has nothing to do with the notation for tetration used in "From Omega to Epsilon-Zero.") If κ and λ are cardinals, we define the cardinal exponent λ^κ to be the cardinality $\overline{\overline{^\kappa\lambda}}$ of the set of all κ-sequences from λ. One motivation for this definition

$$\kappa \text{ lambdas}$$

$$
\lambda^\kappa = \overbrace{(1 + \textcircled{1} + 1 + \ldots)}^{\lambda \text{ ones}} \cdot \overbrace{(\textcircled{1} + 1 + 1 + \ldots)}^{\lambda \text{ ones}} \cdot \overbrace{(1 + 1 + \textcircled{1} + \ldots)}^{\lambda \text{ ones}} \cdot \ldots
$$

is that if we think of λ^κ as the product of κ sums of λ ones, then each of the λ^κ products of κ ones that appears in the final product is obtained by choosing one 1 from each of the κ sets of λ ones. And this process is represented by a member of $^\kappa\lambda$. Another way of putting it is that there should be λ^κ members of $^\kappa\lambda$, since an arbitrary κ-sequence from λ is formed by choosing among λ elements κ times in a row, which can be done $\lambda \cdot \lambda \cdot \lambda \cdot \ldots$ ways, where the product indicated is supposed to have κ members.

On December 7, 1873, Georg Cantor proved the first part of his most famous theorem, now known as Cantor's Theorem: For every cardinal κ, $\kappa < 2^\kappa$. It is quite easy to see that $\kappa \leq 2^\kappa$, since we can get a one-to-one map c from κ into the set $^\kappa2$ of all κ-sequences of zeros and ones. This is done by letting each $a \in \kappa$ correspond to the sequence c_a, which has zeros everywhere except in the ath place. Viewed as a function, the sequence c_a is defined by saying $c_a(b) = 1$ if $b = a$, and $c_a(b) = 0$ if $b \neq a$. Viewed as a κ-tuple, c_2, for instance, would be $<0, 0, 1, 0, 0, \ldots 0, \ldots >$.

The real difficulty in proving Cantor's Theorem is to prove that $\kappa \neq 2^\kappa$. This is done by showing that there can be no map from the set κ *onto* the set $^\kappa2$. Or, put differently, we must show that whenever we have mapped κ onto a set $S = \{s_a; a \in \kappa\}$ contained in $^\kappa2$, there will always be a $d \in {}^\kappa2$ such that $d \notin S$.

$$s_0 = \,<\,\boxed{0}\,,0,0,\ \ldots\ 0,0,\ \ldots\ 0,\ \ldots\ >$$
$$s_1 = \,<1,\,\boxed{1}\,,1,\ \ldots\ 1,1,\ \ldots\ 1,\ \ldots\ >$$
$$s_2 = \,<0,1,\,\boxed{0}\,,\ \ldots\ 0,1,\ \ldots\ 0,\ \ldots\ >$$

$$d(a) = \begin{cases} 1 & \text{if } s_a(a) = 0 \\ 0 & \text{if } s_a(a) = 1 \end{cases}$$

$$s_\omega = \,<1,0,1,\ \ldots\ \boxed{1}\,,1,\ \ldots\ 1,\ \ldots\ >$$
$$s_{\omega+1} = \,<0,1,1,\ \ldots\ 0,\,\boxed{0}\,,\ \ldots\ 1,\ \ldots\ >$$

That is,

$$d(a) = 1 - s_a(a)$$

$$s_\alpha = \,<1,0,1,\ \ldots\ 0,1,\ \ldots\ \boxed{0}\,,\ \ldots\ >$$

$$d = \,<1,0,1,\ \ldots\ 0,1,\ \ldots\ 1,\ \ldots\ >$$

The method of proving that there is such a d is called a *diagonal argument*. Alternatively, we sometimes say that d is found by diagonalizing over S. The method is to make sure that d is different from each member of S by making sure, for each $a \in x$, that d differs from s_a in the ath place. This is done by defining d as the function from κ into 2 such that $d(a) = 1$ if $s_a(a) = 0$ and $d(a) = 0$ if $s_a(a) = 1$. Given that $1 - 0 = 1$ and $1 - 1 = 0$, we can abbreviate this as $d(a) = 1 - s_a(a)$.

We have now shown that $\kappa \leq 2^\kappa$ and that $\kappa \neq 2^\kappa$, so we can conclude that $\kappa < 2^\kappa$. Thus, $\aleph_0 < 2^{\aleph_0}$, $\aleph_1 < 2^{\aleph_1}$, and so on. Note that we now have two ways of passing from a cardinal κ to a larger cardinality. On the one hand, we can mimic the passage from $\aleph_a$ to $\aleph_{a+1}$ by defining κ^+ to be $\{b : \overline{b} \leq \kappa\}$. If κ is a cardinal number, then κ^+ is the first cardinal greater than κ. (It is the *first* since every ordinal before κ^+ has cardinality $\leq \kappa$, and it is *greater* since if it were not, then it would be a member of itself.) On the other hand, we can also obtain a cardinal greater than κ by passing from κ to 2^κ.

We know that 2^κ is greater than κ, and we know that κ^+ is the *least* cardinal greater than κ. So we can conclude that $\kappa^+ \leq 2^\kappa$. Can anything more be said about the relationship of these two cardinals? In his Generalized Continuum Hypothesis (GCH), Cantor made the guess that for all κ, $2^\kappa = \kappa^+$. Another way of putting GCH is: for all a, $2^{\aleph_a} = \aleph_{a+1}$. On the basis of what is presently known about sets, there is no way to prove or disprove GCH.

In order to get a better understanding of this curious state of affairs, we will focus our attention on the special case of GCH called the Continuum Hypothesis (CH). CH is the statement $2^{\aleph_0} = \aleph_1$.

THE CONTINUUM

The continuum problem is the problem of determining which, if any, $\aleph_a$ is equal to $2^{\aleph_0}$. In order really to understand what this problem involves, it is a good idea to look first at a number of different sets that have the cardinality $2^{\aleph_0}$: the power set of omega, the binary tree, the unit interval, the real line, the plane, and three-dimensional mathematical space.

First of all, there is the set $\{x: x \subseteq \omega\}$ of all sets of natural numbers. This set is called the *power-set of omega,* abbreviated $\mathcal{P}\omega$. Typical members of $\mathcal{P}\omega$ would be the empty set $\emptyset$; such finite sets of natural numbers as $\{5\}$, $\{6, 28, 496, 8128, 33550336\}$, $\{n: n \leq 1000\}$, and $\{n:$ there has been a United States presidential election in which one of the candidates received n votes$\}$; such infinite but finitely describable sets of numbers as ω, $\{n \in \omega: n$ is even$\}$, $\{n \in \omega: n > 1000\}$, and $\{n \in \omega:$ a string of ten consecutive sevens appears somewhere in the decimal representation of $n\}$; and probably some completely patternless infinite sets of numbers that do not have any finite description. (The question of whether there actually *are* any such patternless sets was discussed in "Random Reals.")

It turns out that $\overline{\overline{\mathcal{P}\omega}}$ is $2^{\aleph_0}$. This can be proved by constructing a one-to-one map χ from $\mathcal{P}\omega$ onto $^{\omega}2$ as follows. Given any set of natural numbers M, we let χ_M be the ω-sequence that has a one in the nth place if $n \in M$, and a zero in the nth place if $n \notin M$. Thus, if E is the set of all even numbers, then χ_E has the form $<0, 1, 0, 1, 0, 1, \ldots >$. Another example:

$M = \{0, 2, 3, 8, 11, 14, 15, 22, \ldots\}$
$\chi_M = <1, 0, 1, 1, 0, 0, 0, 0, 1, 0, 0, 1, 0, 0, 1, 1, 0, 0, 0, 0, 0, 0, 1, \ldots>$

To prove that χ is *one-to-one,* note that if K and M are different sets of natural numbers, then there has to be some natural number t that is in only one of the two sets K and M. But then $\chi_K(t) \neq \chi_M(t)$, so χ_K and χ_M are different sequences. To prove that χ maps $\mathcal{P}_{\omega}$ onto *all* of $^{\omega}2$, note that if s is any member of $^{\omega}2$, and if S is the set $\{n \in \omega: s(n) = 1\}$ of the numbers of the slots where s has a one, then $\chi_S = s$.

For *any* set x we can form the set $\mathscr{P}x$ of all possible subsets of x and the set $^x2$ of all functions from x into 2. The argument just given can be generalized easily to prove that for any x, $\overline{\overline{^x2}} = \overline{\overline{\mathscr{P}x}}$. I should perhaps mention here that there is some question as to whether the infinite power sets (such as $\mathscr{P}\omega$) really *are* sets, or whether they are perhaps Absolutely Infinite inconsistent multiplicities that canot really exist as a unity. The fact that $\mathscr{P}\omega$ is uncountable certainly makes it difficult to grasp, but such difficulties did not prevent us from accepting the existence of such sets as $\aleph_1$.

It is fairly reasonable to believe that for any κ, $\mathscr{P}\kappa$ is a set. This position can be justified by the following argument. To pick out a subset of κ, you walk from zero out through the ordinals with a pair of brackets in your left hand. Every ordinal that appeals to you is plucked and tossed into the brackets. After κ steps you will have build up one of the possible members of $\mathscr{P}\kappa$. To the extent that we have the idea of *totally free activity,* we have the idea of carrying out such a sequence of κ choices in an arbitrary way. The idea of an *arbitrary sequence of κ choices* serves, then, as the unifying idea that forms the multiplicity $\{y: y \subseteq \kappa\}$ into a unity or set.

Incidentally, this way of thinking of $\mathscr{P}\kappa$ also makes it clear why $\mathscr{P}\kappa$ should be 2^{κ}. The decision for each $\alpha \in \kappa$ on whether or not to include α is a *binary decision,* a choice between two alternatives. How many ways are there to make κ binary decisions in a row? Two times itself κ times, or 2^{κ}. Note that this reasoning also works in finite cases. Thus, $\mathscr{P}3$ has 2^3 elements, for if we regard 3 as being the set $\{0, 1, 2\}$ then $\mathscr{P}3 = \{\emptyset, \{0\}, \{1\}, \{2\}, \{1, 2\}, \{0, 2\}, \{0, 1\}\{0, 1, 2\}\}$.

The diagonal argument proof that for cardinals κ, $\kappa \neq 2^{\kappa}$ can be adapted to prove that for any set x, $\overline{\overline{x}} \neq \overline{\overline{\mathscr{P}x}}$. This is done by showing that no function f can map x *onto* $\mathscr{P}x$, since for any f mapping x into $\mathscr{P}x$, the set $D_f = \{y \in x: y \notin f(y)\}$ is not in the range of f. Why not? Because if we assume that there *is* some member of x, let us say d, such that $f(d) = D_f$, then we can derive a contradiction, leading to the conclusion that the initial assumption that there is such a d is erroneous. And what is the contradiction? Well, if $f(d) = \{y \in x: y \notin f(y)\}$, then it is fairly evident that $d \in f(d) \leftrightarrow d \notin f(d)$.

In any event, we have now more than amply demonstrated that both $^{\omega}2$ and $\mathscr{P}\omega$ have the same cardinality $2^{\aleph_0}$, which we have shown (twice) to be greater than $\aleph_0$. We now examine a way of depicting $^{\omega}2$.

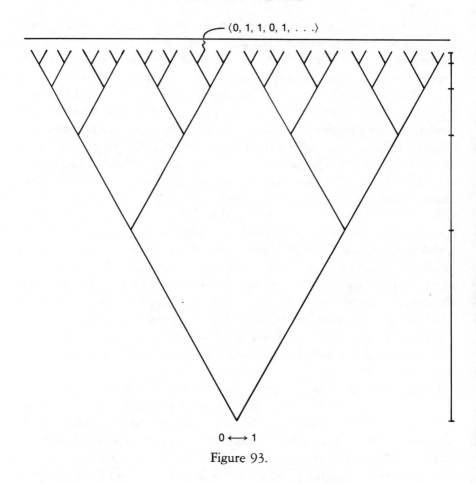

Figure 93.

Depicted in Figure 93 is the infinite binary tree. The tree is constructed by letting a path move upward, forking infinitely many times. By continually halving distances, we fit in ω forks below the horizontal line. We can imagine each point on this line as being a "leaf" that lies at the end of one of the infinitely zigzagged branches up through the tree. A branch that goes all the way up through the tree is given by an ω-sequence of binary decisions. If one takes a pencil and traces a branch up through the tree as drawn, then one can describe one's path by an ω-sequence such as < Left, Right, Right, Left, Right, . . . >. Now, it is clear that if we replace "Left" by "0" and "Right" by "1," then each such path can be identified with a member of $^{\omega}2$ with the set of all branches up through the binary tree.

In a certain sense, the picture of the binary tree is not really a picture of $^\omega 2$, since it is only the various finite initial segments of branches that we see in the tree. The possible paths through the tree exist more conceptually than visually. This is an important distinction. For although the number of paths up through the tree is the uncountable cardinal $2^{\aleph_0}$, the number of nodes in the tree is just $\aleph_0$. A *node* is a fork in the tree, a point where a left-right decision is made. If we start counting the nodes in order of height, we see that there are $1 + 2 + 4 + 8 + 16 + \ldots = \aleph_0$ nodes in all. One can systematically label the nodes by listing all 0-sequences of zeros and ones, all 1-sequences, all 2-sequences, and so on, to get $\emptyset$, $<0>$, $<1>$, $<0, 0>$, $<0, 1>$, $<1, 0>$, $<1, 1>$, $\ldots$, which is readily seen to be a countable list.

The distinction between the countable set of nodes and the uncountable set of branches is sometimes overlooked by philosophers of science who are not mathematicians. I am thinking in particular of Richard Schlegel, who in his otherwise valuable book, *Completeness in Science,* makes two false statements about situations involving binary trees. These two errors concern topics of interest to us, so I will discuss them here.

In the first instance, Schlegel argues that if matter is infinitely divisible, then there will be an *uncountable* number of particles present in each piece of matter.[5] But this is not true at all. For if we think of the infinite binary tree as representing a piece of matter that is infinitely divided, subdivided, etc., then it is clear that the subparticles of the original piece of matter correspond to the *nodes* of the tree, rather than to the *branches* through the tree. For example, if a speck is made of two molecules and each molecule is made of two atoms, and each atom is made of two elementary particles, and so on indefinitely . . . then the total number of particles in the speck is the number of molecules (2), plus the number of atoms (4), plus the number of elementary particles (8), and so on. Schlegel makes the mistake of assuming that there is some ultimate particle at the end of each ω-sequence of splittings, when one would much more reasonably expect that, at the end of an ω-sequence of halving, there will be no matter at all left—only a bare spatial location.

Schlegel makes another (related) blunder in his discussion of the steady-state cosmology (page 139). Recall that the basic assumption of the steady-state cosmology is that new hydrogen atoms tend to appear spontaneously in empty space every so often. Without really changing anything, we can put this assumption differently: Atoms reproduce themselves by "fission" every so often. That is, if one keeps a hydrogen atom locked in a safe for a year and then looks inside, one will find two hydrogen atoms. According to the steady-state cosmology, the universe has no beginning in time. It has been here forever. So, for a past infinity of years, we have had atoms reproducing themselves over and over.

Schlegel jumps to the false conclusion that (assuming we are in such a steady-state universe) we are at the top of a binary tree, and that, therefore, there are uncountably many atoms ($2^{\aleph_0}$ of them). This seems to lead to a contradiction, since there would not be room in a normal Euclidean space for uncountably many atoms of finite size.

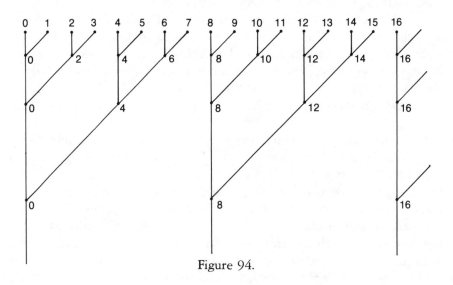

Figure 94.

But in a steady-state universe with splitting atoms and infinite past we would *not* be at the top of a standard binary tree. Instead, we would be at the top of the sort of object drawn in Figure 94. (Note, by the way, that this is basically Figure 18 drawn upside down.) Schlegel falsely claims that a diagram like Figure 94 must have $2^{\aleph_0}$ points on the top line because each point lies above an infinite series of binary forks. The whole difference between this and the true binary tree is that this one is drawn backwards. There will always just be $\aleph_0$ nodes on each line of the reverse tree of Figure 94. In general, the nth line down in the past will contain a node corresponding to each multiple of 2^n.

To return to the main line of discussion, let us investigate a very familiar uncountable set—the real number line. In general, a real number consists of a plus or minus sign, a natural number K and an $e \in {}^{\omega}10$. That is, a real number has the form $\pm K.e_0 e_1 e_2 e_3 \ldots$, where each e_i is a digit between 0 and 9. Sometimes we abbreviate this by writing $\pm K.e$.

An annoying feature of this representation of real numbers is that any real number that ends with an infinite string of nines is identical with some real number ending with an infinite string of zeros. For instance, $10.23999 \ldots = 10.24000 \ldots$ and $.999 \ldots = 1.000. \ldots$ This fact can be justified either algebraically or geometrically.

$$
\begin{array}{rl}
10x = & 9.99 \quad \ldots \\
- \quad x = & .999 \ldots \\
\hline
9x = & 9 \\
x = & 1
\end{array}
$$

The algebraic justification is a bit of sleight of hand that we have already mentioned in "From Pythagoreanism to Cantorism." The trick works because a sequence of $1 + \omega$ nines is no longer than a sequence of ω nines. The geometric justification is that if we start at 0 and repeat-

Figure 95.

edly go $^9/_{10}$ of the remaining distance to 1, then after ω steps we will be at 1. This is just a different version of Zeno's paradox (in which "$^1/_2$" usually appears in place of "$^9/_{10}$"). There are some conceptions of the number lines under which we would *not* choose to identify $.999 \ldots$ with $1.000 \ldots$, instead preferring to say that the former is an infinitesimal amount less than the latter (see "Infinitesimals and Surreal Numbers"). But we will not do this now.

In order to avoid having two notations for the same number, we take R to be the set of all objects of the form $\pm K.e$, where $K \in \omega$, $e \in {}^\omega 10$, and e does not end with an infinite string of nines. We are in the habit of identifying the set R with the set of all points on a line. This is a useful pictorial device, but it should not be taken too literally. As long as we are dealing with an ideal (as opposed to physical) line, there is no difficulty in finding a distinct point to correspond to each distinct real $\pm K.e$. But there is some question as to whether the collection of only these named points really makes up a fully *continuous* line. Let me amplify.

The discrete and the continuous represent fundamentally different aspects of the mathematical universe. One could, perhaps, go so far as to say that it is the left brain that counts up pebbles, but it is the right brain that perceives continuous expanses of space. (I am thinking here of the recent psychological research involving "split-brain" experiments. The brain's left hemisphere controls such digital processes as speech and counting; the right hemisphere handles such analogue processes as singing and space perception.) The left-brain would be on the "Many" side of the table in "Interface Enlightenment," the right-brain with the "One."

Insofar as a set is a collection of distinct elements, it is basically a discrete sort of thing. Because we do allow ourselves to use infinite sets, we can represent the points on a line rather well by these discrete objects. But there is still some question as to whether a continuous line is really just a set of discretely given points.

As was discussed in "Infinitesimals and Surreal Numbers," Zeno's paradox of the arrow hits on this point. Consider an arrow that flies through the air for a minute. Say this continuous stretch of time is really just a set of durationless time-instants. Now, at any one of these instants the arrow is not moving, since motion is not an instantaneous property. Therefore, the arrow is never moving. So how, Zeno asks, did it get from here to there? (There actually *is* a way out of this argument that I have never seen published: according to Special Relativity, an arrow in motion experiences a relativistic length contraction proportional to its speed. So, in fact, the arrow's state of motion *is* instantaneously observable!) The basic question of how a continuous line can be made up of points of no length remains. The radical solution explored in the section "Infinitesimals and Surreal Numbers" is the idea that an *Absolutely Continuous* line can, in fact, never be fully exhausted by *any* set of discrete points, no matter how large such a set might be.

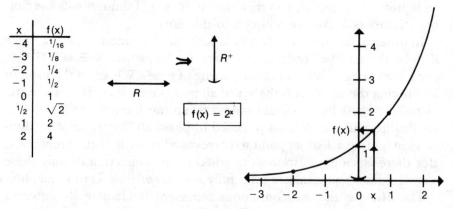

Figure 96.

But we will go ahead and represent R by a line, remembering only not to take this picture too seriously. What is the cardinality of R? Note, first of all, that $\overline{\overline{R}} = \overline{\overline{R^+}}$, where R^+ is the set of all real numbers greater than zero. This fact is demonstrated by the function f given by $f(x) = 2^x$, since this f provides a one-to-one map from R onto R^+. If $(0, 1)$ is

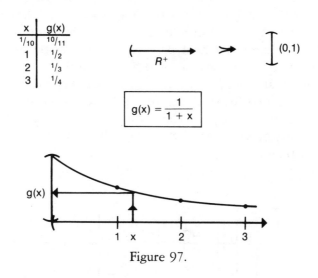

Figure 97.

the set $\{x \in R: 0 \leq x < 1\}$ of all the reals between zero and one, then we can show that $\overline{\overline{R^+}} = \overline{\overline{(0, 1)}}$. This is done by considering a function g that maps R^+ one-to-one onto $(0, 1)$. G is defined by $g(x) = 1/1 + x$.

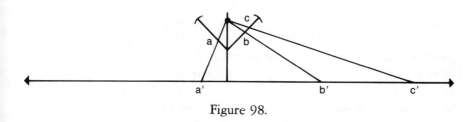

Figure 98.

So, the unit line segment has exactly as many points as does the entire real line. This can also be proved more directly. Suppose that we take a unit line segment and bend it into a right-angled V shape, placing the bottom of the V at the point $1 - \sqrt{1/8}$ on the y-axis. Now, by drawing lines from the point 1 on the y-axis that pass through the V-shaped segment and the infinite real line, we can set up a one-to-one correspondence between the set of points on the unit segment and the set of points on the endless real line. Any two distinct points b and c correspond to distinct points b' and c' and any given point a' on the real line corresponds to a point a on the segment.

One might naturally ask what the cardinality of the real line is. For now, we will use the symbol "c" to stand for $\overline{R}$. It can be shown with certainty that $\aleph_0 < c$. Cantor first proved that $\aleph_0 < c$ on December 7, 1873. We know this because he communicated his proof to his friend Dedekind in a letter the next day.[6] Cantor's first proof of the uncountability of the reals was a bit different from the diagonal argument now used, and this proof will be sketched. The fact that there can be no one-to-one correspondence set up between the sets N and R is the first really interesting fact about the transfinite cardinal numbers, and it can rightly be said that set theory was born on that December day a little more than a century ago.

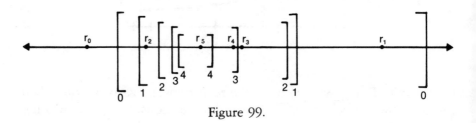

Figure 99.

One must show that there can never be a countable listing $r_0, r_1, r_2, r_3, \ldots$ of real numbers that is *exhaustive*. That is, we must show that given any countable set of real numbers having the form $\{r_n : n \in N\}$, there is some real number d that is different from all of the r_n's. Cantor's first proof of this is simple to present, although a knowledge of the Heine-Borel Theorem is necessary fully to grasp why it works. Cantor's first proof proceeded as follows: Find a closed interval I_0 that fails to contain r_0, then find a closed subinterval I_1 of I_0 such that I_1 misses r_1; continue in this manner, obtaining an infinite nested sequence of closed intervals, $I_0 \supseteq I_1 \supseteq I_2 \supseteq \ldots$, that eventually excludes every one of the r_n; now, let d be a point lying in the intersection of all the I_n's; d is a real number different from all of the r_n.

If c is not $\aleph_0$, then which, if any, alef is it?

The problem of determining where c fits into the hierarchy of alefs is called Cantor's Continuum Problem, and the assertion that $c = \aleph_1$ is known as Cantor's Continuum Hypothesis, or simply CH. Cantor believed strongly that $c = \aleph_1$. Kurt Gödel at one time thought that c should be $\aleph_2$, and a few years ago D. A. Martin wrote a paper that might be interpreted as suggesting that c is $\aleph_3$.[7] (Martin himself denies this interpretation.) But no one really knows. I myself used to think $c = \Omega^+$.

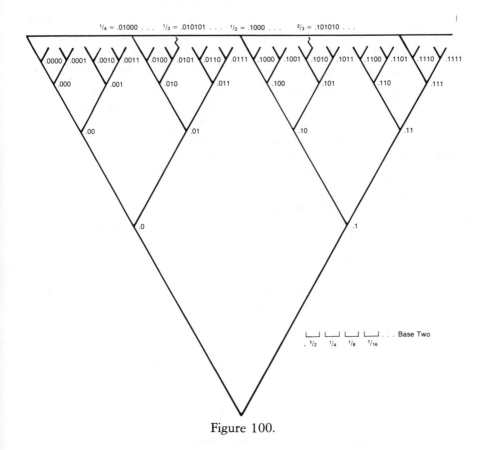

Figure 100.

I would like to postpone discussing the continuum problem just a bit longer. As the reader has probably already suspected, $c = 2^{\aleph_0}$, and I would now like to establish this fact. The easiest way to do this is to look at Figure 100. If we think of that horizontal line segment above the tree as being the segment $[0, 1] = \{x \in R: 0 \le x \le 1\}$, then we can think of each path up through the binary tree as corresponding to a unique point on this segment. It is not too hard to see that, in general, a sequence $s \in {}^\omega 2$ produces a path through the tree leading to the point $s_0/2 + s_1/4 + \ldots + s_n/2^{n+1} + \ldots$. If we go to binary notation, this can be written more simply as $.s_0 s_1 s_2 \ldots {}_{\text{TWO}}$, or $.s_{\text{TWO}}$, (where the "TWO" means that we are to interpret the expansion in terms of powers of two instead of powers of ten.)

Here we have illustrated the way that various members of ${}^\omega 2$ correspond to members of the interval $[0, 1]$. Unfortunately, the map that takes each $s \in {}^\omega 2$ into $.s_{\text{TWO}} \in [0, 1]$ is not one-to-one, since, for example, $.0111 \ldots {}_{\text{TWO}} = {}^1/_2 = .1000 \ldots {}_{\text{TWO}}$ (Zeno again!). But if we ignore the set of all the members of ${}^\omega 2$ ending with an infinite repetition of ones, then the map *is* one-to-one. It turns out that we can harmlessly ignore this set since it is countable, and both ${}^\omega 2$ and $[0, 1]$ are uncountable. So, without going into any more detail, we now claim that $c = 2^{\aleph_0}$.

Cantor originally thought that the cardinality c of the real line would be $\aleph_1$, the cardinality of the set of all points in the plane would be $\aleph_2$, the cardinality of the set of all the points in three-dimensional space would be $\aleph_3$, and so on. But as it turns out, all of these *continua,* or continuous sets of points, have the same cardinality c. Quite formally, we can see rather quickly that the plane should have c points. Why? Well, since the plane is the set of all ordered pairs of real numbers, it has cardinality $c \cdot c = 2^{\aleph_0} \cdot 2^{\aleph_0} = 2^{\aleph_0 + \aleph_0} = 2^{\aleph_0} = c$.

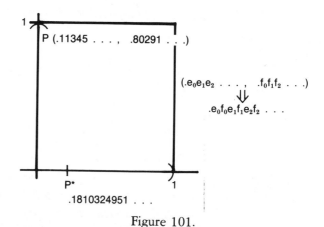

Figure 101.

A meatier proof is obtained if we set up a one-to-one correspondence between the set of points inside the unit square (including the points on the left and bottom edges, but excluding the points on the top and right edges) and the set of points in the half-open unit interval [0, 1). The trick is to match a point P inside the square with the point P^* on the unit interval, where the decimal expansion of P^* is obtained by *shuffling* the expansions of the two coordinates of P. Two members of [0, 1) can be merged to produce a single member of [0, 1) because $2 \cdot \omega = \omega$. Now, it is not too hard to see that the plane is made of $\aleph_0$ of the half-edged unit squares and that the line is made of $\aleph_0$ of the half-open unit segments, so the plane and the line have the same cardinality.

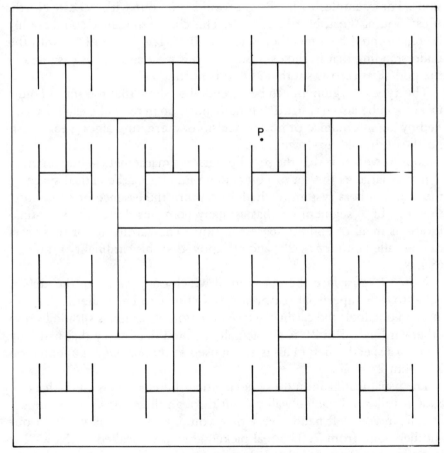

Figure 102.

A more visual and less analytic correspondence between the points of the line segment and the points of the square can be set up as follows.

Start with the line segment. Then pass to the set of all paths through the
binary tree, as indicated in Figure 100. Jiggle the tree around a bit, and
the set of all paths through the tree will actually cover the square. How
do you jiggle the tree? The idea is to pick it up by the root and dangle it
over a unit square. Move down the tree successively rotating each fork
by 90 degrees around the vertical. If you look down from above on
what you then have, it will resemble Figure 102. Another way of thinking of this picture is as the choice tree of a sky diver falling down toward
a square field and making alternate east-west and north-south decisions.
You can reach any point in the square by an ω-sequence of such decisions. For example, to hit P you would start out with $<$East, North,
West, South, West, North, . . . $>$. This choice sequence could readily
be represented by a member $<1, 1, 0, 0, 0, 1, . . . >$ of $^{\omega}2$, with the
understanding that in the even slots zero is west and one is east, and in
the odd slots zero is south and one is north.

This type of argument can be extended to show that the set of points
in a unit cube also has size $2^{\aleph_0}$, for any position in the cube can be specified by an ω-sequence of binary decisions alternating among east-west,
north-south, and up-down.

So now we know that the set of points in a mathematical line segment
is just as large as the set of points in an infinite mathematical space. If
the real numbers system really does capture the essence of continuity,
then this little segment $(-)$ has as many points as there are space-time
locations in all of endless space and time. The cardinality of these sets
can be called either c or $2^{\aleph_0}$, and we know that this cardinality is greater
than $\aleph_0$.

Now we must face the question of which one, *if any,* of the alefs is
equal to c. A superficial reason for expecting c to be an alef is that $c =
2^{\aleph_0}$, and one feels the cardinal exponentiation of cardinals should lead to
other cardinals. But $2^{\aleph_0}$ is not actually defined in any way that naturally
leads to a specific alef. (This is as opposed to the *ordinal* exponentiation
$2^{\omega} = \lim 2^n = \omega$.)

The usual justification for believing that there is *some* $\aleph_a$ such that $c =
\aleph_a$ is as follows. Imagine walking out through the ordinals with the real
line, R, in your left hand. Every time you pass an ordinal, you pick out
another point from R. The real picked at step γ is called r_γ. Now, R is
just a set, and On is Absolutely Infinite, so you will run out of reals long
before you run out of ordinals. In other words, there will be a ρ such
that the set $\{r_\gamma : \gamma \in \rho\}$ exhausts R. It is then evident that $\overline{\overline{R}}$ is the least ρ
such that R can be listed as $\{r_\gamma : \gamma \in \rho\}$. Note that this least ρ must be a
cardinal, i.e., ρ will be an $\aleph_a$ for some a. So we end up with $c = \aleph_a$.

There are two weak points in this argument. First, it could be that the structure of R is so uniform that one simply cannot find a way to keep picking out new members of R indefinitely. This could happen if, for instance, there were a set $I \subseteq R$ of *indiscernibles*—members of R that look so much like each other that you can never manage to find a procedure that will list them all one after another.

To this, one can reply that regardless of whether one has a *rule* for picking out members of R, it is abstractly possible to do so endlessly. The existence of sets does not depend on the existence of rules, names or descriptions, and since it is abstractly possible to pair reals up with ordinals indefinitely, there must be a set that does so. Fine.

What about the second objection to the argument that $c = \aleph_a$ for some a? This is the objection that perhaps one never does run out of reals—perhaps $c \geq \Omega$. It is at least conceivable that the process of discovering new reals never ends, that $\mathscr{P}\omega$ is Absolutely Infinite. The idea here would be that the universe of set theory experiences endless horizontal growth (by adding more reals), as well as endless vertical growth (by adding more ordinals). In this case, $\mathscr{P}\omega$, R, and $^\omega 2$ would not be sets in the usual sense of the word, and $\mathscr{P}R$ would be even worse.

But this seems a little unnatural since one is so inclined to feel that $\{y: y \subseteq \omega\}$ is a set, a "Many that allows itself to be thought of as a One."

In short, we can have either $c = \aleph_a$ for some a, $c \neq \aleph_a$ for any a but $\mathscr{P}\omega$ still a set, or $c \geq \Omega$ and $\mathscr{P}\omega$ not a set. The latter two possibilities are usually formally ruled out by assuming, respectively, the axiom of choice and the power set axiom.

These two axioms make up part of the axioms of Zermelo-Fraenkel set theory, known as ZFC. It will suffice here to say that ZFC codifies the most widespread beliefs of mathematicians about sets. If we agree to believe in the axioms of ZFC, then we can prove that there must be some ordinal a such that $c = \aleph_a$.

c is often called "the cardinal of the continuum," since the word "continuum" is used to denote a continuous region of mathematical space, such as a line, area, or volume. In 1883, Cantor published the remark that he hoped to soon produce a proof that the cardinality of the continuum is the same as that of the second number class.[8] This is the continuum hypothesis, $c = \aleph_1$, also called CH. Cantor never managed to prove CH.

In 1940, Kurt Gödel was able to show that CH is consistent with ZFC. He showed that one can never prove $c \neq \aleph_1$ from the axioms of ZFC.[9] This does not mean that Cantor was *right*—it only means that he was *not provably wrong* on the basis of ZFC.

In 1963, Paul Cohen proved that the negation of CH is consistent with ZFC. He showed that one can never prove $c = \aleph_1$ from the axioms of ZFC.[10] This does not mean that Cantor was wrong—it only means that we *cannot prove that he was right,* using only the axioms of ZFC.

These proofs can be briefly summarized as follows. What Gödel did was to describe a possible universe in which all of the axioms of ZFC hold true and in which $\mathcal{P}\omega = \aleph_1$. This is L, the universe of "constructible sets." Cohen, on the other hand, described various possible universes in which all of the axioms of ZFC hold and $\overline{\overline{\mathcal{P}\omega}} = \aleph_2$, or $\aleph_3$, or $\aleph_{\omega+10}$, or almost anything else. Now, none of their possible universes is believed to be the *real* universe of set theory. But the existence of these possible universes shows that as far as the axioms of ZFC are concerned, either CH or the negation of CH is O.K.

The situation is a little like asking what Scarlett O'Hara did after the end of *Gone with the Wind* . . . one can consistently write a sequel in which she gets back together with Rhett, and one can consistently write a sequel in which she never sees Rhett again. But the book itself does not give us enough information to draw either of these conclusions with certainty. In the same sense, ZFC is not a complete enough description of the universe of set theory to tell us what the power of the continuum is.

Since 1963, there have been a variety of suggestions for new axioms that could be added to ZFC. The only widely accepted new axioms are various axioms of infinity (to be discussed in the next section) and an axiom called Martin's Axiom (after D. A. Martin, although it was R. M. Solovay who first used the axiom). But none of these axioms decides what the value of c is.

In the late 1960s, Kurt Gödel did suggest some new axioms that would decide the size of c. These were the so-called ω-square and $\aleph_1$-square axioms. (The ω-square axiom says that there is a set $S \le {}^\omega\omega$ such that $\overline{\overline{S}} = \aleph_1$, and for every $f \in {}^\omega\omega$ there is a $g \in S$ such that $f <_{\text{bep}} g$, as defined in "Transfinite Numbers"). At first it was thought that these axioms, plus one other, implied that $c = \aleph_2$. But then Gaisi Takeuti showed that on the basis of the two square axioms alone, one can prove that $c = \aleph_1$.[11] At present, there are no very popular axioms that imply that $c \ne \aleph_1$, so it may be that CH will come to be accepted in the future.

One might hope that simply by thinking about c and $\aleph_1$, one could decide whether or not CH is plausible. In 1934, Waclaw Sierpinski did something like this in assembling a book of eighty-eight statements equivalent to CH.[12] But none of these statements is obviously true or false—though in his famous 1947 essay, "What is Cantor's Continuum Problem?," Gödel claimed that some of the equivalents of the continuum hypothesis were "highly implausible."[13]

I have always been intrigued by the following approach to the continuum problem.[14] Define HC to be the set of "hereditarily countable sets." x is in HC iff x is countable and all its members are in HC. HC is what the universe V would look like if there were no such thing as uncountable sets. It is not hard to show that $\overline{\overline{HC}} = c$. Relative to the small universe HC, the class On of all ordinals is $\aleph_1$. So the continuum problem can be phrased this way: "Relative to the universe HC of hereditarily countable sets, is $\overline{\overline{V}} = \overline{\overline{On}}$? Is 'Everything' the same size as 'Absolute Infinity'?" This formulation may seem far-fetched, but it could be that one of the reasons set theory is stalemated by the continuum problem is that we have not yet made enough attempts to identify the problem with some problems outside of pure mathematics.

LARGE CARDINALS

Abstractly speaking, an ordinal is a generalized number that can be "counted up to" by a process of repeatedly taking the next step after all the preceding steps. Usually, an ordinal a is identified with the set $\{b: b < a\}$ of all ordinals less than a. This is because as soon as the collection of steps $\{b: b < a\}$ can be grasped as a unity, the single next step a automatically exists as a definite thing. Thus, conceiving of the first infinite ordinal is really the same thing as forming the multiplicity of all finite ordinals into the unity, or set, $\{0, 1, 2, \ldots\}$. ω exists as a single meaningful concept exactly to the extent to which all the natural numbers exist together in a single unified set.

Imagine all the ordinals lined up one after another like an endless path reaching out toward . . . what? I use the symbol Ω to stand for the end of the path, the last ordinal, the Absolute Infinite, "that than which no greater can be conceived." If Ω is really the greatest ordinal, then the collection of all the steps or ordinals before Ω is just On, the class of all ordinals. Ω exists as a single meaningful concept exactly to the extent that all the ordinals exist together in a single unified entity. But this seems to be a rather small "extent."

For if Ω exists as a single definite step attainable by repeatedly taking "the next step," then Ω is an ordinal, which means that $\Omega < \Omega$, since every ordinal is less than Ω. But no ordinal can be less than itself (since you can never "count up to" something if you have to count up to it before you can count up to it). Another way of putting this is to say that *On* cannot be a set, since there are so many ordinals that it is impossible in principle that they could ever exist together in a single unified entity.

So there is certainly a strong sense in which Ω is not really an ordinal and *On* is not really a unified collction. But yet, but yet . . . I am able to throw around the symbols "Ω" and "*On*" pretty casually, and on the face of it, it seems to make sense to say "Ω has such-and-such a property" (indeed, I just finished saying "Ω is not really an ordinal"). Why not just go ahead and talk about Ω as if it existed as a single definite object, as a sort of *imaginary ordinal*. Strictly speaking, Ω is not an ordinal because it is too big, but it turns out that most of the kinds of things that we say about ordinals seem meaningful when they are said about Ω.

Talking about Ω is an extremely useful and productive thing for set theorists to do. Georg Cantor, the founder of set theory, had quite a bit to say about Ω; and in the last ten or fifteen years this type of discussion has again become respectable. The fact that set theorists are able to talk meaningfully about Ω is a bit surprising, since Ω does not (strictly speaking) exist. The problem of how we *do* talk about such inconceivable things as the Absolute Infinite is an extremely deep and beautiful question in the foundations of set theory. But it is better just to go ahead and start talking about Ω here, leaving aside a deeper analysis of *how* this is done. Let me only remark that this question is really another variation of the One-Many problem. As ungraspable Absolute, Ω is Many, yet as a single guiding idea it is One.

The *large cardinals* we will be discussing in this section are ordinals that share many of the properties of Ω. In general, the more properties of Ω that a cardinal shares, the larger it is.

Recall that an ordinal a is said to be a *cardinal* if we do not have $\bar{\bar{b}} = \bar{\bar{a}}$ for any predecessor b of a. Insofar as Ω is an (imaginary) ordinal, it is also a cardinal. For if b is one of Ω's predecessors, then we do not expect to have $\bar{\bar{b}} = \bar{\bar{\Omega}}$, because if there were a one-to-one function f from b onto Ω, then we would have $On = \{f(a): a < b\}$, implying the contradictory conclusion that On exists as a set, a unity specified by the unity b and the function f. So Ω must be a cardinal.

The same argument can be extended a little to conclude that Ω is a *regular cardinal,* where a cardinal κ is said to be regular if κ cannot be written as the sup of less than κ ordinals less than κ. For if Ω were *not* regular, then we would have some ordinal λ and some set $\{a_b: b < \lambda\}$ of

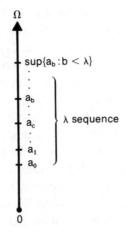

Figure 103.

λ ordinals such that $\Omega = \sup\{a_b: b < \lambda\}$. But this would mean that the Absolutely Infinite could be conceived of in terms of sets, specifically as the supremum of a set of ordinals, which goes against the fundamental assumption that Ω must lie beyond every possible description in terms of sets.

In the ZF set theory, the statement "Ω is regular" is introduced as an explicit assumption called the *Axiom of Replacement*. Informally, the Axiom of Replacement is meant to say that if x is a set and f is a function, then $\{f(y): y \in x\}$ is also a set. The name of this axiom stems from the fact that we can think of the set $\{f(y): y \in x\}$ as being formed by *replacing* each member y of x by its image $f(y)$. For example, it is the axiom of replacement that is invoked to reason from the assumption that $\aleph_1$ is a set to the

$$\{0, 1, 2, \ldots \omega, \ldots \omega + \omega, \ldots \epsilon_0, \ldots \alpha, \ldots\}$$
$$\{\aleph_0, \aleph_1, \aleph_2, \ldots \aleph_\omega, \ldots \aleph_{\omega+\omega}, \ldots \aleph_{\epsilon_0}, \ldots \aleph_\alpha, \ldots\}$$

conclusion that $\{\aleph_a: a \in \aleph_1\}$ is also a set, a conclusion, that enables one to be sure that $\aleph_{\aleph_1} = \sup\{\aleph_a: a \in \aleph_1\}$ is a set as well.

Now, it is evident that the Axiom of Replacement should imply that Ω is regular, since it guarantees that if λ is a set that is an ordinal and if for each $b \in \lambda$, a_b is a set that is an ordinal, then $\{a_b: b \in \lambda\}$ is also a set—so $\sup\{a_b: b \in \lambda\}$ is an ordinal that is also a set, and thus is less than Ω.

The Axiom of Replacement is really a weak form of the *Reflection Principle:* For every conceivable property of ordinals P, if Ω has property P, then there is at least one ordinal $\kappa < \Omega$ that also has property P. The justification of the Reflection Principle is quite simple: If there were some conceivable property P of ordinals such that Ω were the *only* ordinal with property P, then Ω would be conceivable (as the unique ordinal with property P). Therefore, any conceivable property P enjoyed by Ω must also be enjoyed by ordinals less than Ω.

"Conceivable property," incidentally, is supposed to mean a property that is expressible in terms of sets and language of some kind. One does not expect the Reflection Principle to hold for a property that goes " $\square$ is the class of all ordinals," for although "Ω is the class of all ordinals" is true, we do not expect it to be true of any ordinal less than Ω. The point is that the property of *being the class of all ordinals* is not viewed as a *conceivable* property. That is, saying that Ω is the unique thing that is the class of all ordinals does not provide a description of Ω in terms of things simpler than itself.

The fact that Ω is regular can be seen to follow from the Reflection Principle. For given an ordinal $\lambda < \Omega$, and a function determining an ordinal a_b for each $b < \lambda$, then Ω can be seen to satisfy the property "for each $b < \lambda$, $\square$ is greater than a_b." Now, this property is what we call a conceivable property, since it is expressed solely in terms of the given set λ and the given function that finds an ordinal a_b for each $b < \lambda$. Therefore, the Reflection Principle applies, so there must be some ordinal κ such that, the statement "For each $b < \lambda$, κ is greater than a_b" holds true. But then we know that sup $\{a_b : b < \lambda\} \leq \kappa < \Omega$, which implies that Ω cannot be reached by taking the sup of a set of λ ordinals.

Not only is Ω regular, Ω does not have the form κ^+ for any cardinal κ. That is, Ω is not the next cardinal greater than any cardinal. This fact is sometimes expressed by saying that Ω is a *limit cardinal,* rather than a *successor cardinal.* So Ω cannot be reached from below by taking the sup of any sequence shorter than Ω, since Ω is regular; and Ω cannot be reached from below by the process of passing from κ to κ^+ . . . since Ω is a limit cardinal.

A cardinal that has these two properties is called an *inaccessible cardinal.* That is, we say that a cardinal θ is inaccessible if i) θ is regular, and ii) whenever κ is less than θ, κ^+ is less than θ as well. $\aleph_0$ is an inaccessible cardinal since it is regular and since it is not a successor cardinal (because if $\kappa < \aleph_0$, then κ is finite and κ^+ is just $\kappa + 1$). It is difficult to come up with any other inaccessible cardinals. $\aleph_1$ is regular, but it can be described as $\aleph_0{}^+$. $\aleph_\omega$ is not equal to κ^+ for any κ, but it can be described as sup $\{\aleph_n : n \in \omega\}$.

Now, Ω is an inaccessible cardinal greater than $\aleph_0$, so if we apply the Reflection Principle we can conclude that there are other inaccessible cardinals greater than $\aleph_0$. The first one is usually called θ (pronounced "theta"). Since set theorists habitually start counting with zero (rather than one), they often call ω the 0th inaccessible, and call θ the 1st inaccessible.

θ is called *inaccessible* because it is really very hard to get to θ from below. θ cannot be reached by taking the limit of less than θ cardinals, and θ cannot be reached by the operation of taking the next cardinal. If we ask which alef is equal to θ, we get the unhelpful answer that $\theta = \aleph_\theta$.

An interesting way of thinking of θ is to pursue an analogy between the

$$
\begin{array}{ccccc}
0 & 1 & 2 & \omega & \aleph_1 \\
0 & \omega & \aleph_1 & \theta & \rho
\end{array}
$$

two sequences listed above. I have ω corresponding to 1 because these two numbers occur at the two most significant transition points: the transition from nothing (0) to something, and the transition from finite to infinite. Under the first definition of regular (κ is regular if κ is not the sum of less than κ ordinals less than κ) all of the cardinals in the two sequences are regular. $\aleph_1$ corresponds to 2 because $2 = 1^+$ and $\aleph_1 = \omega^+$. That is, 2 and $\aleph_1$ are the first two typical regular successor cardinals. θ corresponds to ω because ω is the first regular limit cardinal after 2, and θ is the first regular limit cardinal after ω. The meaning of the symbol ρ (pronounced "rho") will be explained shortly.

The point of this analogy is that the passage from ω to $\aleph_1$ to θ is something like the passage from 1 to 2 to ω. Just as ω is the first infinite cardinal, θ is the first large cardinal (in the sense that set theorists use the word *large*).

The Reflection Principle says that for any conceivable property P, if Ω has property P, then there is *at least one* ordinal $\kappa < \Omega$ that also has property P. If we examine the justification of the Reflection Principle given above, it becomes evident that the Reflection Principle can actually be strengthened to this: For any conceivable property P, if Ω has property P, then there are Ω ordinals less than Ω that have property P. For otherwise, Ω could be conceived of as the ath ordinal with property P for some (set-sized, and thus, conceivable) ordinal a.

So we see that there are actually Ω inaccessible cardinals less than Ω. Now, apply the reflection principles to: $\boxed{\Omega}$ is inaccessible and there are $\boxed{\Omega}$ inaccessible cardinals less than $\boxed{\Omega}$. This yields a cardinal ν

such that: $\boxed{\nu}$ is inaccessible and there are $\boxed{\nu}$ inaccessible cardinals less than $\boxed{\nu}$.

Such a ν is called *hyperinaccessible.* Another way of defining it is to say that ν is hyperinaccessible if i) and ii) ν is inaccessible, and iii) whenever κ is less than ν, then the first inaccessible greater than κ is less than ν as well.

If we were to define a function θ_a that listed all the inaccessibles and limits of inaccessibles, we would discover that if ν is hyperinaccessible, then $\nu = \theta_\nu$. This is analogous to the fact that for any inaccessible cardinal κ, $\kappa = \aleph_\kappa$. A hyperinaccessible cannot be reached from below by taking the sup of any smaller set of ordinals (since it is regular), nor can it be reached by jumping from cardinal to cardinal (since it is a limit cardinal), nor can it be reached by jumping from inaccessible to inaccessible (since it is the limit of inaccessibles).

We can formulate a notion of *a-hyperinaccessibility* for each ordinal a as follows. The 0-hyperinaccessibles are simply the inaccessibles. The 1-hyperinaccessibles are what we just called the hyperinaccessibles. In general, we will say that ν is $a + 1$-hyperinaccessible if ν is an a-hyperinaccessible that is preceded by ν a-hyperinaccessibles; and for limit ordinals λ, we say that ν is λ-hyperinaccessible if ν is a-hyperinaccessible for every $a < \lambda$.

Note that: $\boxed{\Omega}$ is a-hyperinaccessible for every ordinal $a < \boxed{\Omega}$. If we apply the Reflection Principle to this property of Ω, we discover that there must be inaccessible cardinals μ such that: $\boxed{\mu}$ is a-hyperinaccessible for every $a < \boxed{\mu}$. Such a μ is called *hyper-hyper-inaccessible.* Note that if μ is hyper-hyper-inaccessible, then μ simply cannot be reached from below by using the concept of hyperinaccessibility, since for every $a < \mu$, μ is the μth a-hyperinaccessible.

For convenience, we start writing "hyper²-inaccessible" instead of "hyper-hyper-inaccessible." We can define a notion of a-hyper²-inaccessibility for each a in a way analogous to how a-hyper-inaccessibility was defined. For instance, μ will be 1-hyper²-inaccessible if μ is inaccessible and there are μ hyper²-inaccessibles less than μ. Continuing, we can say μ is hyper³-inaccessible if μ is a-hyper²-inaccessible for all $a < \mu$. Then we can get a notion of hyper^a-inaccessibility for every a; and then (why not) we can define μ to be super-hyper-inaccessible if μ is inaccessible and is preceded by μ hyper^a-inaccessibles for every $a < \mu$. Clearly, there is no end to this process, since for any sequence of Ω degrees of inaccessibility, Ω will be inaccessible and will be preceded by Ω inaccessibles of the ath degree for all $a < \Omega$, and applying the Reflection Principle to this fact always gives inaccessibles of the next level.

What we want to do now is somehow to jump past all of these various degrees of inaccessibility to a whole new level of large cardinals. Informally, the way to do this is to take as a property of Ω the fact that Ω ordinals of each of the degrees of inaccessibility can be found below Ω, apply the Reflection Principle, and get large cardinals ρ that are preceded by ρ inaccessibles, ρ hyperinaccessibles, ρ super-hyper-inaccessibles, and so on. These cardinals are called *Mahlo cardinals,* after Paul Mahlo, who discovered them in 1912.

Formally, we say that ρ is a Mahlo cardinal if ρ is *inaccessible* and *whenever I is a set of ordinals such that there are ρ members of I less than ρ, then* there is an inaccessible cardinal $\kappa < \rho$ such that there are κ members of I less than κ. Putting it another way, ρ is Mahlo if ρ satisfies this *Fixed-Point Reflection Principle:* for every fixed-point property P, if ρ enjoys property P, then there is a $\kappa < \rho$ such that κ enjoys property P. We say that P is a *fixed-point property* if P has the form " $\square$ is inaccessible, and there are $\square$ ordinals less than $\square$ that have property I," where I can be any property of ordinals at all. It is not hard to see that the Fixed-Point Reflection Principle is a weak form of the full Reflection Principle, since the former applies only to *fixed-point* properties, while the latter applies to *all conceivable* properties.

How do we know that any Mahlo cardinals less than Ω exist? By using the Reflection Principle! Given " $\boxed{\Omega}$ satisfies the Fixed-Point Reflection Principle," we can apply the full Reflection Principle to conclude that there must be some $\rho < \Omega$, such that ρ satisfies the Fixed-Point Reflection Principle.

A note of caution must be sounded here. In order for the argument given in the last paragraph to be valid, we must be sure that " $\square$ satisfies the Fixed-Point Reflection Principle" is a conceivable property. We can see that it is by putting it this way: " $\square$ is inaccessible, and if I is a collection of $\square$ ordinals less than $\square$, then there is an inaccessible cardinal $\kappa < \square$ such that there are κ members of I less than κ." The hardest thing one has to do in order to conceive of some cardinal ρ as having this property is to grasp the concept of "arbitrary subcollection I of ρ". This is not hard; insofar as we understand the symbol $\mathscr{P} \square$, we understand the phrase "I is a collection of ordinals less than $\square$." Conceiving of "$\mathscr{P}\rho$" is nothing like trying to conceive of "arbitrary set with ρ as a member," which is impossible.

The last paragraph was, a "note of caution," because one must realize that the Reflection Principle cannot be applied to " $\boxed{\Omega}$ satisfies the full Reflection Principle." Why not? Well, if this were possible, then you could get a $\kappa_0 < \Omega$ such that κ_0 satisfying the full Reflection Principle;

and then applying the same argument to κ_0, you would get a $\kappa_1 < \kappa_0$ satisfying the full Reflection Principle. Continuing through each finite stage, you get a set $K = \{\kappa_n : n \in \omega\}$ of ordinals such that . . . $< \kappa_n < . . . < \kappa_2 < \kappa_1 < \kappa_0$. Now, this is impossible, since every set of ordinals must have a smallest member. So, satisfying the full Reflection Principle is not a conceivable property of ordinals, for if it were, we could get a set K of ordinals that had no least member, and this contradicts the nature of ordinals. But what is the *reason* why satisfying the full Reflection Principle is not a conceivable property of ordinals? The reason is that the notion "arbitrary conceivable property of ordinals" is not itself conceivable.

All the rational thoughts cannot be rationally thought of at once, and all the conceivable properties cannot be conceived of at one stroke. The full Reflection Principle gives a true and understandable statement whenever some particular conceivable property is plugged in for P. But there is a sense in which the full Reflection Principle in its full generality cannot be truly conceived of or understood, for we cannot ever conceive of all the conceivable properties at once. This is as it should be, for Ω may very well be the only ordinal that satisfies the full Reflection Principle, and we know that we cannot conceive of Ω.

The collection of all fixed-point properties *is* conceivable, and that is why the Reflection Principle can be applied to get Mahlo cardinals. There are various other standard conceivable collections of properties for which the same kind of argument can be carried out. Large cardinals obtained in this way are called *indescribable cardinals,* and they are generally quite a bit bigger than the first Mahlo cardinal.

Returning to the Mahlo cardinals for a minute, let us use the symbol ρ_0, or just ρ, to stand for the first Mahlo cardinal. The thing that makes ρ very hard to get to from below is that for no property I can ρ be described as the first inaccessible κ such that there are κ cardinals less than κ having property I.

Let's go back to the two analogous sequences of cardinals introduced a few pages back. We had ρ corresponding to $\aleph_1$ because the process of building up higher and higher degrees of inaccessibility in an effort to get from θ to ρ is so reminiscent of the process of building up higher and higher countable ordinals in an effort to get from ω to $\aleph_1$. The analogy is particularly clear if we think of the former in terms of producing more and more rarified subsets of ρ, and the latter in terms of building up an $\aleph_1$-sequence of functions from $^N N$ under the $<_{\text{bep}}$ ordering. Another analogy between ρ and $\aleph_1$ is that $\aleph_1$ has the "fixed-point" property that whenever I is a set of $\aleph_1$ ordinals less than $\aleph_1$, there is an ordinal $a < \aleph_1$ such that there are a members of I less than a.

All of the definitions of large cardinals given so far could have been strengthened by strengthening the definition of inaccessible cardinals. One defines a *strongly inaccessible* cardinal as a cardinal θ such that i) θ is regular, and ii*) whenever κ is less than θ, then 2^κ is less than θ as well. If 2^κ were always equal to κ^+ (GCH), then of course inaccessibility and strong inaccessibility would be equivalent. But it is, in principle, possible that $2^{\aleph_0}$ could already be as large as ρ, the first Mahlo cardinal. There is, indeed, no upper bound known on the size of the continuum. But since $\kappa < \Omega$ implies $2^\kappa < \Omega$, we can apply the same type of reasoning as before to get strongly inaccessible, strongly hyperinaccessible, and strongly Mahlo cardinals.

There are a great variety of large cardinals after the Mahlo cardinals. First come the indescribable cardinals, then the ineffable cardinals, partition cardinals, Ramsey cardinals, measurable cardinals, strongly compact cardinals, supercompact cardinals, and, finally, the extendible cardinals. Many people are actively engaged in research on these and other types of large cardinals, and one could easily fill a book with facts about large cardinals, I will discuss only the measurable and extendible cardinals here.[15]

$$
\begin{array}{ccccc}
0 & 1 & 2 & \omega & \aleph_1 \\
0 & \omega & \aleph_1 & \theta & \rho \\
0 & \kappa & & \lambda &
\end{array}
$$

The first measurable cardinal is usually called κ. κ is so much bigger than all the cardinals discussed so far that it seems more appropriate to start a new line of analogies for κ. The jump to measurable cardinals is not really analogous to any jump other than the jumps from 0 to 1 and from 0 to ω. As a matter of fact, it turns out that technically speaking, both 1 and ω are measurable cardinals, so we should perhaps call κ the first measurable cardinal *after* ω.

Measurable cardinals were formally defined in 1930 by Stanislaw Ulam, co-inventor of the hydrogen bomb. But it was not until around 1960 that mathematicians realized how big and how strange measurable cardinals really are. The most curious thing about measurable cardinals is that once one knows that they exist, one is forced to the conclusion that there are many more sets of natural numbers than one had previously suspected. It is as if the discovery of some far away galaxy forced us to the conclusion that there are some additional types of microorganisms present in our bodies. Specifically, if a measurable cardinal exists, then there is a set of integers, called $0^{\#}$, that is not in Gödel's universe L of constructible sets.

So what are measurable cardinals? A cardinal κ is said to be measure-able if there is a certain method of deciding which subsets of κ are dense and which subsets of κ are sparse. The "dense" subsets of κ are also known as big subsets, or nodal subsets.[16] The "sparse" subsets of κ are also known as small subsets, or non-nodal subsets.

The method for picking out the dense, big nodal subsets of κ is sim-ply represented by taking the set $\mathcal{N}$ of all the dense, big, nodal subsets of κ. κ is measurable if we can find such an $\mathcal{N}$ that is κ-complete, mean-ing that the intersection of less than κ of the members of $\mathcal{N}$ is also a member of κ.

Formally, we say that κ is measurable if there is a set $M \subset \mathscr{P}\kappa$ such that a) no subsets of κ with cardinality less than κ are in M; b) the inter-section of less than κ members of M is also a member of M; c) for every $A \subset \kappa$, either A or the complement $A' = \{a \in \kappa: a \notin A\}$ of A is in M (but not both). Viewing M as a notion of bigness, a) says that A is big only if A has κ members; b) says that the big subsets of κ are so dense that less than κ of them will always have a big overlap; and c) says that every subset of κ is either big or small. M is called a κ-complete ultrafilter on κ.

To fix the ideas, let us again use κ to mean the first measurable cardi-nal greater than ω; and let us use M to stand for a κ-complete ultrafilter on κ. (There are actually a variety of such ultrafilters on κ. The more natural ones are called *normal* ultrafilters and resemble each other very closely. We assume that M is one of these normal ultrafilters.) What sort of sets are in M? Take, for instance, the set Card of all cardinals less than κ, and the complementary set Card' of all ordinals less than κ that are *not* cardinals. Which of these two sets is to be viewed as big and dense?

In one sense it seems that Card' is the bigger of the two sets, since the infinite cardinals are very far apart and everything in between them is a member of Card'. On the other hand, Card does have the same cardi-nality κ as Card', so it really is not smaller. In fact, Card $\in M$. Set theorists think of Card as consisting of the typical ordinals less than κ, and regard the members of Card' as being just so much packing material (excelsior and styrofoam peanuts) surrounding the valuable Card. Again you might be inclined to say that most ordinals are not inaccessible car-dinals. But as it turns out, the set Inac of all inaccessible cardinals less

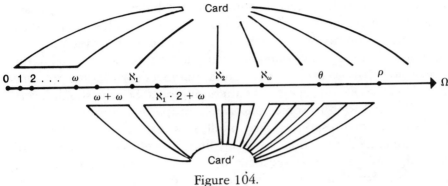

Figure 104.

than κ is a member of M as well, and the cardinals that are not inaccessible are just packing material surrounding Inac.

The guiding rule is that if P is a conceivable property weaker than "☐ is a measurable cardinal," then κ enjoys property P iff the set $\{a \in \kappa : P(a)\}$ of all ordinals less than κ having property P is in M. That is, since a measurable cardinal such as κ is strongly inaccessible, the set of all strongly inaccessibles less than κ is big, dense, nodal—a member of M.

Incidentally, the proofs that a measurable cardinal must be strongly inaccessible and that because of this, the set of strongly inaccessibles less than a measurable cardinal must be nodal . . . these proofs are not easy, and I will not try to give them here. But I will try to explain why measurable cardinals exist.

Basically, it is a matter of convincing oneself that Ω is measurable, and then applying the Reflection Principle to conclude that there are $\kappa < \Omega$ that are measurable as well. The idea is simple. Since Ω does not really exist as a single finished thing, an $X \subseteq \Omega$ can exist only insofar as X corresponds to some property. That is, if $X \subseteq \Omega$, then there must be some conceivable property P of ordinals such that X is the class $\{a \in \Omega : P(a)\}$ of all ordinals less than Ω having property P. Now, we say that this class is *nodal,* or big or dense, if P is a property enjoyed by Ω. And if we let $\mathcal{N}$ be the collection of all the nodal subclasses of Ω, then it turns out that $\mathcal{N}$ is a normal ultrafilter on Ω.

This argument has been tossed around by a number of people over the last ten years or so, and it has still not received universal acceptance.[17] The weak point in the argument is that one feels that there could be some $X \subseteq \Omega$ that were just random, messy classes corresponding to no particular properties of ordinals. Analogy with set-sized ordinals tempts one to believe this. For instance, we know that there are un-countably many subsets of ω; but we are tempted to believe that insofar as a finitely conceivable property of natural numbers has a finite de-scription, there are only countably many such properties, leading to the conclusion that *not* every set of natural numbers corresponds to some finitely conceivable property of natural numbers. So why should every $X \subseteq \Omega$ correspond to some conceivable property or ordinals?

There are two responses. First of all, it is really not so evident that there are indeed only countably many finitely conceivable properties of natural numbers. The whole issue of whether or not there are more properties than there are sets is quite complex, as we saw in Chapter 3. Secondly, one must realize that Ω really is different from any set-sized ordinal. Although we can talk of Ω, it is not really a single finished thing, so one should not expect that there will automatically be a whole lot of subclasses of Ω. If Ω is so big that it can never be thought of as finished, then one cannot really talk about randomly picking out Ω ordi-nals. Instead, one must accept that a subclass of Ω exists only if there is some conceivable property to hold it together.

So, if you swallow that, then you can go ahead and say that Ω is mea-surable, where a class of ordinals is thought of as big, dense, or nodal iff this class corresponds to a property of ordinals that is enjoyed by Ω. The mode of thought employed here (where Ω is thought of one min-ute as a Many, as a bare concept, and is thought of the next minute as a One, as an object that may or may not have a certain property P) is a bit odd, and perhaps even a little shady. Although I find it intelligible, it seems reasonable to call this type of talk *doubletalk*. An extreme exam-ple of this kind of dualistic way of thinking of the universe of set theory occurs in W. N. Reinhardt's justification of the *extendible cardinals*.

The extendible cardinals are the largest cardinals that set theorists know of at present. If one strengthens the definition of extendible car-dinals slightly, one obtains the definition of ∞-extendible cardinals, which are provably inconsistent. This is to say that it is impossible for ∞-extendible cardinals to exist if the class of all sets is indeed a model of ZFC set theory. Because of this, there are some very real doubts as to whether even the extendible cardinals really exist; many people feel that if such large cardinals do exist, they will be the largest large cardi-nals that can ever be found. There is an attractive doubletalk argument for the existence of extendible cardinals, so they probably are really out there. But it would be repeating history to think that with the extendi-ble cardinals we have reached the end of the line. The universe of set theory is infinitely elusive; if there is one thing we can be sure of, it is that the set theories of the future will be vastly more inclusive than anything we have ever dreamed of.

It is interesting to note that the smaller large cardinals have much grander names than the really big ones. Down at the bottom you have the self-styled inaccessible and indescribable cardinals loudly celebrating their size, while above, one of the largest cardinals quietly remarks that it is measurable, and the largest cardinals known simply point out past themselves with the comment that they are extendible.

A precise characterization of extendible cardinals would take us too far afield, but let me attempt a somewhat poetic description of them.

Once again, think of the ordinals as an endless mountain you are climbing. Say that you have climbed as far as the first extendible cardinal, usually called λ. Glancing up, you pick out some landmark that is (necessarily) short of the peak. You are near the lip of a cliff, and walk over to look down at the ordinals you have already climbed past. Far below, you can make out the first inaccessible and the first measurable cardinals, but closer to you there are so many cardinals of this kind that it is hard to pick any one of them out. The part of the cliff nearest to you is rough with cardinalities and interrupted by frequent ledges, but the pattern is so repetitive that there is really no single distinguishing feature near you. There are plenty of ledges as big as the one you are resting on, but it is hard to comprehend how the monotony of the climb could have been endured.

A large eagle floats nearby on the mountain drafts, flexing the fingerlike tips of his wings as he circles. You lie down to rest, merging into an empty peacefulness. You feel yourself snatched up and carried up the mountain, far beyond the landmark you had spotted earlier . . . or is it a dream?

You stand up and look around again, but there is no sign of the eagle. You glance up and see what seems to be the same landmark at the same distance as before; then you go to look down the mountain again. You can still make out the first inaccessible and the first measurable cardinals far below, but now you are trying to tell whether or not you really have been moved up the mountain. There are a number of ledges below, but there is no way to tell whether any of them is the one where you were before the eagle carried you up . . . if indeed he did carry you up at all.

Kurt Gödel

EXCURSION II
GÖDEL'S INCOMPLETENESS THEOREMS

In this excursion we will examine a semiformal proof of Gödel's Incompleteness Theorems. Even at this modest level of precision, the proof takes on a somewhat technical appearance, but this is unavoidable if we are to understand some of the finer points of Gödel's proof.

In the "Formal Systems" section a number of preliminary notions are discussed: the idea of a formal system, the distinction between syntax and semantics, and the notions of consistency and completeness. This is all done with reference to the description of a particular formal system: Peano Arithmetic.

In the "Self-Reference" section we see how to go about generating mathematical sentences that refer to themselves, and in the "Gödel's Proof" section we have the actual proofs of Gödel's theorems, along with a discussion of the precise circumstances under which these theorems apply.

The final section contains an exact mathematical analysis of a vague metaphysical problem, in this case the problem of whether machines can think.

FORMAL SYSTEMS

Gödel's two Incompleteness Theorems state that all formal systems of a certain kind are subject to two limitations. By a formal system we mean a set of mathematical axioms and a set of rules and procedures by which one combines axioms to produce proofs of theorems. Gödel's results apply to any formal mathematical system T that is i) finitely describable, ii) consistent, and iii) as strong as Peano arithmetic. The two Gödel theorems state that any such system T is, firstly, *incomplete* . . . in the sense that there will be some statement about the addition and multiplication of natural numbers that can neither be proved nor disproved by T; and that any such T is, secondly, *unable to establish its own consistency* . . . in the sense that T is unable to prove that no contradiction can be derived from T.

It is the burden of this section to define more precisely the various technical expressions used in the last paragraph. We will begin by developing the notion of a formal system, simultaneously describing a particular formal system: Peano Arithmetic.

Quite generally, a formal mathematical system is a system of symbols together with rules for employing them. A formal system has four components. 1) A basic "alphabet" of *symbols* to be used. Any finite sequence of the fundamental symbols is called a formula. But most of these arbitrary strings of symbols are useless, and one has: 2) A criterion for determining which strings of symbols are "grammatical." These grammatical strings are called *meaningful formulas*. To obtain a formal system, one then takes: 3) A particular set of meaningful formulas as the *axioms* of the system. Lastly, one adopts: 4) A few *rules of inference* that describe the allowable ways of combining axioms to provide proofs of other meaningful formulas.

To be more precise, a *proof from the formal system* T is a sequence of meaningful formulas $M_1, \ldots, M_n$ such that each of the M_i is either an axiom of T or is obtained from some of the previous M_j's by one of the rules of inference. A formula F is said to be *provable from* T if there is such a proof sequence that ends with F.

We will now illustrate these notions by describing the formal system P for arithmetic invented by Giuseppe Peano in 1889. Peano was one of the first to use what we now call symbolic logic. He introduced, for instance, the use of the symbols "$(\exists x)$" to mean "there is an x such that . . ."; and he habitually wrote out all of his lecture notes in his new symbolism. He was teaching at a military academy at the time, and his students were so incensed by his formalistic approach to mathematics that they rebelled (despite his promise to pass them all) and got him fired. Subsequently he found a more congenial setting at the University of Turin.[1]

Anyone who has ever seen a formal system should feel free to skip the following description of P. Anyone who has never seen a formal system will have some difficulty in understanding the description, so he too should skip it, or at most, skim it. The description of P is here primarily as an exhibit.

The basic symbols used in P are of three sorts: logical and punctuation symbols, variable symbols, and special arithmetical symbols. The logical symbols are seven in number: $\sim$, $\lor$, $\&$, $\rightarrow$, $\leftrightarrow$, $\exists$, and $\forall$. These symbols are pronounced, respectively, "not," "or," "and," "implies," "if and only if," "there exists," and "for all." The four punctuation symbols are the round and square parentheses: (,), [, and]. Since the sentences in P may be arbitrarily complicated, one needs an infinite number of variable symbols: $m, n, p, q, x, y, z, v_0, v_1, v_2, \ldots$ The symbols peculiar to the study of arithmetic are five in number: $O, S, +, \times$ and $=$. These symbols are pronounced: "zero," "the successor of," "plus," "times," and "equals."

We have completed phase one of specifying the formal system P by giving the basic symbols used in this system. The only symbols likely to be unfamiliar to the non-mathematician are the successor symbol "S" and the *quantifiers* "$\forall$" and "$\exists$".

In general, if n is some natural number, Sn is supposed to be the *next* natural number. In other words, the set $\{0, 1, 2, 3, \ldots\}$ can be thought of as the set $\{0, S0, SS0, SSS0, \ldots\}$. As we will see, both $+$ and $\times$ can be defined in terms of the extremely basic notion S of taking the next larger natural number. The meaning of the quantifiers will become evident from the following examples of meaningful formulas and their intended meanings.

$(\forall x)\,(\exists y)\,[y = Sx]$	Every natural number has a successor.
$(\forall x)\,(\exists y)\,[x = 0 \lor x = Sy]$	Every natural number is either zero or the successor of some other number.
$SS0 + SS0 = SSS0$	Two plus two is three.
$x + Sy = S(x + y)$	x plus the successor of y is the successor of x plus y.
$(\exists y)\,[x = SS0 \times y]$	There is a y such that x is two times y (i.e., x is even).
$(\exists y)\,[x = SS0 + y]$	X is greater than or equal to two.
$(\forall y)\,(\forall z)\,[x = y \times z \rightarrow (y = x \lor z = x)]$	Whenever x is factored, one of the factors is x itself (i.e., x is prime).
$(\forall n)\,(\exists p)\,(\exists x)\,(\forall y)\,(\forall z)$ $[p = n + x \text{ and } (p = y \times z \rightarrow (y = p \lor z = p))]$	For every n, there is a prime number p that is greater than n (i.e., there are infinitely many primes).

The meaningful formulas of the formal system P are built up in two stages. First we define the notion of a *term,* and then we use this notion to define the meaningful formulas. The definition of *term* has three parts: 1) Every variable symbol is a term. 2) If s and t are terms, then so are $S(t)$, $(s) + (t)$, and $(s) \times (t)$. (The purpose of the parentheses is to avoid ambiguous terms such as $S0 + 0$, which could mean either $S(0 + 0)$, or $(S0) + 0$. In practice, we almost always omit the parentheses, relying on the rule that S, $+$ and $\times$ are to be executed in that order.) 3) A string of symbols is a term only if it can be obtained by repeated applications of 1) and 2). Sx, $SS0$, $v_{13} \times S(SS0 + S0)$ are all terms, but $x0$, SSS, and $S0+$ are not.

Now we can define the notion of *meaningful formula* of P in three steps: 1) If s and t are terms, then $s = t$ is a meaningful formula. 2) If A and B are meaningful formulas, then so are $\sim(A)$, $(A \lor B)$, $(A \& B)$, and $(A \leftrightarrow B)$; and if w is a variable symbol, then $(\exists w) [A]$ and $(\forall w) [A]$ are meaningful formulas as well. 3) A string of symbols is a meaningful formula only if it can be obtained by repeated applications of 1) and 2). (The round parentheses are used to separate "clauses," and the square parentheses are used to indicate the formula to which a quantifier or group of quantifiers applies.) We saw a number of examples of meaningful formulas above.

A variable symbol w is said to be *free* in a formula A if w occurs in A, and if $(\exists w)$ and $(\forall w)$ do not occur before w in A. If a meaningful formula A has no free variables, it is called *a sentence,* and can be regarded as making a statement about natural numbers that is either true or false.

If a meaningful formula A has exactly one free variable w, then we sometimes stress this by writing A as $A(w)$. Given any such $A(w)$ and any term t, we write $A[t]$ to stand for the formula obtained by replacing every occurrence of w in A by t. For instance, if $E(x)$ is the formula $(\exists y)$ $[x = SS0 \times y]$ stating that x is even, then $E[SSSS0]$ is the true sentence $(\exists y) [SSSS0 = SS0xy]$ stating that 4 is even, and $E[SSS0]$ is the false sentence stating that 3 is even.

We are now ready to state the axioms of the formal system P. These axioms fall into four groups: the L group, axioms having to do with the use of the logical symbols; the Q group, which concerns the use of the quantifiers; the E group, involving the symbol "$=$", and the P group, which comprises what is traditionally thought of as Peano's axioms for arithmetic.

Every formal mathematical system includes groups L, Q, and E among its axioms. For this reason, these axioms are often not explicitly mentioned when a formal theory is given. I mention them here for the sake of completeness.

A	B	~A	A V B	A & B	A → B	A ↔ B
T	T	F	T	T	T	T
T	F	F	T	F	F	F
F	T	T	T	F	T	F
F	F	T	F	F	T	T

Figure 105.

If A, B, C, etc., are formulas, we can form various combinations of them using the logical connectives $\sim$, V, &, →, and ↔. The meaning of these logical connectives can be expressed by exhibiting how the truth-values of compound formulas depend on the truth-values of the component parts. (Figure 105). Some combinations will be true regardless of the truth or falsity of the component parts. Such necessarily true combinations are called *tautologies*. A V $\sim A$, $(A \rightarrow B) \leftrightarrow (\sim B \rightarrow \sim A)$, and $(A \text{V} B) \leftrightarrow \sim (\sim A \ \& \sim B)$ are examples of tautological forms. $x = 0$ V $\sim (x = 0)$ and $((\forall y)[\sim (x = Sy)] \rightarrow x = 0) \leftrightarrow (\sim(x = 0) \rightarrow \sim (\forall y) [\sim (x = Sy)])$ are specific instances of the first two of these tautological forms.

A	V	B	↔	~	(~A	&	~B)
T	T	T	T	T	F	F	F
T	T	F	T	T	F	F	T
F	T	T	T	T	T	F	F
F	F	F	T	F	T	T	T

Figure 106.

There is a fixed finite procedure by which one can check if a given string of symbols is a tautology. One simply constructs a truth table and sees if the statement in question is true no matter what the truth values of the component statements are. So there really is no imprecision or hidden complexity in taking the axiom group L to be: L) All tautologies.

Alternatively, one can give eight basic tautologies from which all other tautologies follow according to the Modus Pomens rule of proof to be given below. But I will not state the eight basic tautologies here. Group Q consists of three schemas that clarify the meaning of the quantifiers:

Schema $Q1$) If $A(w)$ is a meaningful formula, w is a variable, and t is a term:
$$(\forall w) [A(w)] \rightarrow A[t].$$
Schema $Q2$) If $A(w)$ is a meaningful formula, w is a variable, and t is a term:
$$A[t] \rightarrow (\exists w) [A(w)].$$
Schema $Q3$) If w is a variable and $A(w)$ is a meaningful formula, then
$$A(w) \rightarrow (\forall w) [A(w)].$$

Schema $Q3$) is the only one that is a bit surprising. $Q3$) is sometimes called the Generalization Rule, and makes the point that if we can prove $A(w)$ holds for a variable w, then, since no special properties of w are used, we have really proved $(\forall w) [A(w)]$ i.e., for all w, A holds). And now for Group E.

Schema $E1$) For any term t, $t = t$.
Schema $E2$) For any terms s and t, $t = s \rightarrow s = t$
Schema $E3$) For any meaningful formula $A(w)$ with a free variable w, and for any terms s and t, $t = s \rightarrow (A[t] \rightarrow A[s])$.

Each of these schemas actually stands, again, for a countably infinite number of axioms. Thus, Schema $E1$) is really all of the axioms $0 = 0$, $0 + 0 = 0 + 0$, $S(S0 + 0) = S(S0 + 0)$, and so on. But rather than writing out each one of these axioms here, we schematically represent them all at once with the formula $t = t$. $E1$) and $E2$) represent what are normally called the *reflexive* and the *symmetric* properties of the relation "=". The *transitive* property, $(s = t \& t = r) \rightarrow s = r$, is derivable from $E3$). $E3$) codifies the principle that equals can be substituated for equals.

The P Group of axioms are as follows:

$P1$) $(\forall x) [\sim(Sx = 0)]$.
$P2$) $(\forall x) (\forall y) [Sx = Sy \rightarrow x = y]$.
$P3$a) $(\forall x) [x + 0 = x]$.
$P3$b) $(\forall x) (\forall y) [x + Sy = S(x + y)]$.
$P4$a) $(\forall x) [x \times 0 = 0]$.
$P4$b) $(\forall x) (\forall y) [x \times Sy = (x \times y) + x]$.
$P5$) For each meaningful formula $A(w)$ with a free variable w,
$$A[0] \rightarrow ((\forall x)[A[x] \rightarrow A[Sx]] \rightarrow (\forall y) [A[y]]).$$

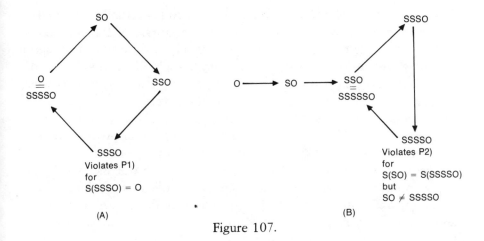

Figure 107.

The purpose of the first two axioms is to guarantee that none of the terms $0, S0, SS0, SSS0, \ldots$ are equal to each other. $P1$) rules out the possibility of the sequence bending around into a circle with $S^n0 = 0$ for some n. $P2$) rules out the possibility of the sequence bending into a loop that hooks in somewhere past 0. $P3$a) and $P3$b) constitute a so-called *recursive* definition of $+$ in terms of S. The definition has the form $n + 0 = n \quad n + Sm = S(n + m)$. To see how it works in action, consider how one uses it to get $3 + 2 = 5$.

$$
\begin{aligned}
SSS0 + SS0 &= S(SSS0 + S0), \text{ by } P3\text{b)} \\
&= S(S(SSS0 + 0)), \text{ by } P3\text{b)} \\
&= S(S(SSS0)), \text{ by } P3\text{a)} \\
&= SSSSS0, \text{ dropping the parentheses.}
\end{aligned}
$$

$P4$a) and $P4$b) define $\times$ recursively in terms of $+$:
$$
\begin{cases} n \times 0 = 0 \\ n \times Sm = (n \times m) + m. \end{cases}
$$
To see this in action, we show $3 \times 2 = 6$.

$$
\begin{aligned}
SSS0 \times SS0 &= (SSS0 \times S0) + SSS0, \text{ by } P4\text{b)} \\
&= ((SSS0 \times 0) + SSS0) + SSS0, \text{ by } P4\text{b)} \\
&= (0 + SSS0) + SSS0, \text{ by } P4\text{a)} \\
&= SSS0 + SSS0, \text{ by } P3\text{a)} \\
&= SSSSSS0, \text{ as can be checked with } P3\text{).}
\end{aligned}
$$

Schema $P5$) is called the *Induction Schema* and is actually a countable infinity of axioms. With the discussion of the Library of Babel in mind, it is not difficult to see that there are actually ω different meaningful formulas with one free variable. For each such $A[w]$, the Induction Schema asserts that *if* 0 has the property A and *if* for any $x, x + 1$ has property A whenever x has property A, *then* every y has property A.

To see a simple example of how this schema is used, consider how one proves $(\forall y)[0 + y = y]$, a fact that is necessary in proving the full commutative law $(\forall x) (\forall y)[x + y = y + x]$. To prove $(\forall y)[0 + y = y]$, we prove i) $0 + 0 = 0$ and ii) $(\forall x)[0 + x = x \rightarrow 0 + Sx = Sx]$, and apply an instance of the Induction Schema. i) follows from $P3$a) by the Q axiom, stating that $(\forall w)A[w] \rightarrow A[t]$ for any formula A, variable w, and term t. ii) is proved by letting x be any number, assuming $0 + x = x$, and then proceeding as follows:

$$0 + Sx = S(0 + x) \text{ by } P3\text{b})$$
$$= S(x) \text{ by hypothesis}$$
$$= Sx \text{ by elimination of parentheses.}$$

There is only one rule of inference used in proofs from the formal system P, called *Modus Ponens*: if A and B are meaningful formulas, one can infer B from $(A \rightarrow B)$ and A. This is a very familiar form of reasoning, since in practice, one very commonly proves a statement B by first proving A and then proving that B is a necessary consequence of A.

We will assume that all of the formal systems we consider have MP as their only rule of inference. This is really no restriction, since in the presence of MP, any rule of the form "Infer B from $A_1, A_2, \ldots, A_n$" can be replaced by the axiom, "$A_1 \rightarrow (A_2 \rightarrow (\ldots (A_n \rightarrow B) \ldots))$."

A proof from P consists of a sequence $M_1, \ldots M_n$ of formulas such that for each M_i with $1 \leq i \leq n$, either a) M_i is an axiom from group L, Q, E, or P; or b) there are j and k such that $j \leq k \leq i$ and M_i follows from M_j and M_k by MP, i.e., M_k is $(M_j \rightarrow M_i)$.

Writing out formal proofs is a very cumbersome process. We write out as an example the formal proof of $(\forall y) [0 + y = y]$.

1. $(\forall x) [x + 0 = x]$ $P3$a.
2. $(\forall x) [x + 0 = 0] \rightarrow 0 + 0 = 0$ Instance of $Q1$.
3. $0 + 0 = 0$ MP (1&2).
4. $0 + x = x \rightarrow (0 + Sx = S(0 + x) \rightarrow 0 + Sx = Sx)$ Instance of $E3$.
5. $(0 + x = x \rightarrow (0 + Sx = S(0 + x) \rightarrow 0 + Sx = Sx))$
   $\rightarrow (0 + Sx = S(0 + x) \rightarrow (0 + x = x \rightarrow 0 + Sx = Sx))$ Instance of L.
6. $0 + Sx = S(0 + x) \rightarrow (0 + x = x \rightarrow 0 + Sx = Sx)$ MP (4&5).

7. $(\forall x) (\forall y) [x + Sy = S(x + y)]$ $P3b.$

8. $(\forall x) (\forall y) [x + Sy = S(x + y)] \rightarrow$
   $(\forall y)[0 + Sy = S(0 + y)]$ Instance of $Q1.$

9. $(\forall y) [0 + Sy = S(0 + y)]$ MP (7&8).

10. $(\forall y) [0 + Sy = S(0 + y)] \rightarrow 0 + Sx = S(0 + x)$ Instance of $Q1.$

11. $0 + Sx = S(0 + x)$ MP (9&10).

12. $0 + x = x \rightarrow 0 + Sx = Sx$ MP (6&11).

13. $(0 + x = x \rightarrow 0 + Sx = Sx) \rightarrow (\forall x)$
    $[0 + x = x \rightarrow 0 + Sx = Sx]$ Instance of $Q3.$

14. $(\forall x) [0 + x = x \rightarrow 0 + Sx = Sx]$ MP (12&13).

15. $0 + 0 = 0 \rightarrow ((\forall x)[0 + x = x$
    $\rightarrow 0 + Sx = Sx] \rightarrow (\forall y) [0 + y = y])$ Instance of $P5.$

16. $(\forall x) [0 + x = x \rightarrow 0 + Sx = Sx] \rightarrow \forall y)[0 + y = y]$ MP (3&15).

17. $(\forall y) [0 + y = y]$ MP (14&16).

This concludes the description of the formal system P.

Fully formalized proofs have a nitpicking, obsessive quality. But by the same token, they are satisfyingly solid and self-explanatory. Nothing is left to the imagination, and the validity of a formal proof can be checked simply by looking at the patterns of symbols. Given the basic symbols, the rules of term and formula formation, the axioms and axiom schemas, and the rules of inference, one can check whether or not a sequence of strings of symbols is a proof in a wholly mechanical fashion.

In their book, *Gödel's Proof,* Nagel and Newman make an interesting comparison between a *calculus* (their word for formal system) and the game of chess:

> Chess is played with 32 pieces of specified design on a square board containing 64 square subdivisions, where the pieces may be moved in accordance with fixed rules. The game can obviously be played without assigning any 'interpretation' to the pieces or to their various positions on the board . . . The pieces and the squares of the board correspond to the elementary signs of the calculus; the legal positions of pieces on the board, to the formulas of the calculus; the initial positions of pieces on the board, to the axioms or initial formulas of the calculus; the subsequent positions of pieces on the board, to formulas derived from the axioms (i.e., to the theorems); and the rules of the game, to the rules of inference (or derivation) for the calculus."[2]

It would greatly simplify things if human discourse could be regarded as the working out of some formal system. One would no longer have to wonder about the meanings of words, but could instead judge the validity of people's arguments by checking them against a fixed, finitely describable set of rules and axioms. Leibniz dreamed of finding such a universal system, the *Characteristica universalis,* and envisioned a day when disagreeing parties would simply get out the rules, saying, "Let us calculate." However, from "What Is Truth?" we know that there can be no finite description of how to generate all true books, for there can be no truth machine.

It is sometimes useful to think of a formal system as a machine, rather than a game. The purpose of a formal system is to generate proofs of theorems, and we can identify a formal system T with a certain machine that prints out the list $T_0, T_1, T_2, \ldots$ of all the theorems proved by the system in question.

One could construct such a machine for P as follows. Build a component that successively prints out the volumes of the Total Library. Build another component that examines each successive volume to see if it is a proof from P. Whenever a proof from P is found, the last formula in the proof is added to the list of theorems. This machine is completely finite, for although there are infinitely many axioms in P, these axioms are schematically described in a finite way.

There are two features that one would like a formal system T to have: completeness and consistency. We say that T is *complete* if for every sentence A in the language of T, either A or $\sim A$ is a theorem proved by T. We say that T is *consistent* if no contradictions are proved by T.

To bring out more vividly the meaning of these two metamathematical concepts, let us think about how these concepts apply to less formal systems. Many novels are primarily collections of statements about some individual. If we take English as the language, and take the usual rules of inference, then a novel can be thought of as a set of axioms concerning some Johnny X. A novel is a *complete* description of Johnny X if it enables one to answer any possible question about him.

Most novels are not complete. Did Raskolnikov get big laughs in Siberia? You'll never find out for sure by reading *Crime and Punishment.* How tall is Vernor Maxwell? No way to know just by reading *Spacetime Donuts.* One might despair of finding or writing a complete novel, but there is one way to do it. Consider a novel whose only sentence is, "Johnny X does not exist at all." In this case we can answer every possible question about Johnny X. Is he five feet nine? No. How many cells are there in his body? None. There are other possibilities of complete descriptions as well. "Johnny X is the three vertex points of an equilateral triangle having no particular space-time-scale location" would seem to serve as a complete description as well.

P (Peano arithmetic) would be a complete theory if every sentence in the language of P could be either proved or disproved by P. As we will see, Gödel proved that P is not complete. Specifically, he proved that there is a sentence having the form $(\forall v_1) \ldots (\forall v_n) [F[v_1, \ldots, v_n] \neq 0]$ that P can neither prove nor refute, F being a polynomial with integer coefficients. This is not entirely a bad thing, for if P were complete, then one would no longer need mathematicians very much.

Why not? If P were complete, then we could build a finite machine that would answer every question we could ask about the natural numbers. If, for instance, we want to know whether Goldbach's conjecture B is true or false, we might build the machine that lists all of P's theorems and wait for either B or $\sim B$ to show up in the list.[3] If P were complete, then one would know with assurance that sooner or later either B or $\sim B$ would be proved by P. But since P is not complete, we cannot rule out the possibility that we might sit watching the printout forever, and that neither B nor $\sim B$ would ever appear as a theorem of P. (A curious sidelight here is that if P cannot prove $\sim B$, then B is in fact true! This follows, since if there *were* an even number not the sum of two primes, then P could exhibit this number, thus proving $\sim B$.) We will have more to say about the significance of the incompleteness of P.

A novel is a *consistent* description of Johnny X if it does not logically lead to a statement of the form A & $\sim A$. A novel containing the sentences "Johnny X was a perfectly normal-looking human being. One morning while he was tying his shoes, he reached under the bed with his third arm and found a glass eye" is not consistent. For insofar as being a normal-looking human being entails having two arms, we can see that this novel leads to the contradiction: "Johnny X has three arms and Johnny X does not have three arms."

Any statement of the form $(A$ & $\sim A) \rightarrow D$ can be seen to be a tautology according to the truth-table definitions of "&," "$\sim$," and "$\rightarrow$." So if a formal system proves a contradiction $(A$ & $\sim A)$, then by Modus Ponens it proves D for any sentence D whatsoever. In other words, if T can prove a contradiction, then the whole system T more or less breaks down and produces "proofs" of every sentence in the language of T. If P were inconsistent, then it would be of no help in deciding if Goldbach's conjecture B is true or not, since P would prove B and P would prove $\sim B$.

The metamathematical statement "P is consistent" is usually abbreviated as "Con(P)", and can be taken to mean "There is no proof from P of the sentence $0 = S0$". (Recall that "$S0$" just means "1".) As mentioned above, Gödel's First Incompleteness Theorem shows that P is not complete. His Second Incompleteness Theorem shows that P cannot prove $\text{Con}(P)$.

It is not immediately evident that $\text{Con}(P)$ is a sentence in the language of P, but we will show in the next section that there is, in fact, a way of coding up formulas by the method called *Gödel numbering*, which enables us to find a sentence in the language of P that expresses the same thing as $\text{Con}(P)$. So although we have every reason to think that P is in fact consistent, it is impossible to prove this fact on the strength of the assumptions that P embodies.

Early in the 1900s a number of paradoxes were found in logic and set theory: the Burali-Forti paradox, the Russell paradox, the Berry paradox, and Richard's paradox. There was a widespread suspicion among mathematicians and philosophers that the use of actual infinites must inevitably lead to contradiction. But nevertheless, mathematical practice seemed to demand the use of such infinite objects as the set of all natural numbers, the set of all real numbers, and the transfinite ordinals.

In this setting, David Hilbert delivered his classic paper of 1925, "On the Infinite." Hilbert was one of the most versatile mathematicians of this century, doing important work in analysis, function theory, number theory, geometry, and the foundations of mathematics. He was far too familiar with the properties of infinite sets to be willing simply to strike out all references to them. Yet he felt that mathematics should somehow be based on completely finite considerations.

Hilbert proposed a *formalist* foundation of mathematics. That is, he suggested that we can view mathematics as the activity of deriving certain strings of symbols from certain other strings of symbols according to certain rules. Thus, although a book on set theory appears to be discussing highly infinite entities, it is possible to maintain that all the book *really* does is to exhibit ways of transforming certain strings of symbols (the axioms of set theory) into certain other strings of symbols (the theorems of set theory.)

The study of how these strings of symbols can be manipulated comprises what Hilbert termed *proof theory*. In order to avoid reintroducing infinities and having to start all over again, Hilbert required that only *finitistic* methods be used in his proof theory (where a method is finitistic only if it involves no infinite searches and can be exhaustively specified in a finite number of words). Hilbert felt that it should be possible to formalize all of mathematics and to find a finitist proof of the consistency of mathematics. This project became known as *Hilbert's Program,* and in 1925 Hilbert felt that the solution was near:

In the present situation, however, this problem of consistency is perfectly amenable to treatment. As we can immediately recognize, it reduces to the question of seeing that '$1 \neq 1$' cannot be obtained as an end formula from our axioms by the rules in force, hence that '$1 \neq 1$' is not a provable formula. And this is a task that fundamentally lies within the province of intuition just as much as does in contentual number theory the task, say, of proving the irrationality of $\sqrt{2}$, that is of proving that it is impossible to find two numerals a and b satisfying the relation $a^2 = 2b^2$, a problem in which it must be shown that it is impossible to exhibit two numerals having a certain property. Correspondingly, the point for us is to show that it is impossible to exhibit a proof of a certain kind.[4]

As mentioned above, we use $\text{Con}(P)$ to stand for the sentence, "There is no proof from P of $0 = S0$." Gödel's Second Incompleteness Theorem shows that $\text{Con}(P)$ *cannot* be proved on the basis of the proof methods embodied in P. This was a real blow for Hilbert's Program, although as Gödel himself pointed out, it is conceivable that there might be finitary proofs of $\text{Con}(P)$ that cannot be represented in P.[5]

Of course, for a Platonist, the consistency of P is evident—one simply notes that all the axioms of P are true when interpreted in the set ω of all natural numbers with the usual definitions of $+$ and $\times$. Now, it is possible to show that the rules of logic preserve truth, so if all the axioms of P are true in the natural numbers, then all the theorems of P must be true in the natural numbers as well. $0 = S0$ is not true of the natural numbers, so it is therefore impossible that the sentence could be a theorem of P. Thus we know that "there is no proof from P of $0 = S0$," is true; so we know $\text{Con}(P)$.

Of course, "seeing" that the Induction Schema is true in the set of natural numbers is anything but a finitary procedure. Around 1940 Gerhart Gentzen was able to establish a result stating that the infinite process just outlined was in some sense equivalent to visualizing all of the ordinals up to the ordinal ϵ_0 that we discussed in "From Omega to Epsilon-Zero."[6]

It is really Gödel's *First* Incompleteness Theorem, however, that deals a death blow to the formalist Hilbert Program. This theorem establishes not only that P is incomplete, but that there is *no* finitely given formalized theory that can correctly answer all questions about the addition and multiplication of natural numbers.

Hilbert was not explicitly opposed to the Platonist view that infinite mathematical objects have a stable existence in the Mindscape. Indeed, he referred to Cantor's class of all sets as a "paradise" from which he refused to be expelled. Nevertheless, Hilbert did believe that talk about actual infinites was a luxury that could be dispensed with. He thought that it was possible to find a complete formal system for mathematics, and that mathematics could then be viewed as a finitary symbol game based on this complete system. And in this he was wrong. No finitely given system can exhaust the riches of the actual infinite. The practicing mathematician's direct intuitions of infinite sets cannot be dispensed with.

For it is by way of this direct perception of the infinite that mathematicians discover new axioms to be added to the old formal system. The working out of the logical consequences of given axioms can be viewed as a wholly finitary process. But deciding which axioms to work out consequences from is a creative, infinitary process that cannot be mechanistically accounted for.

SELF-REFERENCE

Gödel's First Incompleteness Theorem is proved by finding a sentence G_P in the language of P that expresses the metamathematical sentence, "G_P is not provable from P." In other words G is to represent the self-referential sentence, "This sentence is not provable from P." The reader may enjoy already figuring out for himself why G_P must be true but not provable from P.

How are we to express the sentence G in the severely restricted language of P? There are two difficulties. First, we must figure out how to represent the rather complex concept "provable from P" in terms of simple sentences about addition and multiplication. Second, we must find a way of making sentences in the language of P refer to themselves (after all, the locution "this sentence is . . ." is not available in the formal language of P.)

The first difficulty is resolved by means of a trick called *Gödel numbering*. We find a way of assigning a code number to each sentence in the language of P, and then it turns out that the property "n is the Gödel number of a sentence provable from P" is actually representable in the system P. The resolution of the second difficulty is by means of a kind of diagonal argument, which will be explained below.

But first we should describe the Gödel numbering process in some detail. Here, as in everything involving formal systems, it is easy to get bogged down in technical fine points, and the reader who is bored or confused should feel free to skim.

There are infinitely many ways in which one could set up a Gödel numbering. Here we will use a coding like that introduced in "The Library of Babel." We start out by assigning a code number with no zeros in it to each symbol in the language of P.

$\sim$ 1	[11	p 21
$\vee$ 2	] 12	q 22
& 3	S 13	x 23
$\rightarrow$ 4	+ 14	y 24
$\leftrightarrow$ 5	$\times$ 15	z 25
$\exists$ 6	= 16	v_0 26
$\forall$ 7	0 17	v_1 27
(8	m 18	v_2 28
) 9	n 19	: :

Any string A of the basic symbols is coded up by the number $\ulcorner A \urcorner$ obtained by first replacing each symbol by its code number and then separating these symbol code numbers by single zeros. Thus, $\ulcorner SSSO \urcorner$ is 13013013017, since 13 is the code number of S and 17 is the code number of 0. Again, it is not hard to check that the axiom $(\forall x)$ $[x + 0 = x]$ is coded up as 807023090110230140170160230120.

The definitions of the terms, formulas and axioms of P that were given in the last section can all be reformulated so as to apply to the code numbers. For instance, the definition of "term" can be mimicked to define a property $\mathrm{Trm}(x)$ of numbers such that $\mathrm{Trm}(x)$ holds iff x is $\ulcorner t \urcorner$ for some term t. This is done as follows: 1) if $n \geq 17$ and there are no zeros in the decimal expansion of n, then $\mathrm{Trm}(n)$; 2) if $\mathrm{Trm}(n)$ and $\mathrm{Trm}(m)$, then $\mathrm{Trm}(13080n09)$, $\mathrm{Trm}(80n09014080m090)$, and $\mathrm{Trm}(80n09015080m09)$; and 3) $\mathrm{Trm}(x)$ only if x is obtained by a finite sequence of applications of 1) and 2).

In the same vein, we can define a predicate $Fm(x)$ that holds only for the code numbers of meaningful formulae; and we can find a predicate $Ax_P(x)$ such that $Ax_P(x)$ holds if and only if $x = \ulcorner A \urcorner$ for some axiom A of the formal system P.

We can extend the notion of Gödel numbering to *sequences* of sentences in the following way. Say that $A_1, A_2, \ldots, A_n$ is a sequence of sentences in the language of P. We can code up this sequence by first replacing each A_i by its code number $[A_i]$ and then by separating these code numbers by double zeros. Thus, the sequence $A_1, A_2, \ldots, A_n$ is coded up by the natural number $\ulcorner A_1 \urcorner 00 \ulcorner A_2 \urcorner 00 \ldots 00 \ulcorner A_n \urcorner$.

Since one can judge quite mechanically if a given sequence of sentences constitutes a proof from P, there is a predicate $\text{Prov}_P(m,n)$ that holds precisely when m codes up a proof of the sentence coded up by n. At this point we can see how to transform the English sentence, a) "A is provable by the formal system P," into an equivalent statement about natural numbers: b) "$(\exists m)\,[\text{Prov}_P[m, \ulcorner A \urcorner]]$." We can see that a) $\leftrightarrow$ b) as follows. If there is a proof from P of A, then the proof can be coded by some number, so $(\exists m)\,[\text{Prov}_P[m, \ulcorner A \urcorner]]$. Conversely, if $(\exists m)\,[\text{Prov}_P[m, \ulcorner A \urcorner]]$ is true, then there must actually be some natural number M such that $\text{Prov}_P\{M, \ulcorner A \urcorner\}$, and M can be decoded to find a proof of A from P.

Now we wish to take the process a step further. Instead of asking about the *truth* of various number-theoretic statements, we would like just to ask about the derivability of various strings in the formal system P. Keep in mind that although P is *inspired* by the natural numbers, in and of itself P is not really about anything. P is, rather, a set of rules for deriving certain strings of symbols.

It was the hope of Hilbert and other formalists that all reference to "truth" could be replaced by reference to "provability" from P or from some improved formal system T. To a certain extent these hopes are justified.

For instance, not only does $(\forall m)\,(\forall n)\,[m + n = n + m]$ express a true fact about numbers, but it is also a string that is derivable from P. Not only does $2 + 3 = 5$ express a true fact about numbers, but the related string $S^2 0 + S^3 0 = S^5 0$ of P is derivable from P (this string can be written out in full as $SS0 + SSS0 = SSSSS0$). In fact, all of the facts about natural numbers that are familiar to the average person can be converted into strings that can be proved from P. But, as we will see, it is possible with a certain amount of effort to find a string G_P that expresses a true fact about numbers which cannot be proved from P.

I might remark here that even if the formalist program were, *per impossibile,* successful, the slippery notion of "truth" would still not be permanently banished. For to say, instead of "A is true," "A is provable from P," is to say, as noted before, "'$(\exists m)\,[\text{Prov}_P[m, \ulcorner A \urcorner]]$' is true" . . . and we are back where we started. There *would* be some slight gain in this reduction, since to check the truth $(\exists m)\,[\text{Prov}_P[m, \ulcorner A \urcorner]]$ requires only one infinite search through the natural numbers, but still, the dependence on the existence of the set ω as a definite entity has not been avoided.

But let's see how we are to find the sentence G_P that rules out even such a partial reduction in our dependence on the actual infinite. The bulk of Gödel's 1931 paper is devoted to showing that the predicate

$\text{Prov}_P(m, n)$ can be represented in the formal system P. That is to say, Gödel shows indirectly that there must be a formula Prov'_P in the language of P such that for any m and n, $\text{Prov}_P[m, n] \leftrightarrow$ ("$\text{Prov}'_P[S^m0, S^n0]$" is provable from P) and $\sim\text{Prov}_P[m, n] \leftrightarrow$ ("$\sim\text{Prov}'_P[S^m0, S^n0]$" is provable from P). The proof that there is such a formula has to be indirect, for to write Prov'_P out in the limited language of P would take hundreds of pages. Nevertheless, Gödel's proof leaves not the slightest doubt that one could indeed write Prov'_P out, if one really wanted to.

We mentioned before that we can express the English metamathematical statement a) "A is provable from P" by the number-theoretic statement b) "'$(\exists m)[\text{Prov}_P[m, \ulcorner A \urcorner]]$' is true." We know that a) if and only if b). How do these statements relate to c) "'$(\exists m)[\text{Prov}'_P[m, S^{\ulcorner A \urcorner}0]]$' is provable from P"? It is not too hard to see that if a), then c). For if A is provable from P, then there is an M coding up the proof of A, so that $\text{Prov}_P[M, \ulcorner A \urcorner]$. Since Prov'_P represents Prov_P in the sense mentioned above, we know that $\text{Prov}'_P[S^M0, S^{\ulcorner A \urcorner}0]$ is provable from P. Applying an instance of Schema $Q2)$ of the system P and applying the rule of Modus Ponens, we see that '$(\exists m)[\text{Prov}'_P[m, S^{\ulcorner A \urcorner}0]]$' is provable from P as well. Thus we have shown that a) implies c).

The converse implication can conceivably fail. That is, it could happen that P proves $(\exists m)[\text{Prov}'_P[m, S^{\ulcorner A \urcorner}0]]$' even though it also proves '$\sim\text{Prov}'_P[S^M0, S^{\ulcorner A \urcorner}0]$' for each specific natural number M. In this case, P is claiming that there is a proof of A, and at the same time showing that each possible proof of A one considers is no good. There is something wrong with a theory that does such a thing—we will see in the next section that such a theory is called *ω-inconsistent*.

Frequently one can gloss over the distinction between Prov_P and Prov'_P and the distinction between M and S^M0. We will begin to be more casual shortly, but one last formally precise remark should be made. In English we express the consistency of P by, "There is no proof from P of '$0 = S0$'." This can expressed equivalently in number-theoretic terms as "'$\sim(\exists m)[\text{Prov}_P[m, 17016013017]]$' is true." (Recall that $\ulcorner 0 = S0 \urcorner$ is 17018013017.) Corresponding to this number-theoretic sentence, we can form the following string, known as $\text{Con}(P)$, in the language of P: $\sim(\exists m)[\text{Prov}'_P[m, S^{17018030170}0]]$. In the next section we will discover the rather startling result that P *is consistent if and only if P cannot prove* $\text{Con}(P)$!

The result just mentioned is Gödel's Second Incompleteness Theorem, which follows rather easily once his First Incompleteness Theorem is proved. As was promised in the last paragraph, we will no longer maintain the tedious distinctions between Prov'_P and Prov_P, and between M and $S^M 0$, as what we are about to do is already confusing enough as it is.

What we want to do is to find a formula G_P such that $G_P \leftrightarrow \sim (\exists m)$ $[\text{Prov}_P[m, \ulcorner G_P \urcorner]]$ is both true and provable in P. To do this we define a formula $D(n)$ as follows:

$D(n) \leftrightarrow$ (If n is the Gödel number of a formula A with one free variable, then $A[n]$ is not provable from P.) Thus, $D(n)$ holds if *either* n is not the Gödel number of any formula with one free variable *or* if n is $\ulcorner A \urcorner$ for some A with one free variable, and the sentence $A[n]$, obtained by putting the number n in for the variable, is a sentence that is not provable from P. More technically, we could define $D(n) \leftrightarrow \sim (\exists m)$ $[\text{Prov}_P[m, \text{Sub}(n,n)]]$ where $\text{Sub}(n, n)$ is the Gödel number of the sentence, if any, obtained by plugging in n for the free variable in the formula coded up by n.

Now it is possible to represent Prov_P, Sub, and D in the language of P, so there is a number d that is the Gödel number $\ulcorner D(n) \urcorner$ of the formula D that has one free variable n. The sentence G_P that we seek is simply $D[d]$.

Now $D[d]$ says that if $\ulcorner A \urcorner = d$, then $A[d]$ is not provable from P. But, in fact, $\ulcorner D \urcorner = d$, so $D[d]$ says that $D[d]$ is not provable from P. So, G_P asserts its own unprovability from P. If you try to write out G_P in English you get an infinite sentence: "P cannot prove that P cannot prove that P cannot prove that. . . ." If you think of G_P this way it is perfectly clear that $G_P \leftrightarrow (P$ cannot prove that $G_P)$, since all we are doing on the right is putting an extra "P cannot prove" in front of an ω-sequence of them (and recall that $1 + \omega = \omega$).

It was the great achievement of Gödel's 1931 paper to show how to construct such a sentence G_P *wholly within the formal system P*. There are two essential features of P that make this construction possible: i) P is finitely describable, and ii) P is a sufficiently rich theory. It is because P satisfies i) and ii) that the predicate Prov_P can be represented by a formula of P; and it is because of this representation that the predicate D and the sentence G_P can be formed in the language of P.

To elaborate, the import of i) is that it is possible to decide more or less mechanically whether a given sequence of formulas is or is not a valid proof from P. Put differently, the i) says that there is some finitely complex number-theoretic predicate $\text{Prov}_P(m, n)$ that holds if and only if m codes up a proof from P of the sentence coded up by n.

The import of ii) is that the language of P is rich enough, and the axioms of P are strong enough, to ensure that there is a formula Prov$'_P$ that represents Prov$_P$ in the manner mentioned above.

In a nutshell, it is condition i) that enables us to convert the meta-mathematical notion, "provable from P" into a number-theoretic predicate; and it is condition ii) that enables us to represent the truth of this number-theoretic predicate by the provability of a certain string in P.

The whole construction of this section can be carried out for *any* theory T that is i) finitely describable, and ii) as strong as P. So for any such T there will be a formula G_T in the language of T asserting its own unprovability from T in the sense $G_T \leftrightarrow \sim (\exists m) [\text{Prov}_T[m, \ulcorner G_T \urcorner]]$.

GÖDEL'S PROOF

As in the last section, let T be a finitely given theory about some infinitely complex part of the physical or mental universe. We will assume in addition that T is consistent. And again, we let G_T be the formula in the language of T such that $G_T \leftrightarrow \sim (\exists m) [\text{Prov}_T[m, \ulcorner G_T \urcorner]]$.

G_T states that no natural numbers m of a certain kind exist, so it seems legitimate to ask whether G_T is a true or a false statement about natural numbers. Note that G_T is true if it is not provable from T, so it seems that *either* G_T is true and not provable by T, *or* G_T is false and provable by T. Now, if we assume that our theory T *does not prove falsehoods,* then we can already rule out the second option and conclude that it must be that G_T is a true sentence that is not provable by T. As we will see below, this conclusion can be reached under the weaker assumption that T *is consistent.*

In addition to being a true or false statement about natural numbers, G_T can also be represented as a string of symbols in the language of T, and we can ask whether G_T or $\sim G_T$ is provable in the formal system T. Recall from "Formal Systems" that in the case where *both* G_T and $\sim G_T$ are provable from T, we say T is *inconsistent;* and in the case where *neither* G_T nor $\sim G_T$ are provable from T, we say T is *incomplete,* because it does not *decide* G_T (in the sense of proving G_T or proving $\sim G_T$).

There are eight theoretically possible combinations of the truth and the provability of G_T and $\sim G_T$, although only three of these combinations are *actually* possible.

	Both G_T and $\sim G_T$ are provable	Neither G_T nor $\sim G_T$ is provable	G_T alone is provable	$\sim G_T$ alone is provable
G_T is True	IMPOSSIBLE	T is consistent and incomplete	IMPOSSIBLE	T is consistent but ω-inconsistent
G_T is False	T is inconsistent	IMPOSSIBLE	IMPOSSIBLE	IMPOSSIBLE

Why is it completely impossible for G_T alone to be provable? The reason is that if T proves G_T, then T proves $\sim G_T$ as well. For if T proves G_T, then there is a proof from T of G_T, and this proof can be coded up by some number M. Now if M codes up a proof of G_T from T, then $\mathrm{Prov}_T[M, \ulcorner G_T \urcorner]$ is true. Since T obeys conditions i) and ii), Prov_T is represented in T, and since $\mathrm{Prov}_T[M, \ulcorner G_T \urcorner]$ is in fact true, T can prove the string expressing $\mathrm{Prov}_T [M, \ulcorner G_T \urcorner]$ (i.e., P can prove $\mathrm{Prov'}_P[S^M 0, S^{\ulcorner G_P \urcorner} 0]$). Now, we assume that Schema Q2) is part of T, and use it to move from "T proves $\mathrm{Prov}_T[M, \ulcorner G_T \urcorner]$" to "$T$ proves $(\exists m)$ [$\mathrm{Prov}_T[m, \ulcorner G_T \urcorner]]$". But this last phrase says that T proves $\sim G_T$, since $\sim G_T$ is just $(\exists m)$ [$\mathrm{Prov}_T[m, \ulcorner G_T \urcorner]]$.

So now we know that if T proves G_T, then T proves $\sim G_T$. Consider these four equivalent statements of this fact:

1. If T proves G_T, then T proves $\sim G_T$.
2. If T proves G_T, then T is inconsistent.
3. If T is consistent, then T does not prove G_T.
4. If T is consistent, then G_T is true.

Statement 1) implies statement 2), since any theory that proves a statement and its negation is inconsistent. Statement 2) implies statement 1), since an inconsistent theory proves every statement. Statements 2) and 3) are equivalent by contraposition, and statements 3) and 4) are equivalent by the definition of G_T.

We can sum up these facts as follows: *if T is consistent, then G_T is true, but not provable from T.*

In this situation, we can still ask whether $\sim G_T$ is provable from T. If not, then we know that T is incomplete, since it proves neither G_T nor $\sim G_T$ for the particular sentence G_T in the language of T. If T *does* prove $\sim G_T$ then we know that T proves a sentence that is not true . . . so there is something not quite right about T.

Again, we could rule out this possibility by assuming that T does not prove anything that is not true, but "truth" is a very slippery concept. Instead, we can rule out the case where T proves $\sim G_T$ by requiring that T be what is called *ω-consistent*. A theory T is ω-consistent if for no $A(m)$ do we have T claiming that there is an m satisfying A, without T's being actually able to produce one. That is to say, T is ω-consistent if for no $A(m)$ do we have T proving $(\exists m)\,[A(m)]$ *and* proving each of the sentences $\sim A[0]$, $\sim A[1]$, $\sim A[2]$, $\sim A[3]$, . . .

Now, in the case where T is consistent, G_T has no proof from T, so each of the sentences $\sim \mathrm{Prov}_T[0, \ulcorner G_T \urcorner]$, $\sim \mathrm{Prov}_T[1, \ulcorner G_T \urcorner]$, $\sim \mathrm{Prov}_T[2, \ulcorner G_T \urcorner]$, . . . expresses a number-theoretic fact that is provable in T. If T is ω-consistent, then T cannot prove $\sim G_T$ as well, for $\sim G_T$ is the sentence $(\exists m)\,[\mathrm{Prov}[m, \ulcorner G_T \urcorner]]$, each instance of which T has already disproved.

Now we are in a position to state *Gödel's First Incompleteness Theorem:*

If T is a formal system such that

i. T is finitely given,
ii. T extends P,
iii. T is consistent, and
iv. T is ω-consistent;

then T is incomplete.

Let us say a few words about each of the conditions i) through iv) on T. *Condition i)* means, to be precise, that there is a definite algorithmic procedure that can be applied to any number n to determine in a finite amount of time whether or not n codes up an axiom of T. Once we have given the language of T and set up a simple coding such as that used in the last section, we can actually think of T as a set of natural numbers, $\{n: n$ codes up an axiom of $T\}$. Condition i) says that this set is what is commonly called *recursive,* or *computable.*

What would be an example of a theory *not* satisfying condition i)? Suppose that we took Tr to be the set of all the sentences from the language of P that express *true* facts about the natural numbers. Tr is complete because for any given A, either A is true and A is an axiom of Tr and thus provable from Tr, or A is false and $\sim A$ is an axiom of Tr and is thus provable from Tr. So for any A, Tr "proves" (with a proof one sentence long) either A or $\sim A$; and in addition, Tr satisfies conditions ii) - iv). By Gödel's First Incompleteness Theorem we can, therefore, conclude that *the set Tr of all true number-theoretic statements cannot be finitely given.*

Historically, the process of reasoning was just the reverse. That is, Gödel discovered that truth is undefinable, drew the conclusion that there must be a sentence that is true but not provable, and only then set out to construct such a sentence (the G_T we have already seen). I would like to elaborate on this a bit.

Truth is undefinable in the following precise sense: if T is a formal system extending P, then there is no formula $Tru(n)$ in the language of T such that for any sentence A of the language of T, $A \leftrightarrow Tru[\ulcorner A \urcorner]$. This fact is proved by a *reductio ad absurdum* argument. This is, one shows that if the notion "true statement of T" could be expressed in the language of T, then the liar paradox of "What Is Truth?" would arise. For assume that there is a predicate Tru as described above. Then let $E(n) \leftrightarrow \sim Tru[\text{Sub}[n, n]]$ where, as in "Self-Reference," $\text{Sub}[n, n]$ is $\ulcorner A[n] \urcorner$ when $\ulcorner A \urcorner = n$ and A has one free variable. Finally, let $e = \ulcorner E \urcorner$ and form the sentence $E[e]$ of T such that $E[e] \leftrightarrow \sim E[e]$. $E[e]$ is true iff it is false—a contradiction. Therefore we must reject the initial assumption that a truth-definition Tru exists.

Now let Tr be the set of all sentences of T that are *true,* and let Pr be the set of all sentences of T that are *provable from* T. We have just shown that the set Tr cannot be defined by any formula of T, but by the last section we know that the set Pr *can* be defined by a formula of T. In particular, $A \in Pr \leftrightarrow (\exists m) [\text{Prov}_T[m, A]]$. Because of this distinction, we can be sure that $Tr \neq Pr$.

If we assume that *all the axioms and rules of inference of T are true,* then we can conclude that everything provable from T is true, and that $Pr \subseteq Tr$. By the last paragraph we know that this containment is proper, i.e., $Pr \subset Tr$.

This means that there must be some sentence A in the language of T that is true but not provable. Since T does not prove any falsehoods, $\sim A$ cannot be provable either, so A is not decided by T and T is therefore incomplete.

Gödel says that it was this chain of reasoning that led him to discover the First Incompleteness Theorem.[7] This heuristic proof differs from the 1930 version in that a) the heuristic proof depends on the intelligibility of the essentially infinite notion of "truth" in the assumption that the axioms of T are true, whereas the 1930 proof does not use this notion, requiring only that T is consistent and ω-consistent; and b) the heuristic proof shows only that there is *some* A that is undecidable for T, whereas the 1930 proof exhibits the specific undecidable formula G_T.

An example of a theory that *might* not satisfy condition i) is H, the set of all the sentences in Peano arithmetic that human beings will ever learn to be true. Statement ii) certainly holds for H; iii) and probably iv) would seem to hold as well if our methods of reasoning are correct. Thus, it must be that *either* i) fails, and there is no finite description of human mathematical reasoning, *or* H is incomplete, and there are some number-theoretic statements that will never be decided by our mathematics. We will have more to say about H in the next section.

One last remark on condition i) before moving on to conditions ii) - iv). It is not actually necessary to assume that the set of axioms of T is recursive—that is, *computable* in the sense that there is an (idealized) machine that will decide for any given string A whether or not A is an axiom of T. It is enough to assume that the axioms of T are recursively enumerable—that is, *listable* in the sense that there is a machine that will print out all of the axioms of T. The essential thing is that in either case there is a machine that will print out all the theorems of T. This machine can be thought of as working by alternating between two modes. In mode one, the machine prints out the next axiom of T. In mode two, the machine looks over the list of theorems and prints all the statements A for which there is a B such that B and $(B \rightarrow A)$ are in the list.

Condition ii), as it stands, says that the language of T includes all the symbols of the language of P, and that every axiom of P is an axiom or theorem of T. This is obviously true for any of the theories, e.g., analysis or set theory, which set out to encompass large parts of mathematics. But it is nevertheless useful to reduce ii) to three precise subconditions: iia) there must be a sequence $z_0, z_1, z_2, \ldots$ of terms of the language of T such that the relation $k = \ulcorner z_n \urcorner$ is a recursive relation of k and n; iib) there must be a symbol $\sim$ and two symbols x, y in the language of T such that for every recursive relation $R(m, n)$ of two variables there is a formula $R'(x, y)$ of T such that if $R[m, n]$ is true then $R'[z_m, z_n]$ is provable, and if $R[m, n]$ is false, then $\sim R'[z_m, z_n]$ is provable; the formulas of T that are the R' for some recursive relation of one or two variables are called recursive predicates; iic) there must be a symbol $\exists$ such that for any recursive predicate $A(x)$, if $A[z_n]$ is provable from T for any natural number n, then $(\exists x) [A(x)]$ is provable from T as well. In combination with i), iia)-iic) ensure the existence of a formula G_T with the desired properties.

What sorts of theories fail to satisfy ii)? Any theory that has only a finite number of terms in its language fails to satisfy iia). Thus, we can have a complete theory of the mutual relations of the first thousand natural numbers if no numbers greater than one thousand are ever referred to. A theory that has no individual terms at all, only variables, cannot satisfy iia) and iib). Euclidean geometry, for instance, does not speak about any special individual points and lines, but, rather, confines itself to general statements about the existence of points, lines, and circles. Tarski has shown that it is, in fact, possible to extend Euclidean geometry to a complete theory—a theory that proves or disproves every general statement about points, lines, and circles. Again, if DLO is the theory describing the less-than relation, $<$, on the real number line, without mentioning any specific reals, then DLO is complete, in the sense that every statement about the $<$ relation is decided by DLO.

To see an example of a theory that satisfies iia) without satisfying iib), consider the theory P^+. P^+ is the same as P, but with all references to "$\times$" eliminated. It turns out that P^+ is complete in that every statement about the addition of natural numbers can be proved or disproved from P^+. The analogous theory $P^\times$ is also complete. Only when *both* multiplication and addition are present does the theory becomes powerful enough to represent every recursive predicate. After that nothing is gained by extending P to a system that includes exponentiation as a primitive operation as well. Statement iic) simply says that $\exists$ behaves as we expect it to, and is true of any normal theory.

Does Gödel's Incompleteness Theorem have anything to say about our theories of physics? It would seem not, since most theories of physics do not mention infinitely many individuals or qualities that might be used as the z_n. Even if there were a theory U of our universe that singled out an infinite number of particles (or classes of particles or other classes of phenomena) with individual names $z_0, z_1, z_2, \ldots, U$ would still be unlikely to satisfy iib). For there is no reason to suppose that the relations $S(z_k, z_m, z_n)$ and $P(z_k, z_m, z_n)$ that hold if $k = m + n$ and $k = m \times n$, respectively, should be represented in the physical theory U. To give a more concrete example, suppose that U singled out the observer who formulates the theory, and thus the planet Earth; and assume that there are infinitely many planets. We might then call Earth z_0, and for each n let z_n be the term signifying the nth nearest planet to Earth. There is no reason to believe that the set $\{z_n : n$ is prime$\}$ of planets would be singled out by any property of U. Of course, there have been numerologically inclined physicists, such as Kepler and Eddington, and it could be that one of them might construct a physics that deliberately incorporates P. One might also say that insofar as physics includes axioms about measurement, then it must include statements about numbers and again include P. In these cases, Gödel's theorem does apply, but only in a rather uninteresting way.

Condition iii) is quite clear. The minimum requirement here is that for no recursive predicate R' do we have T proving both $R'[z_m, z_n]$ and $\sim R'[z_m, z_n]$. Of course, we never consciously adopt an inconsistent theory since, according to the usual rules of logic, if T proves A and $\sim A$ for some A, then T proves every sentence B in the language of T.

Condition iv) actually can be dispensed with. Given i)-iii), a sentence R_T can be found such that T does not decide R_T (i.e., such that T proves neither R_T nor $\sim R_T$). R_T is constructed more or less like G_T, except that it says "R_T is provable only if there is a shorter proof of $\sim R_T$." Disentangling why T cannot prove R_T or $\sim R_T$ will be left for an exercise.

Now let us say a bit about the sense in which a T satisfying i) - iv) is incomplete. Already in 1930, Gödel was able to show that G_T is equivalent to a certain very simple sort of number-theoretic sentence having to do with the solvability of a certain polynomial over the natural numbers. The recent solution of Hilbert's Tenth Problem[8] has brought the situation to the point where we know, for instance, that there is a polynomial D with integer coefficients, eighty variables, and degree eight such that G_T is equivalent to the statement that the equation $D(x_1, \ldots, x_{80}) = 0$ has no solutions in the natural numbers. This D cannot really be explicitly given, because at least one of the coefficients involved will be thousands of digits long, basically because it must code up a description of the theory T.

Quite recently, work by Harrington, Paris, and Kirby has led to the discovery of a simple and explicit statement Ra about natural numbers that is not provable from P.[9] By thinking about the infinite set ω, one can readily see that Ra is true. Nevertheless, Ra cannot be proved from Peano's axioms for number theory. There is not, however, any obvious method for finding such simple, true, and undecidable sentences for incomplete theories other than P.

Now let us move on to Gödel's other Incompleteness Theorem. In the last section we defined $\text{Con}(T)$ to be the number-theoretic sentence $\sim (\exists m) [\text{Prov}_T[m, \ulcorner 0 = 1 \urcorner]]$.

Göndel's Second Incompleteness Theorem:

If T satisfies conditions i), ii), and iii), then T does not prove $\text{Con}(T)$.

The proof consists, basically, of formalizing within T the argument we gave earlier for the proposition "if T is consistent, then G_T is true." In this way one can show that there is a proof from T of $(\text{Con}(T) \to G_T)$. If

T could prove Con(T) as well, then we could apply Modus Ponens and obtain a proof from T of G_T, which is impossible. Therefore, T cannot prove Con(T).

I should stress that the fact that P cannot prove Con(P) does *not* lead most mathematicians to doubt the consistency of P. What we are encountering here is simply the phenomenon that for any finitely given, sufficiently strong consistent theory T, Con(T) is a true sentence that the machinery of T is not sufficient to arrive at.

Note that Con(T) and G_T both have the form $\sim(\exists m)\,[\mathrm{Prov}_T[m, k]]$ for some number k. As with G_T, Con(T) can be cast in the form of a statement about the non-existence of natural number solutions of a certain polynomial equation. Again, the most complicated thing about this polynomial will be the coefficient or constant term t that codes up a finite description of the theory T. Since the coding process is more explicit than ordinary language, the number t will be rather unwieldy. Still, it is not really unfair to say that an ordinary language description of T constitutes an adequate naming of t, for present-day computer technology is sufficient to build a machine that automatically transforms a description like that given of P in "Formal Systems" into the corresponding code number p. So one is justified in saying that if a) he understands the finite description of a given theory T that is a strong as P, and if b) he knows that T is consistent, then c) he knows true number-theoretic facts (namely, G_T and Con(T)) that T is unable to prove.

A TECHNICAL NOTE ON MAN-MACHINE EQUIVALENCE

Lucas has argued that no machine M can be identical to human mathematical intuition H.[10] For the purpose of this note it will be convenient to let M^* be the set of (Gödel numbers of) theorems listed by the mechanized formal system M. By the same token, H^* is to be the set of (Gödel numbers of) sentences that the human mathematical intuition H will ever be in a position to assert as truths. Lucas claims that for any finitely given M, $M^* \neq H^*$.

His argument can be put as follows: (1) If $M^* \subseteq H^*$, then H can see that M embodies a true formal system. (2) If H knows that M is true, then H knows that M is consistent, and Con(M) $\in H^*$. Gödel's Second Incompleteness Theorem tells us that Con(M) $\notin M^*$. So we can see that if $M^* \subseteq H^*$, then $M^* \neq H^*$. Of course, if $M^* \nsubseteq H^*$, then $M^* \neq H^*$ as well. So no M is equivalent to H.

I would like to formalize the *correct* intuition underlying this *fallacious* argument. It is to be expected that the human mathematical intuition H deals with many extra-mathematical primitives. These primitives function as ideal objects, if you will. I claim that, in particular, H makes use of a certain primitive unary predicate $Tr()$ of natural numbers. $Tr(e)$ is intended to mean that the Turing machine M_e with index e lists a set of (Gödel numbers of) mathematical sentences that are true of the mathematical universe that H believes himself to be thinking of. (By "Turing Machine M_e with index e" I mean a theorem-proving machine M_T, like in "Formal Systems and Machines," whose rules of operation T are coded up by the book B_e, like in "The Library of Babel.")

Using this new predicate, we can formalize the two principles needed for a Lucas-style argument. (1) $M_e^* \subseteq H^* \rightarrow Tr(e) \in H^*$; and (2) $Tr(e) \in H^* \rightarrow \text{Con}(e) \in H^*$. $\text{Con}(e)$ is, of course, the number-theoretic sentence expressing the assertion that M_e formalizes a consistent theory.

The second principle is quite reasonable, and we give it a title: *H-Platonism*) $Tr(e) \in H^* \rightarrow \text{Con}(e) \in H^*$. *H*-Platonism expresses *H*'s belief that his Tr-predicate is based on an objectively existing, and therefore consistently described, universe of mathematical objects.

The first principle is unnecessarily strong. I am willing to grant that all the sentences in H^* are indeed true, and that if $M^* \subset H^*$, then M does indeed list only true theorems. But $Tr(e)$ will actually be in H^* only if H can ever be in a position to see M_e as a whole. And this is possible only if H is capable of naming the large natural number e. So the correct form of (2) is this: $(M_e^* \subseteq H^*$ & e is humanly nameable$) \rightarrow Tr(e) \in H^*$.

In "The Berry Paradox," we discussed the existence of a specific natural number u_H called the human Berry number. u_H is the first number that H cannot find a name for. It may be that there are some humanly nameable numbers scattered here and there beyond u_H. But being less than u_H can serve as a good first approximation to the concept of being humanly nameable.

With this in mind, let us formulate the following version of the first principle necessary for a Lucas-style argument: *H-Consciousness*) $(M_e^* \subseteq H^*$ & $e < u_H) \rightarrow Tr(e) \in H^*$. *H*-Consciousness expresses *H*'s belief that he is not just a formal system, but is, rather, a mathematician discovering mathematical truths.

Insofar as H seems to have evolved as the result of a finitely complex sequence of events, it is not unreasonable to expect that a machine M_h identical to H could also evolve. $H^* = M_h^*$ is compatible with H-Platonism, H-Consciousness and Gödel's Theorem, provided that h is greater than u_H. Presumably h would also satisfy the stronger condition of being humanly unnameable.

What I think I have achieved here is to show that it is possible to formalize and make precise an argument involving such seemingly intractable concepts as "human mathematical intuition," "truth," "the mathematical universe," and "humanly nameable."

ANSWERS TO THE PUZZLES AND PARADOXES

ANSWERS TO CHAPTER ONE

1. No, it is not. We can see this by considering a numerical analogy. Let E be the "universe" of all even numbers. E contains infinitely many numbers, yet it does not contain every possible type of number, to wit, it contains no odd numbers. Although an exhaustive collection of planets would (probably) have to be infinite, an infinite collection of planets need not be exhaustive.

2. The lamp could in fact be either on or off after infinitely many days. Information about its state after any finite number of days is not enough to enable us to extrapolate past infinity. What makes this question interesting is that it is possible give an argument that seems to indicate the lamp will be on, as well as an argument that seems to indicate the lamp will be off. *On*: "The light starts out off, and then we turn it on. Each time we turn it off again, we immediately turn it back on. Therefore it must ultimately be on." *Off*: "Each time we turn the light on, we immediately turn it back off. Therefore it must ultimately be off." This type of lamp is called a Thompson Lamp, and will be discussed again later with reference to the "Grandi Series."

3. No. There is a temptation to say that if everyone thinks the universe is finite, then it really is finite. Indeed, Kant argues something like this in his First Antinomy of Pure Reason. But the argument is fallacious. Just because each natural number is finite does not imply that the set of all natural numbers is finite. Looking at this another way, we can point out that if there are *infinitely many* observers, then the combination of their various finite perceptions can also be infinite.

4. It would mean that you have infinity-plus-one fingers. A curious thing about infinity is that you never count *up* to it. To count an infinite number of fingers is to count through every finite stage, but never to come to any *last* finger. If there is an "∞ finger," then this comes after an infinity of fingers and indicates that you have infinity fingers plus one more.

5. $I + 1$ could not exist, since it would have to be bigger than I, which was assumed to be the largest possible number. But whenever a number x exists, then $x + 1$ exists as well. So if $I + 1$ does not exist then it must be that I does not exist. So we see that we can in fact never find a largest possible number. This is a version of the "Burali-Forti Paradox," which is discussed later. A problem with our conclusion that there can be no largest number is that Ω, the Absolute Infinite, is supposed to be just that: the largest number. One way out of the difficulty is to assert that the largest number Ω exists, but that we can never get our hands on it so as to form $\Omega + 1$.

6. In drawing an arbitrary circle we have three degrees of freedom: the choice of the x-coordinate of the center, the choice of the y-coordinate of the center, and the choice of the radius. So there are ∞^3 circles in the plane. An arbitrary ellipse can be drawn with exactly five degrees of freedom: two for the choice of the center, one for the length of the major axis, one for the length of the minor axis, and one for the angle the major axis makes with the horizontal. So there are ∞^5 ellipses in the plane.

7. Yes, if you say that seven is three plus four—and if you agree that three and four are finite and that the sum of two finite numbers is finite. The kind of argument which is *not* acceptable would be to say: "Seven is the sum of a finite number of ones: $1 + 1 + 1 + 1 + 1 + 1 + 1$." For here you are assuming that the given string of seven ones is finite, and this is just what needs to be proved. By the same token, counting up to seven involves seven steps, and could not be used in a proof of the finiteness of seven. It is abstractly possible to imagine beings that count up to infinite numbers without noticing anything wrong!

8. 6×10^{60} instants so far. Whether one regards as real those numbers which could never physically exist is a debatable topic. My inclination is to say that the world of mathematics exists outside of, and independently of, the physical world and the actions of human beings. The quantum-mechanical lower bound on meaningful time lengths is, by the way, sometimes called a "jiffy," as in, "I'll be back in a jiffy." As mentioned, a jiffy is something like 10^{-44} or 10^{-43} seconds! A discussion of the "jiffy" is found in Paul Davies, *Other Worlds,* (Simon & Schuster, New York 1980).

9. The mathematics of infinity is different from that of ordinary numbers. In a certain sense (the "ordinal" sense) three-times-infinity really is different from infinity. But in the sense intended here (the "cardinal" sense) three-

times-infinity is precisely equal to infinity. Indeed the example given is a proof of this fact.

10. Full equality: "wowowowowo . . . and wowowowowo . . . ," with an infinite number of wo's in each case.

ANSWERS TO CHAPTER TWO

1. Do ω miles in the first half hour and ω miles in the second half hour. To do ω miles in one half hour, do the nth mile in the time interval between $1/2 - 1/2^n$ and $1/2 - 1/2^{n+1}$.

2. $1 + 1/2 + 1/4 + 1/8 + 1/16 + \ldots = 2$

   $1 - 1/2 + 1/4 - 1/8 + 1/16 - \ldots = 1/(1 - 1/2) = 1/(3/2) = 2/3$

   For $a = 1/3$, the total is $2/3$, for $a = 2/3$, the total is 3. For $a = .1$, the total is $10/9$, which can also be written $1.111111. \ldots$ If we substitute $a = 1$ we get $1 + 1 + 1 + 1 + \ldots = 1/0$, which is sometimes called ∞. So this looks fairly reasonable. If we use $a = -1$, we get $1 - 1 + 1 - 1 + 1 - \ldots = 1/2$. This sum has an interesting relation to the Thompson Lamp of Chapter One, Problem 2). We can think of "adding one" as corresponding to turning the lamp on, and "subtracting one" as corresponding to turning the lamp off. Now a running total of 1 arises when the lamp is in fact on, and a running total of 0 arises when the lamp is in fact off. The series for $a = -1$ is called the Grandi Series, and one can argue that it sums to 1, or one can argue that it sums to 0. It is amusing that our formula compromises with a "sum" of $1/2$. If we put values of a below -1 or above $+1$ into our formula, the result is complete nonsense. Thus, if we let $a = 2$, we get the "equation," $1 + 2 + 4 + 8 + 16 + \ldots = -1$.

3. This is a bit tricky. One way to do it is to use the two facts that i) there are infinitely many prime numbers p, (recall that a prime is a number with no divisors other than unity and itself,) and ii) if p and q are distinct primes, then for any natural numbers m and n, p^n and q^m are distinct. Now what we do is put the first ω guests in the ω rooms whose numbers are powers of 2, put the next ω guests in the ω rooms whose numbers are powers of 3, put the next ω guests in the ω rooms whose numbers are powers of 5, . . . , put the nth ω guests in the ω rooms whose numbers are powers of the nth prime, and so on. Note that using this scheme leaves a lot of rooms empty (such as room number 6), but all we wanted was to fit the guests in at all. A mapping that leaves *no* room empty can be constructed by putting the mth person from the nth group of omega people into the room with number $1/2(n^2 + m^2 + 2mn - 3m - n + 2)$.

4. The register at Hilbert's Hotel must have $\aleph_1$ pages. Recall that $\aleph_1$ is the first uncountable ordinal. For any countable ordinal a, it is possible to fit a guests into the hotel's ω rooms. So to be sure that there is always room to sign in, we need a number of pages greater than every countable a, that is, we need $\aleph_1$ pages.

5. The pages of the book would be ordered like the numbers in the set of integers: $\{. . . , -3, -2, -1, 0, 1, 2, 3, . . .\}$. Condition ii) rules out the possibility of having uncountably many pages. Condition iii) indicates that after each page there is a next page (for otherwise one would not be able to count from illustration to illustration). This rules out the possibility of having a dense ordering like, for instance, the set of rational numbers. Condition i) implies that the book has infinitely many pages, and that is infinite in either direction. The set of integers is in fact the only countable, nowhere dense ordering, with no first or last element. In a footnote on p. 58 of his *Labyrinths* (New Directions, New York 1964), Borges describes a book "containing an infinite number of infinitely thin leaves," ordered like the rational numbers. In *White Light,* I describe a book with as many pages as there are real numbers.

6. $\omega \cdot \aleph_1$ is equal to $\aleph_1$, but $\aleph_1 \cdot \omega$ is not. The former statement is true, since any initial segment of an ordering of $\aleph_1$ copies of ω will have the form $\omega \cdot a + n$, for some countable a and some finite n. But such ordinals are countable and thus less than $\aleph_1$. So $\omega \cdot \aleph_1$ reaches as far as $\aleph_1$, but never gets past it. $\aleph_1 \cdot \omega$, on the other hand, is an arrangement of ω copies of $\aleph_1$, and thus reaches well past $\aleph_1$.

7. Under this definition ω is even, as $2 \cdot \omega = \omega$. $\omega + 4$ is also even, as $2 \cdot (\omega + 2) = \omega + 4$. Note, by the way, that the sort of "distributivity" just used does not work if the order is reversed. That is, $(\omega + 2) \cdot 2 = \omega \cdot 2 + 2$. Note also that if we were to change the definition slightly and say that an ordinal number a is even* if for some ordinal b, $b \cdot 2 = a$, then $\omega + \omega$ would be the first transfinite ordinal which is even*. An even number can be gotten by summing up some number b of two's; an even* number can be gotten by summing up two of some number b. The former concept is much more useful, but a confusion between the two concepts led some early thinkers to say that ω is both even and odd.

8. 0, 1, and 2 are the regular finite numbers. This is discussed in the second section of Excursion I.

9. Let e be any infinitesimal quantity, and consider the line with equation $y = 1 - e \cdot x$. This line differs from the horizontal line $y = 1$ in that it drops an infinitesimal amount lower with each unit of motion to the right. If it were to strike the x-axis at some point $(I, 0)$, then we would have $0 = 1 - e \cdot I$, and thus $I = 1/e$. But now, since e is infinitesimal, we have $e < 1/n$ for every natural number n, and this implies that $1/e > n$ for

every natural number n. So if $I = 1/e$, then I is infinite. Thus we see that the line $y = 1 - e \cdot x$ satisfies the conditions of i) passing through the point $(0,1)$, ii) being different from the line $y = 1$, and iii) being parallel to the x-axis in the sense of never intersecting it at any finite point.

10. No. Consider a string of length ω. Just as there is no last natural number, there is no last point on a string of length ω. Alternatively, one might consider a string whose length corresponds to the half-open interval $[0,1) = \{x: 0 \le x < 1\}$. The problem here is that there is no real number (or surreal number) which comes *just before 1*.

ANSWERS TO CHAPTER THREE

1. $^410 = 10^{(10^{(10^{10})})} > 10^{(10^{(10^2)})} = 10^{(10^{(10^{100})})} = 10^{\text{googol}} = \text{googolplex}$.

2. 1 is interesting as it is the first number. 2 is interesting for many reasons, one of which is that 2 is the only number x such that $x + x = x \cdot x = x^x = {}^x x$. 3 is interesting as the only number which is the sum of all the numbers less than itself. 4 is interesting as the first non-trivial perfect square. 5 is interesting as it is the first number that is the sum of an even and an odd (other than one). 6 is the first number that is the sum of its proper divisors. 7 is the first number n so that we cannot construct a regular n-gon with ruler and compass. 8 is the first non-trivial perfect cube. 9 is interesting because when we move from 8 to 9 we go from 2^3 to 3^2. 10 is interesting since it is equal to $1 + 2 + 3 + 4 + 5$. And so on. There is a book by Philip J. Davis, *The Lore of Large Numbers,* (Random House, New York, 1961), which includes a list of properties that make all the numbers up through 100 more or less interesting. You can, for instance, cheer up a friend turning 36 by consulting this list to learn that $36 = 1^3 + 2^3 + 3^3$.

Are *all* the numbers interesting? Assume that there are some uninteresting numbers somewhere up the line. Let U be the first uninteresting number. The *very first* uninteresting number. How strange U is to be the *first* uninteresting number, how interesting a number it is! But now U is both interesting and uninteresting, a contradiction. Therefore we were mistaken to assume that there are any uninteresting numbers.

This argument is very much in the style of the Berry paradox, particularly if we realize that for large numbers, "being interesting" is akin to "having a short description."

Since we do not in fact believe that all numbers are interesting, how are we to avoid the argument given above? There is no easy answer, but one approach might be to say that i) the property of "being interesting" is vague and in fact finitely undefinable, so ii) the property of "being uninteresting" is also finitely undefinable, so iii) we cannot in fact construct the set of all uninteresting numbers in order to find its least member U.

3. The length of the nth hypotenuse is the square root of $n + 1$.

4. Yes. You say, "In exactly one minute I am going to shout one word, either 'YES', or 'NO'." You then look in the book to see what it predicts you will do. The book will predict either i) that you will shout, "YES," or it will predict ii) that you will shout, "NO," shout nothing, or do something different. In case i) you shout, "NO," and in case ii) you shout "YES," thereby disproving the book's prediction.

One could of course argue that you will never get around to carrying out this refutation, or that, even worse, when you attempt it the book correctly predicts that you drop dead before shouting anything! And of course, if you never saw the book at all, you certainly could not refute it.

But the point being argued here is not that there could never be a correct "Book of Your Life," but rather that if you are shown any purported such book you can, in principle, refute it. A dramatization of this appears in Chapter 13 of my *White Light*. There is also a (slightly dishonest) discussion of the problem in Chapter 6 of Alvin I. Goodman, *A Theory of Human Action* (Prentice-Hall, Englewood Cliffs, N.J., 1970). I call Goodman's discussion "slightly dishonest," for in the illustrative story he tells, the protagonist makes only one attempt to refute his Book of Life, and this one try is spoiled by the protagonist's disingenuous misreading of the time for which the prediction is made.

5. Carroll continues as follows, "'You will devour it!' cried the distracted Mother. 'Now,'" said the wily Crocodile, 'I *cannot* restore your Baby: for if I do, I shall make you speak *falsely:* and I warned you that, if you spoke *falsely,* I would *devour* it.' 'On the contrary,' said the yet wilier Mother, 'you cannot *devour* my Baby: for if you do, you will make me speak *truly,* and you promised me that, if I spoke *truly,* you would *restore* it!'" So what happens then? Given that the Crocodile's "sense of honour outweighed his love of Babies," the Crocodile can neither *restore* nor *devour* the Baby. He thus enters a state of indecision similar to a Truth Machine which has been fed a book like in our "What is Truth" section. Perceiving the Crocodile's state of dazed inaction, the courageous Mother slides into the Nile and snatches her Baby from the beast's dreadful Jaws. Whether she makes it back out of the water is, of course, another question.

6. This sentence is equivalent to our sentence A) THIS SENTENCE IS FALSE. For any phrase P, we can imagine forming a longer phrase "P"P by putting P in quotes and then putting P again. If, as Hofstadter suggests, the phrase is HUBBA, then we get "HUBBA" HUBBA. (What a wonderful phrase, by the way, to shout! The first HUBBA is spoken ironically, in quotes, but on the second HUBBA you are ecstatic, carried away, and quoteless!) If the phrase is IS A SENTENCE FRAGMENT, then you get

a true sentence: "IS A SENTENCE FRAGMENT" IS A SENTENCE FRAGMENT. If the phrase is IS A BEAUTIFUL HUMAN BEING, then you get a false sentence: "IS A BEAUTIFUL HUMAN BEING" IS A BEAUTIFUL HUMAN BEING. And if the phrase is APPENDED TO ITS OWN QUOTATION IS FALSE, then you get a sentence Q) "AP-PENDED TO ITS OWN QUOTATION IS FALSE" APPENDED TO ITS OWN QUOTATION IS FALSE, which actually is asserting that itself, Q, is false.

7. There are numerous possibilities. The best I've thought of is: . . . AND ENDED QUITE SUDDENLY AND ENDED QUITE SUDDENLY AND ENDED QUITE SUDDENLY.

8. The best explanation of this regress is Carroll's dialogue itself. Formally, what is going on is that we start with A and with $A \rightarrow C$. Now if we could just convert the logical implication symbol "$\rightarrow$" into the imperative phrase "forces us to conclude that", then we would have C. But if this is blocked by a sceptic, we try to assert it by saying, $(A \ \& \ (A \rightarrow C)) \rightarrow C$. But the same problem arises again. "$\rightarrow$" refers to formal logic, but "forces us to conclude that" refers to human behavior, and there is, in the last analysis, no reason why symbols on paper can force *any* kind of behavior! As the desperate logician tries harder and harder to do this, he is forced ever deeper into a regress, the next step of which would be, $(A \ \& \ (A \rightarrow C) \ \& \ ((A \ \& \ (A \rightarrow C)) \rightarrow C)) \rightarrow C$.

ANSWERS TO CHAPTER FOUR

1. As Smullyan points out, the paradox results because "provable" is not de-finable in any absolute and finite sense. In this respect the paradox is quite similar to the liar paradox, which is escaped by pointing out that "truth" is undefinable. The only precise notion of "provable" which we have is "prov-able by the theory T", where T is some specific, finitely specified formal system. And there is no paradox in the fact that "THIS SENTENCE CAN NEVER BE PROVED BY T" is at the same time not provable within T, yet provable *outside T*, given the additional assumption that T is consistent.

2. This is a difficult question, and is to some extent treated in Douglas Hof-stadter's dialog, "A Conversation with Einstein's Brain," in Douglas Hof-stadter and Daniel Dennett, *The Mind's I*, (Basic Books, New York, 1981). Numerous of the other selections in this collection are, by the way, rele-vant to the questions raised here.

The problem is that one tends to think of the *self* as being something above and beyond one's hardware and software. Yet if the *self* is, as I have argued, nothing more than pure existence, then one is also immortal—for pure existence continues independently of individuals' deaths.

Another objection to the idea of immortality as an abstract software pattern is that a living person changes as time goes on, yet a coded-up simulation would seem to be static, unchanging thing. There is a two-step answer to this. Firstly, one could, in principle, imagine a simulation of a person which continues changing and interacting with a simulation of the real world. Secondly, such a time-dependent pattern is in fact a fixed object in four-dimensional spacetime. This pattern can in turn be coded up as a set. So not only would *you* be coded up in the universe *V* of all sets, but all of the possible continuations of your life would be coded up there as well, even those continuations of your life which extend out to transfinite length.

3. Unless we believe that the soul is an actual physical component of the body (and some people have actually tried to *weigh* souls by setting dying people on delicate scales), then it would seem that an exact replica of yourself is as much "you" as is a "you" resurrected from a hundred years of frozen sleep. The physical continuity of the body is not really so important, in view of the fact that all of our cells are replaced every dozen years or so. But the first question posed is not a matter of an exact replica of you, but rather of an exact replica of one of your states of mind. Would this make a difference to you?

It would certainly be comforting, in advance of death, to reflect on this sort of artistic immortality. And the fact is, that this is one of the only two types of physical immortality we can be certain of, the other being genetic immortality, in the sense of living in one's descendants. Of course, in neither case do you have a ghost of yourself moving around and thinking "I am," but if every "I am" is the same, then what's the difference?

I might add, by way of explaining this line of thought, is that I wrote these questions shortly after the murder of John Lennon. Sitting, mourning him by listening to "Day in the Life," mouthing the words, twisting my face to resemble Lennon's, it occurred to me that in that instant, I *was* John Lennon, as was anyone else who listened to his music with the same attention.

This idea is expressed in Thomas Pynchon's *Gravity's Rainbow,* (Viking Press, New York 1973), p. 516.

"John Dillinger, at the end, found a few seconds' strange mercy in the movie images that hadn't quite yet faded from his eyeballs–Clark Gable going off unregenerate to fry in the chair, voices gentle out of the deathrow steel *so long, Blackie* . . . there was still for the doomed man some shift of personality in effect–the way you've felt for a little while afterward in the real muscles of your face and voice, that you *were* Gable, the ironic eyebrows, the proud, shining, snakelike head–to help Dillinger through the bushwhacking, and a little easier into death."

Who, leaving a favorite movie has not momentarily felt like the star? Perhaps, momentarily, you really were.

4. The state which reoccurs is the playing of the record. Yet in and of itself, this state cannot be called self-reproducing: it is parasitic upon the behavior of a person. We might perhaps compare the record to a virus which reproduces by taking over a cell's genetic material to turn the cell into a "virus factory." Of course, listening to the song does not destroy a person, it only reduces his supply on nickels. The complete system that is reproduced is song plus eager listener.

   This example suggests a related notion: that ideas are independently "alive" patterns who perpetuate themselves by being thought. Zeno's Paradox, for instance, would be viewed as a sort of mind parasite which, by now, you have been infected with.

5. Here are three solutions which I have found. Perhaps the second two can be shortened. COLD, CORD, CARD, WARD, WARM. BEER, BEAR, BEAD, BEND WEND, WIND, WINE. FISH, DISH, DASH, BASH, BASS, BOSS, BOWS, BOWL, FOWL.

6. Let U be a Universal Truth Machine. We wish to use U as a component in a Truth Sorting Machine T_U. The method is this. You put U inside a large box, and let it begin printing out true statements, some of which will be book-length. There is a slot in the box. When someone shoves a book B in the slot, a robot picks it up and begins comparing it to the true books which U has printed out. If U is truly universal, then U will eventually print out as true either B or a book B' which states that B is not true. As soon as one of these events occurs, the watchful robot announces the decision by throwing B out the True door or the Not-True door. Thus the combination of U, box, and robot acts as a Truth Sorting Machine T_U.

   Now let T be a Truth Sorting Machine. We wish to use T as a component in a Universal Truth Machine U_T. The method is this. You put T inside a large box, right next to a machine which mechanically prints out all possible books, one after the other. Note that to print out all possible books is a purely mechanical task: first all the one-letter books, then all the two-letter books, etc. For each of these books, T decides whether or not it is true. The true books are ejected from a slot as print-out. Eventually any true book will be printed deemed true, and issued. Thus the combination of U, box, and book printer acts as a Universal Truth Machine U_T.

   Note, by the way, that if we were concerned not with "truth" but with "provability from P," where P is some fixed consistent formal system, then the equivalence would not hold. That is, if UPM is a machine which lists all theorems of P, and if PSM is a machine which decides for any given sentence whether or not it is a theorem of P, then PSM is essentially *stronger* than UPM. The reason is that there could some sentence S such that S is not a theorem of P, and S' is not a theorem is P either, (where S' would be, as above, a sentence negating S). If we tried to decide whether or not S is a theorem of P by watching the UPM, we might begin to suspect that it is *not* a theorem, but we would never be given the theorem S' which definitely settles the question. This fact is expressed in technical language by saying that the set of theorems of P is "recursively enumerable" (meaning mechanically listable), but not "recursive" (meaning that membership in the set can be mechanically decided.)

7. Computers reproduce with our aid. Even if a particular program has arisen somewhat randomly, we can have it printed out and passed on to some other machine. The basis for selection is simple: we only reuse given hardware and software configurations which work accurately, rapidly and interestingly. The mutation of programs is carried out primarily with humans as agents. Perhaps a given program works, but then someone sees how to improve an algorithm and reduce its running-time. The "evolution" view of computers is in fact so prevalent that it is common for people to speak of "generations" of computers. It is also interesting to note that the day of programs too complex for any individual to understand is already here. Very large programs, such as the one in charge of launching the space-shuttle, have been assembled piecemeal by many people, and are now so large that it is doubtful if any one person knows the whole program. Still, this seems to an incomprehensibility brought on more by sheer bulk and multiplication of special cases than by actual complexity.

8. Begin with this tautologous disjunction: *Either* the human mind is able to solve more problems than any given machine M, *or* there is some particular machine M_H which the human mind is not able to surpass. The second alternative breaks into two subcases: M_H is actually equivalent to the human mind, or M_H proves some things which the human mind cannot prove. In the first subcase, $\text{Con}(M_H)$ is a humanly unsovable number-theoretic question. In the second subcase, we take any of the number-theoretic questions which M_H solves, and humans do not, as an example of something humanly unsolvable. So now we have established Godel's claim.

It is hard to decide just how much content this remark of Gödel's really has. The problem is that "the human mind" is not a very well-defined concept. Is it to mean an average person's mind? Gödel's mind? The collective minds of all who have existed so far? The collective minds of all humans who will ever exist? The collective minds of all humans who *might* ever exist?

Suppose that we imagine that the human race will never die out, and that we take human mathematical knowledge to be the sum total of all the mathematical facts that will ever be known. If we call this sum total H, then we have H arising as the limit of an increasing sequence of sets $H_1, H_2, \ldots,$ $H_n, \ldots$. Perhaps H will be infinite: not just schematically infinite, but infinite in the strong sense of having not finite description. In this case, Gödel's first alternative could perhaps be said to hold. Someone might object that for very large values of n, H_n will in fact be too large for anyone to know. But the knowledge in H_n could be distributed over a very large number of people. Alternatively, we can imagine the science-fictional example of large computers used as cerebral prosthetic devices. (This idea is explored in my novel *Spacetime Donuts,* where certain future thinkers enlarge their mental capacity by plugging their brains in to a giant computer.)

If H were indeed essentially infinite would that make us better than machines? No, it would not. For one can equally well imagine generation after generation of larger and smarter robots evolving to produce a sequence of levels of robot mathematical knowledge: $R_1, R_2, \ldots, R_n, \ldots$ which approaches in the limit some essentially infinite body of knowledge R. The sum-total H of human mathematical knowledge would be greater than the knowledge of any one machine, but it would not necessarily be greater than the sum-total R of all machine knowledge.

ANSWERS TO CHAPTER FIVE

1. It could be said that what the various large things have in common with the form Largeness is that both are instances of still higher-order concept of LARGENESS. This can of course be continued indefinitely, just as with the Bradley-style regress described in the "What is Truth?" section. The source of the difficulty seems to be that any definite, named form is somehow too small to capture the extended and intuitively felt concept which it is to embody. There is something about overarching concepts which makes them resist being treated as manageable and limited things. We have a primitive concept of what it is to be a set, but if we try to freeze this concept into the collection of all sets, we get only some large set which can be transcended. This does not necessarily mean that there is no class of all sets, or that there is no form Largeness. It means only that these classes and forms are essentially beyond rational comprehension. In mathematical philosophy, we describe this situation by saying that such large concepts can be known *intensionally,* but not *extensionally*. To know a collection "intensionally" is to know what is the criterion for membership in it. To know a collection "extensionally" is to be able to visualize it as a completed whole.

2. If all possible universes exist–and in a sense they must, at least as abstract possibilities coded up in V–then it becomes hard to grasp them as a One. But we can try, as John Wheeler does in Misner, Thorne & Wheeler, *Gravitation* (W.H. Freeman, San Francisco 1973). The idea is to postulate a higher-order Superspace in which each of the possible universes is repre-

sented as a point. If the world has alef-null degrees of freedom, then Super-space will be of countably infinite dimension. If the world has continuum many degrees of freedom, then Superspace will be uncountably infinite dimension. The idea is that we think of each of the dimensions as corresponding to one of the choices that must be made in putting together a possible spacetime. But now we have already imagined two different sorts of Super-spaces, and we can go on to think of many other sorts: Absolutely Infinitely many, in fact. There could be Superspaces in which all of the universes fail to have gravity, Superspaces in which each of the universes has time-length of alef-two years, and so on. Each of these Superspaces is now a point in a Supersuperspace which is as big as the class of all sets.

3. $3' = \{\{\},\{\{\}\},\{\{\},\{\{\}\}\}\}$. $4' = \{\{\},\{\{\}\},\{\{\},\{\{\}\}\},\{\{\},\{\{\}\},\{\{\},\{\{\}\}\}\}\}$. Note the similarity of these patterns to the thought-balloon patterns shown in the "Absolute Infinity" section of Chapter One.

4. If a and b are in V_n, then $\{a\}$ and $\{a,b\}$ are subsets of V_n and thus in V_{n+1}. So now $\{\{a\},\{a,b\}\}$ is a subset of V_{n+1} and is thus an element of V_{n+2}. Each of the pairs in the set representing a given rational number lies in the set V_ω. The set of all these pairs is in $V_{\omega+1}$. So we would say the "rational numbers" lie in $V_{\omega+1}$. A set of "rational numbers" is thus a subset of $V_{\omega+1}$, and therefore lies in $V_{\omega+2}$. From the first part of this question we know that if U and L lie in $V_{\omega+2}$, then the ordered pair $<U,L>$ appears two steps higher up—at, that is, $V_{\omega+4}$. So one could say that the standard representations of the real numbers occur first in the set $V_{\omega+4}$.

5. "One of the disciples came out of the ranks, took the stick away from the master, and breaking it in two, exclaimed, 'What is this?' "—D.T. Suzuki, *An Introduction to Zen Buddhims*, (Grove Press, New York 1964), p. 66. You, of course, may find your own way of jumping up into the interface. This is a sort of One/Many problem, for if you single out the given object as "a short staff," you are dividing it from the rest of reality and, in a sense, opposing its essential union with the One. If, on the other hand, you say it is *not* a short staff, then you denying the rational analysis of the world into parts with the Many makes possible.

NOTES

The footnotes are broken up by chapter. Normally I have given complete bibliographical information on a reference only the first time that it occurs. The same information is also to be found in the Bibliography. In referring to classical works existing in many editions, I have tried always to give a chapter and section number, as well as the page number for the specific edition cited.

NOTES ON CHAPTER ONE

1. The work was John Wallis' *Arithmetica Infinitorum* of 1656. See J. F. Scott, *The Mathematical Work of John Wallis* (London: Taylor and Francis, 1938).
2. Blaise Pascal, *Pensées et Opuscules,* Pensée No. 205 (Leon Brunschvicg, ed., Paris: Classiques Hachette, 1961), p. 427.
3. Aristotle, *Physics,* III.7.208a (Richard Hope, trans., University of Nebraska Press, 1961), p. 56.
4. Georg Cantor, *Gesammelte Abhandlungen* (Abraham Fraenkel and Ernst Zermelo, eds., Berlin: Springer Verlag, 1932), p. 404.
5. Plotinus, *Enneads,* V.5.11 (Boston: C. T. Branford, 1949).
6. Saint Augustine, *City of God,* XII. 18 (New York: E. P. Dutton, 1947).
7. Saint Thomas Aquinas, *Summa Theologiae,* Ia, 7, 2–4 (London: Blackfriars, 1944).
8. Galileo Galilei, *Two New Sciences* (Henry Crew and Alfonso De Salvio, trans., New York: Macmillan, 1914), p. 26.

9. *Ibid.*, p. 31.

10. *Ibid.*, p. 32.

11. *The Analyst* is reprinted in Volume III of Alexander Campbell Fraser, ed., *The Works of George Berkeley* (Oxford: Clarendon Press, 1901), p. 1. The "infidel mathematician" is the famous astronomer Edmund Halley. Also reprinted in this volume is Berkeley's curious work, *Siris: A chain of Philosophical Reflexions and Inquiries concerning the virtues of Tar-water, and divers other subjects connected together and arising one from another.* Tar-water is exactly what it sounds like: a tonic produced by mixing water with the tarry resin obtained from pines. Berkeley believed the substance to be a panacea.

12. Benoit Mandelbrot, *Fractals: Form, Chance and Dimension* (San Francisco: W. H. Freeman, 1978), p. 36. This is a fascinating and wide-ranging book. The chief technical tool is a method of assigning a fractional dimension-value to objects such as the Koch curve, which is said to have dimension log 4/log 3, which is approximately 1.26.

13. Georg Cantor, *Gesammelte Abhandlungen,* p. 378.

14. The whole of the section on physical infinities was published previously as the paper, "Physical Infinities," *Speculations in Science and Technology* 1, (April, 1978), pp. 43–58. I am very thankful to William M. Honig, the editor of this journal, for his encouragement.

15. Immanuel Kant, *The Critique of Pure Reason,* First Antinomy (Norman Kemp Smith, trans., New York: St. Martin's Press, 1964), pp. 396–398.

16. Lucretius, *On the Nature of the Universe* (Ronald E. Latham, trans., Harmondsworth, England: Penguin Books, 1951), p. 55.

17. Aristotle, *Physics,* III.8.208a, p. 57.

18. Jorge Luis Borges, *Labyrinths* (New York: New Directions, 1962), pp. 189–192. Borges is the writer of fiction who has written most profoundly about infinity. Most of the relevant stories and essays are in this anthology. Of particular interest is his essay, "Avatars of the Tortoise," pp. 202–208. This essay begins with a paragraph that could serve as the epigraph for *Infinity and the Mind:* "There is a concept which corrupts and upsets all others. I refer not to Evil, whose limited realm is that of ethics, I refer to the infinite."

19. Rudolf v.B. Rucker, *Geometry, Relativity and the Fourth Dimension* (New York: Dover Publications, 1977), p. 39. See also Rudy Rucker, "On Hyperspherical Space and Beyond," *Isaac Asimov's Science Fiction Magazine* (November, 1980), pp. 92–106.

20. Dante Aligheieri, *The Divine Comedy,* Paradisio, Canto 33 (Charles S. Singleton, trans., Princeton University Press, 1975), pp. 380–381.

21. This reading of Dante was first suggested by Andreas Speiser, and is presented in J. J. Callahan's excellent article, "The Curvature of Space in a Finite Universe," *Scientific American* (August, 1976).

22. Giordano Bruno, *On the Infinite Universe and Worlds* (Dorothy Singer, translator, New York: Greenwood Press, 1968).
23. David Hume, *A Treatise of Human Nature* (L. A. Selby-Bigge, ed., Oxford: Clarendon Press, 1896), p. 27.
24. In the summer of 1975 I wrote a science-fiction novel about circular scale. The novel was partially serialized in *Unearth* magazine in 1977 and 1978 and is now published. Rudy Rucker, *Spacetime Donuts* (New York: Ace Books, 1981).
25. Josiah Royce, *The World and the Individual, First Series,* Appendix: The One, the Many and the Infinite (New York: Macmillan, 1912), pp. 504–507.
26. Georg Cantor, *Gesammelte Abhandlungen,* p. 204.
27. *Ibid.,* p. 374.
28. Saint Thomas Aquinas, *Summa Theologiae,* Ia.7.1
29. Quoted in Allen Wolter, "Duns Scotus on the Nature of Man's Knowledge of God," Review of Metaphysics (I.2, 1941), p. 9.
30. This use of Hegel's terminology was suggested to me by William Small, "A Note on Dialectics in Mathematics," Iowa Academy of Science, Vol. 67, 1960, pp. 389–393.
31. Richard Dedekind, *Essays on the Theory of Numbers* (New York: Dover Publications, 1963), p. 64. Originally appeared in 1872.
32. Bernard Bolzano, *Paradoxes of the Infinite* (London: Routledge and Kegan Paul, 1950), pp. 84–85. This book first appeared in 1851, three years after Bolzano's death. Bolzano came close to creating modern set theory, although he used a different definition of transfinite numbers than Cantor.
33. Georg Cantor, *Gesammelte Abhandlungen,* p. 443. This important letter is translated by Stefan Bauer-Mengelberg in: Jean van Heijenoort, ed., *From Frege to Gödel* (Cambridge, Mass., Harvard University Press, 1967), p. 114.
34. Despite what I will argue in this section, it is at least abstractly possible to hold any of eight possible views regarding the existence or nonexistence of the three sorts of infinity (mathematical, physical, and Absolute). It is

	MATHEMATICAL INFINITIES	PHYSICAL INFINITIES	ABSOLUTE INFINITE
Abraham Robinson	No	No	No
Plato	No	Yes	No
Thomas Aquinas	No	No	Yes
L. E. J. Brouwer	No	Yes	Yes
David Hilbert	Yes	No	No
Bertrand Russell	Yes	Yes	No
Kurt Gödel	Yes	No	Yes
Georg Cantor	Yes	Yes	Yes

an interesting pastime to try to find representatives of each of the eight possible standpoints. Here is one I worked out in Heidelberg. I should caution that most of these thinkers have not *explicitly stated* an opinion on all three kinds of infinity, but each of their standpoints is at least consistent with their slot.

35. Georg Cantor, *Gesammelte Abhandlungen,* pp. 378–439.
36. Saint Thomas Aquinas, *Summa Theologiae,* Ia.7.4.
37. David Hilbert and S. Cohn-Vossen, *Geometry and the Imagination* (New York: Chelsea, 1952), p. 159.
38. This example appears in Louis Couturat, *De l'Infini Mathématique,* (Paris: Baillière & Co., 1896), p. 474. This book is primarily a defense of the new Cantorian theory of transfinite numbers. Cantor's theory was, in those years, under heavy attack from a number of philosophers who did not really grasp the mathematical subtleties of the theory. Another strong defense of Cantor is to be found in Constantin Gutberlet, *Das Unendliche, metaphysisch and mathematisch betrachtet* (Mainz: G. Faber, 1878). Gutberlet was a Jesuit priest who corresponded with Cantor about the infinite. This book is remarkable for its number of proofs of God's infinitude.

NOTES ON CHAPTER TWO

1. Aristotle, *Metaphysics,* 985b and 987a, in Richard McKeon, ed., *The Basic Works of Aristotle* (New York: Random House, 1941), pages 698 and 700.
2. *The Thirteen Books of Euclid's Elements,* Vol. 1 (Thomas Heath, ed., New York: Dover Publications, 1956), p. 409.
3. Georg Cantor, *Gesammelte Abhandlungen,* pp. 395–396.
4. Rudy Rucker, *White Light, or, What Is Cantor's Continuum Problem?* (New York: Ace Books, 1980). The subtitle is actually the title of one of Kurt Gödel's papers.
5. Heinz Bachmann, *Transfinite Zahlen* (Heidelberg: Springer-Verlag, 1967), probably has the best and most comprehensive description of different procedures for naming countable ordinals. Georg Cantor, *Contributions to the Founding of the Theory of Transfinite Numbers* (New York: Dover, 1955), is also valuable, as it contains the translation of Cantor's clearest description of his transfinite ordinals, which was originally published in 1897.
6. G. H. Hardy, *Orders of Infinity, the 'Infinitärcalcul' of Paul DuBois-Reymond,* (Cambridge, England: Cambridge University Press, 1910).
7. See Erik Ellentuck, "Gödel's Square Axioms for the Continuum," *Mathematische Annalen* 216 (1975), pp. 29–33.
8. The Polish science fiction writer Stanislaw Lem once wrote a short story

about Hilbert's Hotel. This storylet appears in N. Ya. Vilenkin, *Stories About Sets* (New York: Academic Press, 1968).

9. See, for instance, Bertrand Russell's discussion of Zeno and continuity in his dusty tome, *The Principles of Mathematics* (Cambridge, England: Cambridge University Press, 1903). This book also contains interesting presentations of the immediately post-Cantorian views of infinity and infinitesimals.

10. This is from one of Cantor's letters quoted in Joseph W. Dauben, *Georg Cantor, His Mathematics and Philosophy of the Infinite,* (Cambridge, Mass.: Harvard University Press, 1979), p. 131. Dauben's book contains a deep and balanced account of the interrelationships between Cantor's mathematical and theological thinking. I would, with Dauben, specifically like to warn the reader away from Eric Temple Bell's lurid and inaccurate description of Cantor's life in *Men of Mathematics* (New York: Simon & Schuster, 1937). A different biography, closer in spirit to that of Dauben, is H. Meschkowski, *Probleme des Unendlichen: Werk und Leben Georg Cantors* (Braunschweig: Vieweg-Verlag, 1967).

11. Abraham Robinson, "The Metaphysics of the Calculus," reprinted in J. Hintikka, ed., *The Philosophy of Mathematics* (London: Oxford University Press, 1969), pp. 153 and 163. An expansion of these ideas appears in the last chapter of Robinson's now classic book, *Non-Standard Analysis* (Amsterdam: North-Holland, 1974).

12. Almost no one discusses the philosophy of mathematics without discussing Zeno's paradoxes. A basic overview of the literature can be found in Gregory Vlastos' entry on "Zeno" in *The Encyclopedia of Philosophy* (Paul Edwards, ed., New York: Macmillan, 1967), V.8, pp. 369–378. This beautifully put together encyclopedia is a logical starting point for further research into almost any of the topics I discuss. With regard to Zeno, note also Wesley Salmon, ed., *Zeno's Paradoxes* (New York: Irvington, 1970). This book of essays presents most of the possible views.

13. Quoted in Hao Wang, *From Mathematics to Philosophy* (New York: Humanities Press, 1974), p. 86. Wang spent several years discussing this book with Gödel, and it includes some of Gödel's most mature thought. Rather than adopting doctrinaire positions, Wang lays out with an entomological precision the exact and objective morphologies of key issues in the philosophy of mathematics.

14. See Joseph Dauben, "C. S. Peirce's Philosophy of Infinite Sets," *Mathematics Magazine* 50 (May, 1977), pp. 123–135.

15. This is a modification of Theorem 2 on page 336 of K. Kuratowski and A. Mostowski, *Set Theory* (Amsterdam: North-Holland, 1968). This book also contains a straightforward presentation of Cantor's transfinite numbers.

16. J. H. Conway, *On Numbers and Games* (New York: Academic Press,

1976). A description in dialogue of how to work with Conway's numbers appears in Donald E. Knuth, *Surreal Numbers* (Reading, Mass.: Addison-Wesley, 1974). Martin Gardner also discusses these numbers in the "Nothing" chapter of his *Mathematical Magic Show* (New York: Vintage Books, 1978). The "Everything" chapter of the same book is also to be recommended as well as the "Alef-one" chapter in Gardner's *Mathematical Carnival* (New York: Knopf, 1975).

17. See footnote 11) for Robinson's book.

18. See footnote 11) for Robinson's interesting historical surveys of calculus. Another very good essay on the history of mathematics is Abraham Robinson, "Some Thoughts on the History of Mathematics," *Compositio Mathematica* 20 (1968), pp. 188–193.

19. Charles Howard Hinton, *Speculations on the Fourth Dimension: Selected Writings of C. H. Hinton* (Rudolf v.B. Rucker, ed., New York: Dover Publications, 1980), p. 81. Hinton believed that our space has a slight hyper-thickness, so that the smallest particles in our brains are actually four-dimensional. Therefore, he reasoned, it is in fact possible to form four-dimensional images in one's brain. Adapting this line of thought, one can argue that if matter is infinitely divisible, then we can actually form infinite images in our brains.

20. H. Jerome Keisler, *Elementary Calculus* (Boston: Prindle, Weber & Schmidt, 1976), and Henle and Kleinberg, *Infinitesimal Calculus,* (Cambridge, Mass.: MIT Press, 1978).

21. Rudy Rucker, *White Light,* pp. 71–73.

22. W. Kaufmann, *Cosmic Frontiers of General Relativity* (Boston: Little, Brown, 1977).

23. This idea and many other relevant topics are discussed in José Benardete, *Infinity* (Oxford: Clarendon Press, 1964). Benardete takes great pains to counter the claim that "the universe has alef-one stars," is unverifiable, and therefore (in the eyes of logical positivism), meaningless. He also has some highly original thoughts on Zeno's paradoxes.

24. See Cantor's *Gesammelte Abhandlungen,* pp. 275–277, or my translation of this passage: Rudy Rucker, "One of Georg Cantor's Speculations on Physical Infinities," *Speculations in Science and Technology* (October, 1978), pp. 419–421. This notion of Cantor's is also discussed in my *White Light,* particularly in Chapter 23.

25. Gottfried Leibniz, *The Monadology and Other Philosophical Writings* (Robert Latta, trans., London: Oxford University Press, 1965).

26. Jorge Luis Borges, *The Book of Sand* (New York: E. P. Dutton, 1977). In a footnote to his story "The Library of Babel" (reprinted in *Labyrinths*), Borges describes a different fantastic book, one with infinitely thin pages, densely packed like the rational numbers. Here, each time one thinks one has a page, it turns out, on closer inspection, to be collection of thinner

pages. In *White Light,* I describe a book with c pages and alef-null words per page.

27. This construct is due to Max Dehn, and is described in David Hilbert, *The Foundations of Geometry* (Chicago: Open Court, 1902), p. 129. The curious thing about the Dehn plane is that, although the Fifth Postulate fails, the sum of the angles in any triangle is still 180 degrees. (The proof works only because Dehn takes the non-standard reals for his lines, yet keeps standard reals for the measures of the angles.)

28. Robert Anton Wilson, *Schrödinger's Cat: The Universe Next Door* (New York: Pocket Books, 1980), p. 103.

NOTES ON CHAPTER THREE

1. This section was published previously as the paper, "The Berry Paradox," *Speculations in Science and Technology* (June, 1979), pp. 197–208.

2. Bertrand Russell, "Mathematical Logic is Based on the Theory of Types," *American Journal of Mathematics* 30 (1908), p. 223. In this paper Russell proposes introducing a hierarchy of languages L_n, where "nameable in language L_n" is always a concept expressible in L_{n+1}, but not in L_n. If we think of L_ω as the union of all the L_n, then we can pass on to $L_{\omega+1}$, and so, eventually, through all the ordinals. Gödel has proved that there is a sense in which nothing new is added after the level $\aleph$.

3. Jorge Luis Borges, "Avatars of the Tortoise," in *Labyrinths,* (New York: New Directions, 1962), p. 208.

4. Rene Daumal, *Mount Analogue* (San Francisco: City Lights Books, 1959), p. 63. Actually more recent psychological tests seem to indicate that the average person can hold up to seven things in his field of consciousness. *Mount Analogue* is an interesting, but unfinished, book about a climb up a mysterious and possibly infinite mountain on earth that no one has ever noticed because the space near it is so curved that one tends to see "around" it. This book was one of the inspirations for my *White Light.*

5. *Op. cit.,* p. 64.

6. The essay is reprinted in J. Newman, ed., *The World of Mathematics, Vol. 1* (New York: Simon and Schuster, 1956), pp. 420–429.

7. Strictly speaking, we have only shown that TRANS cannot be described in less than one billion minus K words, where K is the relatively small constant number of words needed to turn a description of TRANS into a description of u_0-TRANS. K would be something like one thousand.

8. The "earlier" and "later" Wittgenstein are Wittgenstein as the author of the earlier *Tractatus Logico-Philosophicus* (London: Routledge and Kegan Paul, 1961) and of the later *Philosophical Investigations* (Oxford: Black-

well, 1953) respectively. The *Tractatus* first appeared in 1921 and was the only work Wittgenstein actually published during his lifetime (1889–1951). A posthumous work of Wittgenstein that is not as polished as the *Investigations* is his *Remarks on the Foundations of Mathematics* (Oxford: Blackwell, 1956). One of the oddest things in the book is the segment (pp. 49–63) in which Wittgenstein makes an attack of sorts on Gödel's Theorem and Cantor's Theorem simply by pretending that (as a finitist) he cannot even understand what these theorems are supposed to mean! Several thoughtful essays on Wittgenstein's *Remarks* appear in Benacerraf and Putnam, eds., *Philosophy of Mathematics: Selected Readings* (Englewood Cliffs, N.J.: Prentice-Hall, 1964).

9. Gregory Chaitin, "Randomness and Mathematical Proof," *Scientific American* (May, 1975), pp. 47–52. See also the article by Chaitin's colleague at the T. J. Watson Research Center of IBM, Charles Bennett, "Mathematical Games," *Scientific American* (November, 1979), pp. 20–34. Both articles are contributions to the new field of "algorithmic information theory."

10. The Berry paradox argument is only needed to prove these results for Turing machines, i.e., computers with potentially infinite memories. A much simpler argument proves the same results for finite-state machines, i.e., computers with a fixed finite memory capacity. See Marvin Minsky, *Computation: Finite and Infinite Machines* (Englewood Cliffs, N.J.: Prentice-Hall, 1967).

11. If we regard the successive digits of a sequence as the numbers that come up on a ten-slotted wheel of fortune, then a statistically random sequence is one for which there is no finitely given "betting strategy" which guarantees winning more than a tenth of the time. See Richard Von Mises, *Probability, Statistics and Truth* (Hilda Geiringer, trans., New York: Macmillan, 1957).

12. Thomas L. Heath, *A History of Greek Mathematics* (Oxford: Clarendon Press, 1921).

13. I can imagine generalizing i) and ii) to n-dimensional space for each n; and I would like to know if nth roots can be geometrically extracted in n-dimensional space if one is allowed the appropriate one through n dimensional forms of i) and ii). The first interesting case would be the extraction of 5th roots by a construction in five-dimensional space.

14. For historical details on these developments, see Moris Kline, *Mathematical Thought from Ancient to Modern Times* (New York: Oxford University Press, 1972), pages 763 and 980. This book is very comprehensive and well organized. See also Howard Eves, *An Introduction to the History of Mathematics* (New York: Holt, Rinehart and Winston, 1964).

15. See G. H. Hardy, *Divergent Series* (Oxford: Clarendon Press, 1949).

16. See José Benardete, *Infinity,* and Wesley Salmon, ed., *Zeno's Paradoxes.*

17. Reprinted in Clifton Fadiman, ed., *Fantasia Mathematica* (New York: Simon and Schuster, 1958), pp. 237–247.

18. Jorge Luis Borges, *Labyrinths*, pp. 51–58.

19. *Ibid.*, p. 54.

20. See "The Continuum," in Excursion I, for an explanation of what c, the cardinality of the continuum, is.

21. I have thought about Richard's paradox for so many years that, I am sorry to say, this particular subsection has become quite technical. Most readers will probably want to skip it, or to save it until after reading "The Continuum" in Excursion I.

22. Jules Richard communicated his paradox to the world by means of a letter, "The Principles of Mathematics and the Problem of Sets," which is reprinted in Jean van Heijenoort, *From Frege to Gödel*, pp. 142–144.

23. I do not know of a paper where Wette makes a completely clear statement of this idea. An example of his work is: Eduard Wette, "Definition eines (relativ vollständigen) formalen Systems konstruktiver Arithmetic," in Jack J. Bulloff, Thomas C. Holyoke, and S. W. Hahn, eds., *Foundations of Mathematics: Symposium Papers Commemorating the Sixtieth Birthday of Kurt Gödel* (New York, Springer-Verlag, 1969).

24. Arthur S. Eddington, *Fundamental Theory* (Cambridge, England: Cambridge University Press, 1946). This posthumously published work was put together from an unfinished manuscript.

25. Recently there has been talk among physicists that protons may be unstable, with an average lifetime of but 10^{30} years each. If this is true, then atoms don't last forever. (The question of whether any physical object has an infinite lifetime is apparently distinct from whether the universe itself has infinite lifetime.) See Steven Weinberg, "The Decay of the Proton," *Scientific American* (June, 1981), pp. 64–75.

26. See, for instance, Paul A. Benioff, "On the Relationship between Mathematical Logic and Quantum Mechanics," Journal of Symbolic Logic 38, p. 547.

27. This idea, due to Hugh Everett, is described in Brian S. DeWitt and Neill Graham, eds., *The Many-Worlds Interpretation of Quantum Mechanics* (Princeton University Press, 1973). See also Chapter Four of Rudolf v.B. Rucker, *Geometry, Relativity and the Fourth Dimension* and, most recently, Douglas R. Hofstadter, "Metamagical Themas," *Scientific American* (July, 1981), pp. 18–30.

28. Herman Melville, *Moby Dick,* Chapter 119 (New York: New American Library, 1961), p. 477.

29. I discovered this in Alan Ross Anderson, "St. Paul's Epistle to Titus," in: Robert L. Martin, ed., *The Paradox of the Liar* (New Haven: Yale University Press, 1970), pp. 1–11.

30. There is an interesting chapter on the traditional role of infinite regresses

in John Passmore, *Philosophical Reasoning* (New York: Basic Books, 1969). See also Borges, "Avatars of the Tortoise," in *Labyrinths,* pp. 202–212.

31. Francis Herbert Bradley, *Appearance and Reality* (New York: Macmillan, 1899). Josiah Royce, *The World and the Individual, First Series,* contains a detailed discussion of Bradley's ideas.

32. Tarski's proof is, in fact, simply a refinement of the proof of Gödel's Incompleteness Theorem of 1930 to be discussed in the next chapter.

33. Sam Loyd, *Mathematical Puzzles of Sam Loyd* (Martin Gardner, ed., New York: Dover, 1959), pp. 116–117.

34. See Lewis Carroll, *Lewis Carroll's Symbolic Logic* (William Bartley, ed., New York: Clarkson Potter, 1977), pp. 425, 426–438. This long book includes Carroll's system for solving "sorites problems." These problems are lists of ten or twenty related statements that the solver is to combine to get a single conclusion. (*Example.* Given: (1) None but goblins are tidy; (2) No fat pets of mine are noisy; (3) All lobsters are pets of mine; and (4) All goblins are fat and noisy. What interesting conclusion can be drawn? No lobsters are tidy!)

35. This statement of the Liar paradox is due to W. V. O. Quine, the eminent contemporary philosopher. An entertaining discussion of this appears in Douglas Hofstadter, *Gödel, Escher, Bach: An Eternal Golden Braid* (New York: Basic Books, 1979), pp. 431–437. Hofstadter's stimulating book discusses many of the topics I treat here. He finds an excellent analogy for the biting-their-own-tail quality of logical paradoxes in Escher's brilliant lithographs of "impossible" scenes.

36. John Barth, "Frame-Tale," in: *Lost in the Funhouse* (New York: Grosset and Dunlap, 1969), pp. 1–2.

37. The model for this paradox is Lewis Carroll's "What the Tortoise said to Archilles," in *Lewis Carroll's Symbolic Logic,* pp. 431–434. Douglas Hofstadter reprints this dialogue in *Gödel, Escher, Bach,* pp. 43–45.

NOTES ON CHAPTER FOUR

1. I have drawn the details of Gödel's life from Georg Kreisel's biographical memoir, "Kurt Gödel, 1906–1978," which is to be published by the Royal Society of London.

2. This quote is taken from John Passmore's article on "Logical Positivism," in *The Encyclopedia of Philosophy,* Vol. 5, pp. 52–57.

3. Ludwig Wittgenstein, *Tractatus,* 7, p. 151.

4. This is from paragraph 6.52 of the *Tractatus,* p. 149. The whole of the book's section 6 is more or less mystical in tone, speaking continually of

things that can be *known* but not *rationally expressed*. A maddening thing about Wittgenstein's influence on modern philosophy is that his soaring section 6 is ignored, and the book's closing, *"Wovon man nicht sprechen kann, darüber muß man schweigen,"* is taken as a valid and permanent injunction against just the sort of mystical ruminations that Wittgenstein himself presents in section 6. Bertrand Russell, in his introduction to the English translation of the *Tractatus,* comments dryly on this split between what Wittgenstein proposes and what he actually does: "After all, Mr. Wittgenstein manages to say a good deal about what cannot be said," p. xxi.

5. Bertrand Russell and Albert North Whitehead, *Principia Mathematica* (New York: Cambridge University Press, 1910–1913).

6. Kurt Gödel, "An Example of a New Type of Cosmological Solution of Einstein's Field of Equations of Gravitation," *Reviews of Modern Physics* 21 (1949), pp. 447–450.

7. Kurt Gödel, *The Consistency of the Continuum Hypothesis* (Princeton University Press, 1940).

8. Both of these essays are reprinted in P. Benacerraf and H. Putnam, eds., *Philosophy of Mathematics* (Englewood Cliffs, N.J.: Prentice-Hall, 1964), on pp. 211–232 and pp. 258–273, respectively. These papers first appeared in 1944 and 1947.

9. *Ibid.,* p. 272.

10. See Stanislaw Ulam, *Adventures of a Mathematician* (New York: Charles Scribner's Sons, 1976). Here Ulam quotes von Neumann as saying, "How can any of us be called professor when Gödel is not?"

11. Kurt Gödel, "A Remark on the Relationship Between Relativity Theory and Idealistic Philosophy," in: Paul Schilpp, ed., *Albert Einstein: Philosopher Scientist, Vol. II* (New York: Harper & Row, 1959), pp. 557–562. A good modern discussion of Gödel's "rotating universe cosmology" can be found in S. W. Hawking and G. F. R. Ellis, *The Large Scale Structure of Space-Time* (Cambridge, England: Cambridge University Press, 1973), pp. 168–170.

12. Kurt Gödel, "Über eine Bisher Noch Nicht Benutzte Erweiterung des Finiten Standpunktes," *Dialectica* 12, (1958), pp. 280–287.

13. A slightly different version of this section is to be published by *Science 81* in April 1982.

14. The speaker was Simon Kochen. See, "In Memoriam Kurt Gödel," *The Mathematical Intelligencer* (July, 1978), pp. 182–185.

15. Franz Kafka, *The Castle* (Willa and Edwin Muir, trans., New York: Knopf, 1976). I spent a lot of time in Heidelberg reading *The Diaries of Franz Kafka* (Max Brod, ed., New York: Schocken Books, 1949), to the point where I wrote a story in his style: Rudy Rucker, "The Fifty-Seventh Franz Kafka," *The Little Magazine* (Summer, 1982).

16. The manuscript, called "Some Considerations Leading to the Probable Conclusion that the True Power of the Continuum is Alef-Two," was given to me by my thesis adviser, Erik Ellentuck. See the reference in footnote 4) of Chapter Two for more information on this manuscript.

17. See note 11) above.

18. For more on time-paradoxes, see Rudolf v.B. Rucker, "Faster than Light, Slower than Time," *Speculations in Science and Technology* 4, (Oct.1981).

19. The idea in question was Circular Scale, as discussed in "Infinities in the Small."

20. This section is based on a talk I gave at the Thomas J. Watson Research Center of the International Business Machines Corporation and was previously published as a paper under the same title in *Speculations in Science and Technology* 3 (June, 1980), pp. 205–217. My thanks to Gregory Chaitin and Charles Bennett for inviting me to speak, and to Cobb Anderson, who shared some of these ideas with me.

21. Quoted in Hao Wang, *From Mathematics to Philosophy* (New York: Humanities Press, 1974), p. 324. Wang took the quotation from the unpublished text of a Josiah Willard Gibbs Lecture delivered by Gödel in Providence, Rhode Island, on December 26, 1951.

22. Many relevant essays appear in Alan R. Anderson, ed., *Minds and Machines* (Englewood Cliffs, N.J.: Prentice-Hall, 1964). See also the annotated bibliography in Howard DeLong, *A Profile of Mathematical Logic* (Reading, Massachusetts: Addison-Wesley, 1971), and above all, Chapter X of Hao Wang's *From Mathematics to Philosophy*.

23. A very interesting attempt at a completely self-explanatory language appears in Hans Freudenthal, *LINCOS: Design of a Language for Cosmic Intercourse* (Amsterdam: North-Holland, 1960).

24. If one objects to the use of the self-referential phrase, "this statement," in G, then G can be rephrased to avoid it. We do this by letting G have a text that runs like this: *"appended to its own quotation yields a statement that the UTM based on (WELL, AXIOM, RULE) never prints out" appended to its own quotation yields a statement that the UTM based on (WELL, AXIOM, RULE) never prints out.* This trick is due to W. V. O. Quine. See note 23, Chapter Three, for a reference.

25. More details on the Gödel proof can be found in Excursion II. For an utterly precise treatment see C. Smorynski, "The Incompleteness Theorems," in J. Barwise, ed., *Handbook of Mathematical Logic* (Amsterdam: North-Holland, 1977), pp. 821–865. One of the first attempts at a semipopular treatment of the Incompleteness Theorems appears in E. Nagel and J. Newman, *Gödel's Proof* (New York: New York University Press, 1958).

26. See the sections "Infinities in the Mindscape" and "Conversations with Gödel" and my paper, "The Actual Infinite," *Speculations in Science and Technology* 3 (April, 1980), pp. 63–76. It should be stressed that many

mathematicians and philosophers question the existence of such a nonrationalizable mathematical intuition. Unless one believes in the existence of this intuition, the Gödel argument sketched in this section will not go through.

27. John von Neumann, *Theory of Self-Reproducing Automata* (Urbana: University of Illinois Press, 1966). Von Neumann also discusses the themes of competition, mutation, and evolution taken up here. In the process of studying these ideas I became so interested in them that I wrote a novel about robot evolution: Rudy Rucker, *Software* (New York: Ace Books, 1982). An excerpt from *Software* appears in: Douglas Hofstadter and Daniel Dennett, *The Mind's I* (New York: Basic Books, 1981). The theme of robot self-reproduction has also been treated in: Edward F. Moore, "Artificial Living Plants," *Scientific American* (October, 1956), pp. 118–126. This article proposes setting self-reproducing robot refineries afloat on the open sea. Periodically the extra refineries would be harvested and used for their raw materials.

28. Georg von Tiesenhausen and Wesley A. Darbro, "Self-Replicating Systems—A Systems Engineering Approach," *NASA Technical Memorandum TM-78304,* (Marshall Space Flight Center, Alabama, 1980).

29. A convincing dramatization of this notion can be found in John Varley, *The Ophiuchi Hotline* (New York: Dell, 1978). Here the heroine's brain patterns are recorded and transferred onto the brain of a new body cloned from her old. A different method of hardware replacement is involved in the science-fictional concept of *matter-transmission*. Here a precise description of a person's body is extracted (destroying the body in the process), coded up, sent via radio (or tachyon) beam, decoded, and used as the blueprint for a new and identical body. See Robert Weingard, "On Travelling Backward in Time," *Synthese* 24 (1972), pp. 117–132.

30. D. T. Suzuki, *The Field of Zen* (New York: Harper & Row, 1970), p. 37. The actual phrasing goes like this: "Question: 'I am told that one reality moistens all beings. What is one reality?' Answer: 'It is raining.'"

31. See Turing's 1936 technical paper, "On Computable Numbers, with an Application to the *Entschiedungsproblem*," reprinted in M. Davis, ed., *The Undecidable* (Hewlett, N.Y.: Raven Press, 1965), pp. 116–151, with special attention to pp. 135–138. See also Turing's brilliant and readable 1950 paper, "Computing Machinery and Intelligence," reprinted in Anderson's *Minds and Machines*. The Davis anthology also contains a translation of Gödel's original paper on the Incompleteness Theorem, as well as many other important papers. Douglas Hofstadter, "Metamagical Themas," *Scientific American* (May, 1981), dramatizes Turing's imitation game in terms of a conversation among three individuals, any one of whom could be male, female, or robot. The reader must guess which is which!

32. This formulation is taken from Wang's *From Philosophy to Mathematics,* p. 326. In this passage Wang reports that Gödel rejected the first assertion as "a prejudice of our time, which will be disproved scientifically (perhaps by the fact that there aren't enough nerve cells to perform the observable operations of the mind)."

33. C. G. Jung, *Synchronicity* (Princeton, New Jersey: Princeton University Press, 1973). See also Jung's foreword to *The I Ching* (Richard Wilhelm and Cary Baynes, trans., Princeton, New Jersey, Princeton University Press, 1967), pp. xxi–xxxix.

34. But see E. Wigner's classic paper, "Remarks on the Mind-Body Question," in I. J. Good, ed., *The Scientist Speculates* (New York: Basic Books, 1962), pp. 284–302. See also the (much less responsible) essays in A. Puharich, ed., *The Iceland Papers* (Amherst, Wisconsin: Essentia Research Associates, 1979).

35. Bernard d'Espagnat, "The Quantum Theory and Reality," *Scientific American* (November, 1979), pp. 158–181.

36. Georg Cantor, *Gesammelte Abhandlungen,* p. 374.

37. Lewis Carroll, *Through the Looking Glass,* Chapter V (New York: Random House, 1946), p. 76.

38. This example is from Raymond Smullyan, *What Is the Name of This Book?* (Englewood Cliffs, New Jersey: Prentice-Hall, 1978), p. 240. This book constitutes perhaps the greatest collection of logic puzzles ever assembled. Enough of Smullyan's perverse, quirky personality comes through in the puzzles to prevent dryness. There is a great picture of Smullyan on the back dust-jacket, in which he looks as if he really *doesn't* know what the name of this book is!

39. Douglas Hofstadter, *Gödel, Escher, Bach,* p. 500.

40. See note 21) for the source of this quote. An ambiguity in this remark of Gödel's arises because of a certain unclarity in the phrase, "the human mind." If we take this to mean the collective minds of all humans who will exist, then it may very well be that this "mind" has no finite description— especially if there are infinitely many generations of people who each have a brain that has experienced certain random mutations.

It can be argued that any single mind can also be made indescribably complex by incorporating random changes. This is the moral of the cartoon in Figure 108. Keep in mind that R is the robot's program, W is Wheelie Willie's program (if he has one), and that for any program T, $Con\ (T)$ is a number-theoretic sentence coding up the assertion that T does not lead to contradictions. Wheelie Willie is asserting here that he has no definite program, as he allows his mind to change randomly under environmental influences. Yet at each instant he *does* have a program, and, what is worse, there is a sort of master program governing the way in which he responds

Figure 108.

to randomization. When the robot seeks to one-up Wheelie Willie by self-randomization it may be that, although it still has a program, the new program is too complex for W. W. to grasp.

NOTES ON CHAPTER FIVE

1. This subsection appeared in my paper, "The One/Many Problem in the Foundations of Set Theory," in *Logic Colloquium 76* (Amsterdam: North-Holland, 1977), the proceedings of a conference held in Oxford, England. I thank Dana Scott for inviting me. I learned of the distinction between monism of kinds and monism of substance in Roland Hall's essay, "Monism and Pluralism," in the *Encyclopedia of Philosophy*.

2. Although Cantor himself hints at the connection with a reference to Plato on page 204 of his *Gesammelte Abhandlungen*, it is the philosopher Josiah Royce who first makes explicit the connection between set theory and the One/Many problem. The reference is Royce's essay, "The One, the Many and the Infinite," which appears as an appendix to his *The World and the Individual, First Series*. The main point of this difficult essay seems to be that an infinite set is a good model of the Absolute in that it is both One and Many . . . One by virtue of its finite definition (e.g., "the natural numbers"), yet Many by virtue of the human inability to grasp every member at once. The idea of "self-representative system" discussed in "Infinities in the Mindscape" stems from this essay of Royce.

3. Arthur Lovejoy, *The Great Chain of Being* (Cambridge, Massachusetts: Harvard University Press, 1953), p. 12.

4. William James, *A Pluralistic Universe* (New York: Longmans, Green & Co., 1909), p. 34. James had a sort of student or friend a man called Benjamin Paul Blood, a gentleman-farmer in Utica, New York. Blood was, to my knowledge, America's first chemical mystic, a turn-of-the-century Tim Leary, if you will. The drug that brought him his revelations was the anaesthetic substance, ether. Blood attracted James's attention with a pamphlet called "The Anaesthetic Revelation and The Gist of Philosophy." A correspondence followed, and as is well-known, James made his own experiments with ether. The really unusual thing about Blood is that, unlike most mystics, he was not a monist. His lifework is a weird and wonderful book called *Pluriverse* (Boston: Marshall Jones, 1920). A good description of the Blood-James relationship is found in Hal Bridges, *American Mysticism: From William James to Zen* (Lakemont, Georgia: CSA Press, 1977).

5. Plato, *The Dialogues of Plato, Vol. 2*, Philebus 15 (B. Jowett, trans., New York: Random House, 1937), pp. 347–348.

6. Georg Cantor, *Gesammelte Abhandlungen*, p. 204: "*Unter einer 'Mannigfal-*

tigkeit' oder 'Menge' verstehe ich nämlich allgemein jedes Viele, welches sich als Eines denken lässt." In 1895, Cantor restated this definition as follows: "By a 'set' we mean any gathering into a whole M of distinct perceptual or mental objects m (which are called the 'elements' of M)" (*ibid.,* p. 282). A very good discussion of these and other definitions of "set" appears in Chapter VI of Wang's *From Mathematics to Philosophy.* The word "set" also has, of course, very many non-mathematical meanings. As a matter of fact, "set" has the longest definition of any word appearing in the *Oxford English Dictionary!*

7. The quote is taken from the translation published as "Letter to Dedekind" in Jean van Heijenoort's anthology, *From Frege to Gödel,* p. 114. This valuable anthology contains some other important material on the orgins of the One/Many problem in set theory, including Cesare Burali-Forti's 1897, "A Question on Transfinite Numbers," pp. 104–112. This paper is the first to point out that the order type Ω of the class of all ordinals is a problematic notion since, on the one hand, Ω should be the largest possible number; but on the other, if we actually have Ω, what is to stop us from forming $\Omega + 1$, showing that Ω was *not* the largest possible number? The only way out of this bind is to assert that the Absolute Infinity Ω exists only as an "inconsistent multiplicity," so that we never really do *have* Ω as a definite, conceivable number. (But what, then, am I talking about when I say, "Ω"?)

Another important paper in the van Heijenoort anthology is Russell's 1902 "Letter to Frege," pp. 124–125. Frege was a pre-Cantorian set theorist of logician who constructed a foundation of mathematics based on the assumption that for any property P one could form a definite set, called $\hat{x}P$, of all the objects with property P. Thus, for instance, the number "two" was defined as "$\hat{x}$ (there are distinct objects y and z such that x is the set with exactly y and z as members)." Now, in 1902, Russell discovered that such an unlimited set formation principle leads to the "Russell paradox" of the set R of all sets that are not members of themselves $(R = \hat{x}(x \notin x))$, with R being a member of itself if and only if it is not a member of itself. He wrote this to Frege, who was on the point of publishing the second of his two volumes of *Grundgesetze der Arithmetik.*

Frege's response to Russell also appears in the van Heijenoort anthology (pp. 126–129), and is worth quoting as an example of fine scientific detachment. One should keep in mind that the Russell paradox, coming when it did, essentially destroyed a good part of Frege's lifework. Responding in 1902, Frege wrote:

Your discovery of the contraction caused me the greatest surprise and, I would almost say, consternation, since it has shaken the basis on which I intended to build arithmetic. . . . It is all the more serious since not only the foundations of my arithmetic, but also the sole possible foundations of arithmetic, seem to vanish. . . . In any case, your discovery is very remarkable and will perhaps result in a great advance in logic, unwelcome as it may seem at first glance. . . . The second volume of my *Grundgesetze* is to appear shortly. I shall no doubt have to add an appendix in which your discovery is taken into account. If only I already had the right point of view for that!

8. For an extended and lucid exposition of intuitionism, see Michael Dummett, *Elements of Intuitionism* (Oxford: Clarendon Press, 1977). Some further comments of mine on the old formalism-intuitionism-Platonism puppet-show can be found in "The Actual Infinite," *Speculations in Science and Technology* 3 (April, 1980), pp. 63–76.

9. See the essays "Why," by Paul Edwards, and "Nothing," by P. L. Heath in the *Encyclopedia of Philosophy,* V.8, pp. 296–302, and V.5, pp. 524–525, respectively.

10. Is it truly legitimate to regard the various individual objects in the world as sets? Certain monists of substance would say not, arguing that in order to express fully all the aspects of any given thing, it is necessary to bring in everything, so that no indivdual thing would embody a *conceivable* form after all.

There are two other weak points in the view that everything is a set. First of all, the experienced fact that things are one way and not another, that I am myself and the world is this world—this sort of particularlity does not seem to be provided for by saying that I am a certain point in a certain complex relational system. That is, there does not seem to be any way to represent set-theoretically the fact that it is *this* world that really exists. This objection could be countered with the claim that *every* possible world really exists.

A second, related, objection to the view that everything is a set is that the set-theoretic model does not seem to account for the fact that the world is *going on.* John Wheeler speaks of this difficulty as it relates to an imagined room full of equations intended to represent the physics of the universe: "Stand up, look back on all those equations, some perhaps more hopeful than others, raise one's finger commandingly, and give the order 'Fly!' Not one of those equations will put on wings, take off, or fly. Yet the universe 'flies.'" [Misner, Thorne & Wheeler, *Gravitation* (San Francisco: W. H. Freeman, 1973), p. 1208.] This objection could perhaps be met by the assertion that there is nothing more to the "life" of the world than the various forms and formations that occur.

11. This would be a person of strong formalist learnings who denies the objective existence of sets and identifies them simply with states of the human brain. He could say that "thinking of such-and-such a set" is simply a certain finite neuronal pattern that occurs occasionally in the physical

world. But, and this is the weakness in formalism, he would be unable to explain why our discussions of sets seem *meaningful,* and why certain facts about sets force themselves upon us as being *true.* The argument against formalism is akin to the argument against solipsism: If everything is just a dream of mine, then why do things exhibit such an obstinate lack of concern for my desires and preconceptions?

The best spokesman for formalism in recent times was Abraham Robinson: see his "Formalism 64," in Y. Bar-Hillel, ed., *Logic, Methodology and Philosophy of Science* (Amsterdam: North-Holland, 1964), pp. 228–246. See also Paul J. Cohen, "Comments on the Foundations of Set Theory," in Dana S. Scott, ed., *Axiomatic Set Theory, Proceedings of Symposia in Pure Mathematics, XIII, Part 1* (Providence, Rhode Island: American Mathematical Society, 1971), pp. 9–15.

Generally, any thoughtfully written discussion of these foundational questions is of value. Happily, the time when philosophers of mathematics would become bitter and angry over the issues seems to be past. Of course, *Infinity and the Mind* could change that!

12. The quote appears in Fr. Allen Wolter, "Duns Scotus on the nature of man's knowledge of God," *Review of Metaphysics,* I: 2 (1941), p. 9.

13. This is a translation from the last page of Ernst Zermelo, "Über Grenzzahlen und Mengenbereiche," *Fundamenta Mathematica* 16 (1930), pp. 29–47.

14. Section 6.522 of Wittgenstein's *Tractatus,* pp. 149–151.

15. Rudolf Otto, *Mysticism East and West* (New York: Macmillan, 1960), pp. 57–72. The book first appeared in 1932. It is primarily a comparison between the thinking of the thirteenth-century German priest, Meister Eckhart, and the ninth-century Indian teacher, Sankara. The great Zen master Daisetz Teitaro Suzuki also discusses Eckhart in a book called *Mysticism: Christian and Buddhist* (Westport, Connecticut: Greenwood, 1976).

16. Rudy Rucker, *Spacetime Donuts* (New York: Ace Books, 1981), Chapter 5.

17. Erwin Schrödinger, *What is Life? & Mind and Matter* (Cambridge, England: Cambridge University Press, 1969), p. 93. The essay was first published in 1944. Schrödinger is best known for his work in quantum mechanics.

18. D. T. Suzuki, *The Field of Zen* (New York: Harper & Row, 1970), pp. 21–27.

19. *Ibid.,* p. 22.

20. *Ibid.,* p. 24.

21. See Note 4).

22. Benjamin Paul Blood, "The Anaesthetic Revelation and the Gist of Philosophy" (Privately printed in Amsterdam, New York, 1874), p. 34. Most libraries will have a microfilm of the pamphlet in their Americana collec-

tions. This quote impresses me so much that I also use it in *White Light* (p. 221) to separate a scene in which the hero merges into the White Light (i.e., the Absolute) from the next scene, where he is back to his ordinary life in a small upstate New York town. A quote similar to Blood's but made by Xenas Clark can be found in William James, *The Varieties of Religious Experience* (New York: Macmillan, 1961), pp. 306–307.

23. *Op. cit.,* p. 23.

24. It is interesting to note that several sorts of One-Many problems arise in quantum mechanics. One of the most difficult philosophical issues there is to distinguish between an observer and the system he observes. By way of pointing out this difficulty, Niels Bohr gives the example of someone who picks up a walking stick in a dark room. At first the stick feels like something outside of you—part of the system. But if you then begin feeling around the room by moving the stick, you begin to think of the stick as an extension of your arm—part of the observer. The quotes are from Niels Bohr, *Atomic Theory and the Description of Nature* (Cambridge, England: Cambridge University Press, 1934), pp. 56–91.

The notorious quantum-mechanical "collapse of the wave-function" can be viewed as a passage from the One of a causally developing state in superspace, to a choice between Many different strands of definite realities. I wrote a story about this: Rudy Rucker, "Schrödinger's Cat," *Analog* (March 30, 1981), pp. 70–84.

25. Plato, *The Dialogues of Plato, Vol. 2,* Parmenides 132, pp. 92–93. See also Chapter Two of John Passmore, *Philosophical Reasoning* (New York: Basic Books, 1969) and the discussion of Aristotle's "Third Man" argument in "Avatars of the Tortoise," in Jorge Luis Borges, *Labyrinths* (New York: New Directions, 1962).

26. Gyomay Kubose, *Zen Koans* (Chicago: Henry Regnery, 1973), p. 5.

NOTES ON EXCURSION I

1. L. E. J. Brouwer, *Collected Works* (Amsterdam: North-Holland, 1975), p. 133.

2. This way of illustrating the proof is taken from Alexander Abian, *The Theory of Sets and Transfinite Arithmetic* (Philadelphia: Saunders, 1965).

3. The letter is printed in the appendix to Herbert Meschkowski, *Probleme des Unendlichen: Werk und Leben Georg Cantors* (Braunschweig: Vieweg, 1967).

4. This comment appears in Cantor's "Mitteilungen zur Lehre vom Transfiniten," reprinted in his *Gesammelte Abhandlung,* pp. 378–439.

5. Richard Schlegel, *Completeness in Science* (New York: Appleton-Century-Crofts, 1967), p. 223.

6. The letter appears in *Briefwechsel Cantor-Dedekind* (Paris: Hermann, 1937), edited by E. Noether and J. Cavailles. I was one of the only people who noticed or observed what one might call the centennial of set theory on December 7, 1973. See my letter in the *Notices of the American Mathematical Society* 20 (November, 73), p. 362.

7. David Anthony Martin, "Hilbert's First Problem: The Continuum Hypothesis," in F. Browder, ed., *Proceedings of Symposia in Pure Mathematics XXVIII* (Providence, Rhode Island: American Mathematical Society, 1976), pp. 81–92.

8. *Gesammelte Abhandlungen,* p. 192.

9. Kurt Gödel, *The Consistency of the Continuum Hypothesis* (Princeton, New Jersey: Princeton University Press, 1940).

10. Paul J. Cohen, *Set Theory and the Continuum Hypothesis* (New York: Benjamin, 1966).

11. Gaisi Takeuti, "Gödel Numbers of Product Space," in Gert H. Müller and Dana S. Scott, eds., *Higher Set Theory* (Heidelberg: Springer Lecture Notes #669, 1978).

12. Waclaw Sierpinski, *Hypothèse du Continu,* (Warsaw, Poland: Monografie matematyczne, 1934).

13. See Benacerraff and Putnam, eds., *Philosophy of Mathematics,* p. 267.

14. Rudy Rucker, "On Cantor's Continuum Problem," *Journal of Symbolic Logic 41,* p. 551. This is not a full-length paper; only a short research note. Another description of my ideas on the continuum problem can be found in *White Light,* pp. 34–36.

15. Frank Drake, *Set Theory: An Introduction to Large Cardinals* (Amsterdam: North-Holland, 1974).

16. The "nodal class" terminology is from Gaisi Takeuti, "The Universe of Set Theory," in Bulloff, Holyoke and Hahn, eds., *Foundations of Mathematics* (Springer-Verlag, 1969), pp. 74–128.

17. See, in particular, William Reinhardt, "Remarks on Reflection Principles, Large Cardinals, and Elementary Embeddings," in: Thomas Jech, ed., *Axiomatic Set Theory, Proceedings of Symposia in Pure Mathematics XIII, Part 2* (Providence, Rhode Island: American Mathematical Society, 1974), pp. 189–205. See also Hao Wang, "Large Sets," in Butts and Hintikka, eds., *Logic, Foundations of Mathematics and Computability Theory* (Dordrecht, Holland: Riedel, 1977), p. 309.

NOTES ON EXCURSION II

1. There are many books that describe the building up of formal logical systems. A very good semipopular presentation can be found in Howard De-Long, *A Profile of Mathematical Logic* (Reading, Massachusetts: Addison

Wesley, 1971). A somewhat more technical treatment is found in Joseph R. Shoenfield, *Mathematical Logic* (Reading, Massachusetts: Addison-Wesley, 1967).

2. Ernest Nagel and James R. Newman, *Gödel's Proof* (New York: New York University Press, 1958), 34–35. This little book, expanded from a 1956 article in *Scientific American,* was for many years the only nontechnical account of Gödel's Incompleteness Theorem. The book by DeLong mentioned above is now to be preferred, as the book by Nagel and Newman is in some ways too naive. A detailed and readable account of the Gödel results is also to be found in Douglas Hofstadter's *Gödel, Escher, Bach.*

3. Goldbach's conjecture is the statement that every even number greater than two is the sum of two prime numbers. (Technically, 1 is not thought of as prime, but 2 is, as it has exactly two divisors.)

4. Reprinted in Jean von Heijenoort, *From Frege to Gödel,* p. 383.

5. Kurt Gödel, "On Formally Undecidable Propositions of Principia Mathematica and Related Systems," in: Martin Davis, ed., *The Undecidable,* p. 37. This is the text of Gödel's original 1931 paper.

6. There was no popular accounts of Gentzen's work. A technical account with references is Gaisi Takeuti, *Proof Theory* (Amsterdam: North-Holland, 1975).

7. See Hao Wang, *From Mathematics to Philosophy,* p. 9, and Gödel's "On Undecidable Propositions of Formal Mathematical Systems," pp. 63–65, as reprinted in Martin Davis, ed., *The Undecidable.*

8. Martin Davis, Yu. Matijacevic, and Julia Robinson, "Hilbert's Tenth Problem. Diophantine Equations: Positive Aspects of a Negative Solution," in F. Browder, ed., *Proceedings of Symposia in Pure Mathematics XXVIII* (Providence, Rhode Island: American Mathematical Society, 1976), pp. 223–378.

9. Jeff Paris and Leo Harrington, "A Mathematical Incompleteness in Peano Arithmetic," in Jon Barwise, ed., *A Handbook of Mathematical Logic* (Amsterdam: North-Holland, 1977), pp. 1133–1142.

10. J. R. Lucas, *The Freedom of the Will* (Oxford: Clarendon Press, 1970). A good listing of articles about the Lucas argument can be found in the annotated bibliography of DeLong's *Profile of Mathematical Logic.*

BIBLIOGRAPHY

Abian, Alexander: The Theory of Sets and Transfinite Arithmetic. Philadelphia: Saunders 1965

Alighieri, Dante: The Divine Comedy, Paradisio, Canto 33. Translated by Charles S. Singleton, Princeton N.J.: Princeton University Press 1975

Anderson, Alan R. (ed.): Minds and Machines. Englewood Cliffs, N. J.: Prentice-Hall 1964

Anderson, Alan Ross: St. Paul's Epistle to Titus. In: The Paradox of the Liar, ed. R. L. Martin, pp. 1–11. New Haven: Yale University Press 1970

Aquinas, Saint Thomas: Summa Theologiae. London: Blackfriars 1944

Aristotle: Metaphysics. In: The Basic Works of Aristotle, ed. R. McKeon. New York: Random House 1941

Aristotle: Physics. Translated by Richard Hope. Lincoln, Nebr.: University of Nebraska Press 1961

Augustine, Saint: City of God. New York: E. P. Dutton 1947

Bachmann, Heinz: Transfinite Zahlen. Heidelberg: Springer-Verlag 1967

Barth, John: Frame-Tale. In: Lost in the Funhouse. New York: Grosset and Dunlap 1969

Bartley, Willaim (ed.): Lewis Carroll's Symbolic Logic. New York: Clarkson Potter 1977

Bell, Eric Temple: Men of Mathematics. New York: Simon & Schuster 1937

Benacerraf, P. and Putnam, H. (eds.): Philosophy of Mathematics. Englewood Cliffs, N. J.: Prentice-Hall 1964

Benardete, José: Infinity. Oxford: Clarendon Press 1964

Benioff, Paul A.: On the Relationship between Mathematical Logic and Quantum Mechanics. Journal of Symbolic Logic. 38, p. 547

Bennett, Charles: Mathematical Games. Scientific American. 20–34 (November, 1979)

Berkeley, George: Siris: A Chain of Philosophical Reflections and Inquiries Concerning the Virtues of Tar-Water, and Divers Other Subjects Connected Together and Arising One from Another. In: The Works of George Berkeley, Vol. III, ed. A. C. Fraser. Oxford: Clarendon Press 1901

Blood, Benjamin Paul: The Anaesthetic Revelation and the Gist of Philosophy. Amsterdam, N. Y.: Privately printed 1874

Blood, Paul: Pluriverse. Boston, Mass.: Marshall Jones 1920

Bohr, Niels: Atomic Theory and the Description of Nature. Cambridge, England: Cambridge University Press 1934

Bolzano, Bertrand: Paradoxes of the Infinite. London: Routledge and Kegan Paul 1950

Borges, Jorge Luis: Labyrinths. New York: New Directions 1962

Borges, Jorge Luis: The Book of Sand. New York: E. P. Dutton 1977

Bradley, Francis Herbert: Appearance and Reality. New York: Macmillan 1899

Bridges, Hal: Ameican Mysticism: From William James to Zen. Lakemont, Ga.: CSA Press 1977

Brouwer, L. E. J.: Collected Works. Amsterdam: North-Holland 1975

Bruno, Giordano: On the Infinite Universe and Worlds. Translated by Dorothy Singer, New York: Greenwood Press 1968

Bulari-Forti, Cesare: A Question of Transfinite Numbers. In: From Frege to Gödel, ed. Jean van Heijenoort, pp. 104–112. Cambridge, Mass.: Harvard University Press 1967

Callahan, J. J.: The Curvature of Space in a Finite Universe. Scientific American. (August, 1976)

Cantor, Georg: Gesammelte Abhandlungen, eds. A. Fraenkel and E. Zermelo. Berlin: Springer-Verlag 1932

Cantor, George: Contributions to the Founding of the Theory of Transfinite Numbers, ed. P. Jourdain. New York: Dover 1955

Cantor, Georg: Letter to Dedekind. In: From Frege to Gödel, ed. J. van Heijenoort, pp. 113–117. Cambridge, Mass.: Harvard University Press 1967

Cantor, George and Dedekind, Richard: Briefwechsel Cantor-Dedekind. Noether, E. and Cavailles, J. (eds): Paris: Hermann 1937

Carroll, Lewis: Through the Looking Glass. New York: Random House 1946

Carroll, Lewis: What the Tortoise Said to Archilles. In: Lewis Carroll's Symbolic Logic, ed. W. Bartley, pp. 431–434. New York: Clarkson Potter 1977

Chaitin, Gregory: Randomness and Mathematical Proof. Scientific American. 47–52 (May, 1975)

Cohen, Paul J.: Set Theory and the Continuum Hypothesis. New York: Benjamin 1966

Cohen, Paul J.: Comments on the Foundations of Set Theory. In: Axiomatic Set Theory, Proceedings of Symposia in Pure Mathematics, XIII, Part 1, ed. D. S. Scott, pp. 9–15. Providence, R. I.: American Mathematical Society 1971

Conway, J. H.: On Numbers and Games. New York: Academic Press 1976

Couturat, Louis: De l'Infini Mathématique. Paris: Baillière & Co. 1896

Dauben, Joseph: C. S. Peirce's Philosophy of Infinite Sets. Mathematics Magazine. 50, 123–135 (May 1977)

Dauben, Joseph W.: Georg Cantor, His Mathematics and Philosophy of the Infinite. Cambridge, Mass.: Harvard University Press 1979

Daumal, Rene: Mount Analogue. San Francisco: City Lights Books 1959

Davies, Paul: Other Worlds. New York: Simon & Schuster 1980

Davis, Martin, Matijacevic, Yu. and Robinson, Julia: Hilbert's Tenth Problem. Diophantine Equations: Positive Aspects of a Negative Solution. In: Proceedings of Symposia in Pure Mathematics XXVIII, ed. F. Browder, pp. 223–378. Providence, R. I.: American Mathematical Society 1976

Davis, Philip J.: The Lore of Large Numbers. New York: Random House 1961

Dedekind, Richard: Essays on the Theory of Numbers. New York: Dover Publications 1963

DeLong, Howard: A Profile of Mathematical Logic. Reading, Mass.: Addison-Wesley 1971

d'Espagnat, Bernard: The Quantum Theory and Reality. Scientific American. 158–181 (November, 1979)

DeWitt, Brian S. and Graham, Neill (eds.): The Many-Worlds Interpretation of Quantum Mechanics. Princeton, N. J.: Princeton University Press 1973

Drake, Frank: Set Theory: An Introduction to Large Cardinals. Amsterdam: North-Holland 1974

Dummett, Michael: Elements of Intuitionism. Oxford: Clarendon Press 1977

Eddington, Arthur S.: Fundamental Theory. Cambridge, England: Cambridge University Press 1946

Edwards, Paul: Why? In: The Encyclopedia of Philosophy, Vol. 8, ed. P. Edwards, pp. 296–302. New York: Macmillan 1967

Ellentuck, Erik: Gödel's Square Axioms for the Continuum. Mathematische Annalen. 216, 29–33 (1975)

Eves, Howard: An Introduction to the History of Mathematics. New York: Holt, Rinehart and Winston 1964

Fadiman, Clifton (ed.): Fantasia Mathematica. New York: Simon & Schuster 1958

Freudenthal, Hans: LINCOS: Design of a Language for Cosmic Intercourse. Amsterdam: North-Holland 1960

Galilei, Galileo: Two New Sciences. Translated by Henry Crew and Alfonso De Salvio. New York: Macmillan 1914

Gardner, Martin (ed.): Mathematical Puzzles of Sam Loyd, New York: Dover 1959

Gardner, Martin: Mathematical Carnival. New York Knopf 1975

Gardner, Martin: Mathematical Magic Show. New York: Vintage Books 1978

Gödel, Kurt: The Consistency of the Continuum Hypothesis. Princeton, N. J.: Princeton University Press 1940

Gödel, Kurt: An Example of a New Type of Cosmological Solution of Einstein's Field Equations of Gravitation. Reviews of Modern Physics. 21, 447–450 (1949)

Gödel, Kurt: Über eine Bisher Noch Nicht Benutzte Erweiterung des Finiten Standpunktes. Dialectica. 12, 280–287 (1958)

Gödel, Kurt: A Remark on the Relationship Between Relativity Theory and Idealistic Philosophy. In: Albert Einstein: Philosopher Scientist, Vol. II., ed. Paul Schilpp, pp. 557–562. New York: Harper & Row 1959

Gödel, Kurt: On Formally Undecidable Propositions of Principia Mathematica and Related Systems. In: The Undecidable, ed. M. Davis, pp. 5–38. Howlett, N. Y.: Raven Press 1965

Goodman, Alvin I.: A Theory of Human Action. Englewood Cliffs, N. J.: Prentice-Hall 1970

Gutberlet, Constantin: Das Unendliche, Metaphysisch und Mathematisch Betrachtet. Mainz: G. Faber 1878

Hall, Roland: Monism and Pluralism. In: The Encyclopedia of Philosophy, Vol. 5, ed. P. Edwards, pp. 363–365. New York: Macmillan 1967

Hardy, G. H.: Orders of Infinity, the 'Infinitärcalcul' of Paul DuBois-Reymond. Cambridge, England: Cambridge University Press 1910

Hardy, G. H.: Divergent Series. Oxford: Clarendon Press 1949

Hawking, S. W. and Ellis, G. F. R.: The Large Scale Structure of Space-Time. Cambridge, England: Cambridge University Press 1973

Heath, P. L.: Nothing. In: The Encyclopedia of Philosophy, Vol. 5. ed. P. Edwards, pp. 524–525. New York: Macmillan 1967

Heath, Thomas L.: A History of Greek Mathematics. Oxford: Clarendon Press 1921

Heath, Thomas (ed.): The Thirteen Books of Euclid's Elements, Vol. 1. New York: Dover Publications 1956

Henle, James and Kleinberg, Eugene: Infinitesimal Calculus. Cambridge, Mass.: M. I. T. Press 1978

Hilbert, David: The Foundations of Geometry. Chicago: Open Court 1902

Hilbert, David and Cohn-Vossen, S.: Geometry and the Imagination. New York: Chelsea 1952

Hofstadter, Douglas: Gödel, Escher, Bach: An Eternal Golden Braid. New York: Basic Books 1979

Hofstadter, Douglas: Metamagical Themas. Scientific American. (May, 1981)

Hofstadter, Douglas R.: Metamagical Themas. Scientific American. 18–30 (July, 1981)

Hofstadter, Douglas and Dennett, Daniel: The Mind's I. New York: Basic Books 1981

Hume, David: A Treatise of Human Nature, ed. L. A. Selby-Bigge. Oxford: Clarendon Press 1896

James, William: A Pluralistic Universe. New York: Longmans, Green & Co., 1909

James, William: The Varieties of Religious Experience. New York: Macmillan 1961

Jung, C. G.: Forward. in: The I Ching. Translated by Richard Wilhelm and Cary Baynes, pp. xxi–xxxix. Princeton, N. J.: Princeton University Press 1967

Jung, C. G.: Synchronicity. Princeton, N. J.: Princeton University Press 1973

Kafka, Franz: The Castle. Translated by Willa and Edwin Muir. New York: Knopf 1976

Kafka, Franz: The Diaries of Franz Kafka, ed. M. Brod. New York: Schocken Books 1949

Kant, Immanuel: The Critique of Pure Reason. Translated by Norman Kemp Smith. New York: St. Martin's Press 1964

Kaufmann, W.: Cosmic Frontiers of General Relativity. Boston: Little, Brown 1977

Keisler, H. Jerome: Elementary Calculus. Boston: Prindle, Weber & Schmidt 1976

Kline, Morris: Mathematical Thought from Ancient to Modern Times. New York: Oxford University Press 1972

Knuth, Donald E.: Surreal Numbers. Reading, Mass.: Addison-Wesley 1974

Kochen, Simon and Wang, Hao: In Memoriam Kurt Gödel. The Mathematical Intelligencer. 182–185 (July, 1978)

Kreisel, Georg: Kurt Gödel, 1906–1978. To be published by the Royal Society of London

Kubose, Gyomay: Zen Koans. Chicago, Ill.: Henry Regnery 1973

Kuratowski K. and Mostowski, A.: Set Theory. Amsterdam: North-Holland 1968

Leibniz, Gottfried: The Monadology and Other Philosophical Writings. Translated by Robert Latta. London: Oxford University Press 1965

Lovejoy, Arthur: The Great Chain of Being. Cambridge, Mass.: Harvard University Press 1953

Lucas, J. R.: The Freedom of the Will. Oxford: Clarendon Press 1970

Lucretius: On the Nature of the Universe. Translated by Ronald E. Latham. Harmondsworth, England: Penguin Books 1951

Mandelbrot, Benoit: Fractals: Form, Chance and Dimension. San Francisco: W. H. Freeman 1978

Martin, David Anthony: Hilbert's First Problem: The Continuum Hypothesis. In: Proceedings of Symposia in Pure Mathematics XXVIII, ed. F. Browder, pp. 81–92. Providence, R. I.: American Mathematical Society 1976

Melville, Herman: Moby Dick, Chap. 119. New York: New American Library 1961

Meschkowski, Herbert: Probleme des Unendlicken: Werk und Leben Georg Cantors. Braunschweig: Vieweg 1967

Minsky, Marvin: Computation: Finite and Infinite Machines. Englewood Cliffs, N. J.: Prentice-Hall 1967

Misner, C., Thorne, K. and Wheeler, J.: Gravitation. San Francisco: W. H. Freeman 1973

Moore, Edward F.: Artificial Living Plants. Scientific American. 118–126 (October, 1956)

Nagel, Ernest and Newman, James R.: Gödel's Proof. New York: New York University Press 1958

Newman, J. (ed.): The World of Mathematics, Vol. 1. New York: Simon & Schuster 1956

Otto, Rudolf: Mysticism East and West. New York: Macmillan 1960

Paris, Jeff and Harrington, Leo: A Mathematical Incompleteness in Peano Arithmetic. In: A Handbook of Mathematical Logic, ed. J. Barwise, pp. 1133–1142. Amsterdam: North-Holland 1977

Pascal, Blaise: Pensées et Opuscules, Pensée No. 205, ed. L. Brunschvicq. Paris: Classiques Hachette 1961

Passmore, John: Logical Positivism. In: The Encyclopedia of Philosophy, Vol. 5, ed. P. Edwards, pp. 52–57. New York: Macmillan 1967

Passmore, John: Philosophical Reasoning. New York: Basic Books 1969

Plato: The Dialogues of Plato. Translated by B. Jowett. New York: Random House 1937

Plotinus: Enneads. Boston: C. T. Branford 1949

Puharich, A. (ed.): The Iceland Papers. Amherst, Wisc.: Essentia Research Associates 1979

Pynchon, Thomas: Gravity's Rainbow. New York: Viking Press 1973

Reinhardt, William: Remarks on Reflection Principles, Large Cardinals, and Elementary Embeddings. In: Axiomatic Set Theory, Proceedings of Symposia in Pure Mathematics XIII, Part 2, ed. T. Jech, pp. 189–205. Providence, R. I.: American Mathematical Society 1974

Richard, Jules: The Principles of Mathematics and the Problem of Sets. In: From Frege to Gödel, ed. Jean van Heijenoort, pp. 142–144. Cambridge, Mass.: Harvard University Press 1967

Robinson, Abraham: Formalism 64. In: Logic, Methodology and Philosophy in Science, ed. Y. Bar-Hillel, pp. 228–246. Amsterdam: North-Holland 1964

Robinson, Abraham: Some Thoughts on the History of Mathematics. Compositio Mathematica. 20, 188–193 (1968)

Robinson, Abraham: The Metaphysics of the Calculus. In: The Philosophy of Mathematics, ed. J. Hintikka. London: Oxford University Press 1969

Robinson, Abraham: Non-Standard Analysis. Amsterdam: North-Holland 1974

Royce, Josiah: The World and the Individual, First Series, Appendix: The One, the Many and the Infinite, pp. 504–507, New York: Macmillan 1912

Rucker, Rudolf v.B.: Notices of the American Mathematical Society. 20, p. 362 (November 1973)

Rucker, Rudolf v.B.: On Cantor's Continuum Problem. Journal of Symbolic Logic. 41 (June, 1976)

Rucker, Rudolf v.B.: Geometry, Relativity and the Fourth Dimension. New York: Dover Publications 1977

Rucker, Rudolf v.B.: The One/Many Problem in the Foundations of Set Theory. In: Logic Colloquium 76, eds. R. O. Gandy and J. M. E. Hyland. Amsterdam: North-Holland 1977

Rucker, Rudolf v.B.: Physical Infinities. Speculations in Science and Technology. 1, 43–58 (April, 1978)

Rucker, Rudolf v.B.: One of Georg Cantor's Speculations on Physical Infinities. Speculations in Science and Technology 1. 419–421 (October, 1978)

Rucker, Rudolf v.B.: The Berry Paradox. Speculations in Science and Technology. 2, 197–208 (June, 1979)

Rucker, Rudolf v.B.: The Actual Infinite. Speculations in Science and Technology. 3, 63–76 (April, 1980)

Rucker, Rudolf v.B.: Towards Robot Consciousness. Speculations in Science and Technology. 3, 205–217 (June 1980)

Rucker, Rudolf v.B. (ed.): Speculations on the Fourth Dimension: Selected Writings of C. H. Hinton. New York: Dover Publications 1980

Rucker, Rudolf v.B.: Faster than Light, Slower than Time. Speculations in Science and Technology. 4, 375–383 (October 1981)

Rucker, Rudy: On Hyperspherical Space and Beyond. Isaac Asimov's Science Fiction Magazine. 92–106 (November, 1980)

Rucker, Rudy: White Light, or, What is Cantor's Continuum Problem? New York: Ace Books 1980

Rucker, Rudy: Spacetime Donuts. New York: Ace Books 1981

Rucker, Rudy: Schrödinger's Cat. Analog. (March 30, 1981)

Rucker, Rudy: Software. New York: Ace Books 1982

Rucker, Rudy: The Fifty-Seventh Franz Kafka. New York: Ace Books 1982

Russell, Bertrand: The Principles of Mathematics. Cambridge, England, Cambridge University Press 1903

Russell, Bertrand: Mathematical Logic as Based on the Theory of Types. American Journal of Mathematics. 30, (1908)

Russell, Bertrand and Whitehead, Albert North: Principia Mathematica. New York: Cambridge University Press 1910–1913

Russell, Bertrand: Letter to Frege. In: From Frege to Gödel, ed. Jean van Heijenoort, pp. 124–125. Cambridge, Mass.: Harvard University Press 1967

Salmon, Wesley (ed.): Zeno's Paradoxes. New York: Irvington 1970

Schlegel, Richard: Completeness in Science. New York: Appleton-Century-Crofts 1967

Schrödinger, Erwin: What is Life? & Mind and Matter. Cambridge, England: Cambridge University Press 1969

Scott, J. F.: The Mathematical Work of John Wallis. London: Taylor and Francis 1938

Shoenfield, Joseph R.: Mathematical Logic. Reading, Mass.: Addison-Wesley 1967

Sierpinski, Waclaw: Hypothèse du Continu. Warsaw, Poland: Monografie Matematyczne 1934

Smorynski, C.: The Incompleteness Theorems. In: Handbook of Mathematical Logic, ed. J. Barwise, pp. 821–865. Amsterdam: North-Holland 1977

Smullyan, Raymond: What is the Name of This Book? Englewood Cliffs, N. J.: Prentice-Hall 1978

Suzuki, D. T.: An Introduction to Zen Buddhism. New York: Grove Press 1964

Suzuki, D. T.: The Field of Zen. New York: Harper & Row 1970

Suzuki, D. T.: Mysticism: Christian and Buddhist. Westport, Conn.: Greenwood 1976

Takeuti, Gaisi: The Universe Set Theory. In: Foundations of Mathematics, eds. J. Bulloff, T. Holyoke and S. Hahn, pp. 74–128. New York: Springer-Verlag 1969

Takeuti, Gaisi: Proof Theory. Amsterdam: North-Holland 1975

Takeuti, Gaisi: Gödel Numbers of Product Space. In: Higher Set Theory, eds. G. H. Müller and D. S. Scott, Lecture Notes No. 669. Heidelberg: Springer-Verlag 1978

Turing, Alan M.: Computing Machinery and Intelligence. In: Minds and Machines, ed. A. R. Anderson. Englewood Cliffs, N. J.: Prentice-Hall 1964

Turing, Alan M.: On Computable Numbers, with an Application to the *Entschiedungsproblem.* In: The Undecidable, ed. M. Davis, pp. 116–151. Hewlett, N. Y.: Raven Press 1965

Ulam, Stanislaw: Adventures of a Mathematician. New York: Charles Scribner's Sons 1976

Varley, John: The Ophiuchi Hotline. New York: Dell 1978

Vilenkin, N. Ya.: Stories About Sets. New York: Academic Press 1968

Vlastos, Gregory: Zeno. In: The Encyclopedia of Philosophy, Vol. 8, ed. P. Edwards, pp. 369–378. New York: Macmillan 1967

Von Mises, Richad: Probability, Statistics and Truth. Translated by Hilda Geiringer. New York: Macmillan 1957

von Neumann, John: Theory of Self-Reproducing Automata. Urbana, Ill.: University of Illinois Press 1966

von Tiesenhausen, Georg and Darbro, Wesley A.: Self-Replicating Systems— A Systems Engineering Approach. NASA Technical Memorandum TM-78304. Marshall Space Flight Center, Alabama: 1980

Wang, Hao: From Mathematics to Philosophy. New York: Humanities Press 1974

Wang, Hao: Large Sets. In: Logic, Foundations of Mathematics and Computability Theory, eds. Butts and Hintikka, p. 309. Dordrecht, Holland: Riedel 1977

Weinberg, Steven: The Decay of the Proton. Scientific American. 64–75 (June, 1981)

Weingard, Robert: On Travelling Backward in Time. Synthese. 24, 117–132 (1972)

Wette, Eduard: Definition eines (Relativ Vollständigen) formalen Systems Konstruktiver Arithmetic. in: Foundations of Mathematics: Symposium Papers Commemorating the Sixtieth Birthday of Kurt Gödel, eds. J. J. Bulloff, T. C. Holyoke and S. W. Hahn. New York: Springer-Verlag 1969

Wigner, E.: Remarks on the Mind-Body Question. In: The Scientist Specu-
    lates, ed. I. J. Good, pp. 284–302. New York: Basic Books 1962
Wilson, Robert Anton: Schrödinger's Cat: The Universe Next Door. New
    York: Pocket Books 1980
Wittgenstein, L.: Philosophical Investigations. Oxford: Blackwell 1953
Wittgenstein, L.: Remarks on the Foundations of Mathematics. Oxford: Black-
    well 1956
Wittgenstein, L.: Tractatus Logico-Philosophicus. London: Routledge and
    Kegan Paul 1961
Wolter, Allen: Duns Scotus on the Nature of Man's Knowledge of God. Re-
    view of Metaphysics (1941)

Zermelo, Ernst: Über Grenzzahlen und Mengenbereiche. Fundamenta Math-
    ematica. 16, 29–47 (1930)

INDEX